Demystifying the Complexities of Supramolecular Systems:

Real World Applications

Keyur D. Bhatt
Ganpat University, India

IGI Global
Scientific Publishing
Publishing Tomorrow's Research Today

Vice President of Editorial	Melissa Wagner
Director of Acquisitions	Mikaela Felty
Director of Book Development	Jocelynn Hessler
Production Manager	Mike Brehm
Cover Design	Jose Rosado

Published in the United States of America by
IGI Global Scientific Publishing
701 East Chocolate Avenue
Hershey, PA, 17033, USA
Tel: 717-533-8845 | Fax: 717-533-7115
Website: https://www.igi-global.com E-mail: cust@igi-global.com

Library of Congress Cataloging-in-Publication Data

Names: Bhatt, Keyur Dinesh, 1986- editor.
Title: Demystifying the complexities of supramolecular systems : real world applications / edited by Keyur Bhatt.
Description: Hershey, PA : Engineering Science Reference, an imprint of IGI Global, [2026] | Includes bibliographical references and index. | Summary: "This book bridges the gap between theory and application by providing a clear and accessible understanding of the theoretical foundations of supramolecular chemistry and demonstrating how these concepts translate into practical applications. It further facilitates education and learning by serving as a valuable educational tool for students at various levels, offering a structured and insightful guide to supramolecular chemistry with real-world relevance"-- Provided by publisher.
Identifiers: LCCN 2024038203 (print) | LCCN 2024038204 (ebook) | ISBN 9798369327746 (h/c) | ISBN 9798369349168 (s/c) | ISBN 9798369327753 (eISBN)
Subjects: LCSH: Supramolecular chemistry.
Classification: LCC QD878 .D46 2024 (print) | LCC QD878 (ebook) | DDC 541/.22--dc23/eng/20241114
LC record available at https://lccn.loc.gov/2024038203
LC ebook record available at https://lccn.loc.gov/2024038204

British Cataloguing in Publication Data
A Cataloguing in Publication record for this book is available from the British Library.

Table of Contents

Detailed Table of Contents

This chapter explores the intersection of molecular docking and nanotechnology, highlighting how computational methods drive innovations at the nanoscale. The fundamental principles and algorithms of molecular docking are highlighted before moving on to its pivotal role in nanotechnology. The applications are diverse, from improving drug delivery systems through enhanced interactions between nanoparticles and drugs to guiding the precise design of nanomaterials through protein-nanomaterial interactions. The chapter includes case studies that provide real-world examples of molecular docking's applications. It also addresses challenges and offers a balanced perspective on integrating molecular docking in nanotechnology. The chapter explores future directions and emerging trends, envisioning advancements in docking algorithms and their potential impact on future nanotechnology innovations. In conclusion, the chapter outlines implications for nanotechnology advancements and encourages the exploration of molecular docking in nanotechnology.

Materials that are biological, nonbiological, biomimetic, or hybrid are employed in a variety of nanomedicine research applications, such as high-throughput screening platforms, wearable devices, antibacterial, antiviral, and diagnostic instruments. The systematic, biomimetic, and rational process of creating the optimal therapeutic combinations has included NPs-based drug delivery. Numerous noteworthy developments in cellular, preclinical, and clinical nanomedicine research have been made, encompassing both fundamental and applied aspects. It is anticipated that in the years to come this excessive emphasis will be rectified, with research efforts shifting to other medical problems including antibiotic resistance and artificial organs. Because of the diversity of the design of the nanomaterial platform, nanomedicine is well-positioned to be advantageous in many domains, whether through multimodal beneficial approach or exceedingly specialized multifunctional design for pertinent natural application.

The field of supramolecular chemistry includes dynamic covalent chemistry as well as the investigation of all forms of intermolecular non-covalent bond formation in engineered molecular systems. Reversible non-covalent interactions, a key component of supramolecular chemistry, include hydrogen-bonding, π-π stacking, hydrophobic, van der Waals, and electrostatic interactions. Advances in supramolecular chemistry in recent times have largely aided biomedical applications. For an understanding of how pharmaceuticals interact with binding sites in biological systems, supramolecular chemistry is crucial. Therapeutic compounds should be momentarily contained within the carrier vehicles and released at a predetermined location or time in order to create effective drug delivery systems (DDS). This chapter summarizes the current trends and notable developments in supramolecular systems with specific functions, along with their many biomedical uses in drug delivery, gene transfection, protein delivery, molecular recognition, and biomimetic chemistry.

In the pharmaceutical sciences, developing efficient, sustainable, and biocompatible drug delivery systems is crucial. Traditional methods face issues such as poor solubility and low bioavailability. Supramolecular chemistry, which involves the formation of complex structures through noncovalent interactions, offers solutions to these challenges. This chapter discusses the use of supramolecular systems for green drug delivery, aiming to enhance efficacy, specificity, and sustainability. The authors examined how these structures can improve drug solubility, target specificity, and controlled release, reducing environmental impact. This chapter covers fundamental supramolecular chemistry principles; the various structures, such as cyclodextrins, dendrimers, and micelles and their drug delivery applications. This chapter concludes with the future prospects of supramolecular approaches in green drug delivery, highlighting ongoing research, challenges, and directions such as integrating nanotechnology and personalized medicine for better drug delivery systems.

Sulphur-containing heterocycles, including thiazoline, thiophane, and thiazole, have garnered significant attention in medicinal chemistry due to their diverse biological activities and crucial roles in the biosynthesis of essential compounds. This review highlights the pivotal roles of these sulfur-containing heterocycles' structural diversity and biological significance. The synthesis section delves into the methods and mechanisms behind the formation of these heterocycles. Thiazoline plays essential role in the biosynthesis of thiamine (vitamin B1) and is synthesized in living organisms. Thiophane is a key structural motif in various biologically active natural products, while thiazole is widely prevalent in antimicrobial peptides and antibiotics. The biological evaluation aspect explores the diverse pharmacological activities of these compounds. Thus, the chapter features insights into synthesis and biological activities of these moieties, provides a valuable resource for researchers in medicinal chemistry and chemical biology.

The growing use of pesticides in agriculture poses significant risks to food safety and environmental health. Traditional detection methods are accurate but require complex instruments and skilled personnel. Biosensing approaches using functionalized gold nanoparticles (AuNPs) have emerged as rapid, cost-effective, and portable alternatives. This mini-review explores the synthesis, surface modification, and characterization of AuNPs, emphasizing their surface plasmon resonance (SPR) properties for colorimetric biosensing. Pesticide exposure induces aggregation of ligand-functionalized AuNPs, leading to visible color changes detectable without advanced equipment. Recent studies demonstrate the successful detection of pesticides such as paraquat, glyphosate, and deltamethrin across diverse matrices. These biosensors offer advantages like visual output, simplicity, and fast response. Despite some limitations, functionalized AuNPs show strong potential for developing efficient biosensing systems, supporting safer agriculture and environmental monitoring.

This chapter delves into the diverse applications of nanotechnology, focusing on its role in nanomedicines and its potential to revolutionize healthcare. It covers nanocarriers in drug delivery, diagnostic applications, and therapeutic approaches, highlighting nanotechnology's impact on regenerative medicine, as well as its industrial and environmental applications. The chapter also discusses challenges and considerations, such as biocompatibility, toxicity, regulatory frameworks, and ethical implications, providing a comprehensive perspective on nanomedicine complexities.

Preface

In today's rapidly advancing world, the integration of supramolecular systems and nanotechnology is leading to remarkable breakthroughs across multiple scientific and industrial domains. These innovative fields are unlocking solutions to some of the most critical global challenges, from revolutionizing healthcare through targeted drug delivery systems to developing sustainable materials for environmental and industrial applications. The need for continued research and innovation in these areas is more pressing than ever before, as the complexities of these technologies offer both vast potential and significant hurdles.

Demystifying the Complexities of Supramolecular Systems: Real World Applications aims to illuminate the intricate and transformative role that supramolecular chemistry and nanotechnology play in shaping the future. This book not only highlights the cutting-edge developments in these fields but also addresses the fundamental problems that have hindered progress—problems that the chapters within this volume seek to solve. As editors, we believe that the key to overcoming these challenges lies in deepening our understanding and enhancing our ability to apply these complex systems to real-world scenarios.

In Chapter 1, "Molecular Docking and its Applications in Nanotechnology," authors Popat Mohite, Shubham Munde, Sudarshan Singh, and Bhupendra Prajapati delve into the essential intersection of molecular docking and nanotechnology. The chapter outlines the principles and algorithms of molecular docking, offering insight into its pivotal role in the development of nanomaterials. By highlighting case studies, the authors explore how molecular docking enhances drug delivery systems by improving nanoparticle-drug interactions and guides the precise design of nanomaterials. The chapter also addresses current challenges and presents future directions, emphasizing the need for refined docking algorithms to unlock further advancements in nanotechnology.

Chapter 2, "Applications in Nanomedicine and Beyond," authored by Selvakumar P, Seenivasan S, Sumanta Bhattacharya, Sudha D, and Akila A, explores the significant strides in nanomedicine, with a focus on the application of nanomaterials in therapeutic fields. From antibacterial to diagnostic instruments, the chapter highlights the diverse roles of biological, nonbiological, and hybrid materials in nanomedicine. The authors advocate for expanding research to tackle emerging medical challenges, including antibiotic resistance and the creation of artificial organs. With an emphasis on multifunctional nanomaterial platforms, this chapter sheds light on how nanomedicine continues to evolve as a multifaceted and dynamic field.

Chapter 3, "Supramolecules as Efficient Tools for Modern Targeted Drug Delivery Approaches," by Nikita Patel, Miral Vora, Asha Joshi, and Bhupesh Bhatt, explores the applications of supramolecular chemistry in drug delivery. The chapter outlines how non-covalent interactions, such as hydrogen bonding and electrostatics, enable precise control over drug release. The authors highlight the growing importance of supramolecular systems in creating effective drug delivery systems (DDS), emphasizing their role in the targeted delivery of therapeutic compounds, gene transfection, and protein delivery. This chapter is a crucial overview of how supramolecular chemistry is transforming the landscape of biomedicine.

In Chapter 4, "Supramolecular Approaches to Green Drug Delivery," Bancha Yingngam focuses on the growing need for environmentally friendly and biocompatible drug delivery solutions. The chapter discusses how supramolecular systems, such as cyclodextrins, dendrimers, and micelles, offer significant improvements in drug solubility, specificity, and controlled release. Yingngam emphasizes the role of supramolecular chemistry in addressing challenges like low bioavailability and environmental sustainability, ultimately positioning these systems as critical to the future of green chemistry in pharmaceutical sciences.

Chapter 5, "Exploring Thiazoline, Thiophane, and Thiazole Advances in Sulphur-Based Heterocyclic Compounds," authored by Nandan Chamanlal Pomal and Keyur D. Bhatt highlights the pivotal roles of these sulfur-containing heterocycles' structural diversity and biological significance. The synthesis section delves into the methods and mechanisms behind the formation of these heterocycles. Thiazoline plays essential role in the biosynthesis of thiamine (vitamin B1) and is synthesized in living organisms. Thiophane is a key structural motif in various biologically active natural products, while thiazole is widely prevalent in antimicrobial peptides and antibiotics. The biological evaluation aspect explores the diverse pharmacological activities of these compounds. Thus, articles features insights into synthesis and biological activities of these moieties, provides a valuable resource for researchers in medicinal chemistry and chemical biology.

In Chapter 6, "Functionalized Gold Nanoparticles Enable Colorimetric Biosensing of Pesticides for Advanced Applications," Keyur Bhatt, Nandan C. Pomal, Dhruvansh Patni, and Nihal Patel explore the synthesis, surface modification, and characterization of AuNPs, emphasizing their surface plasmon resonance (SPR) properties for colorimetric biosensing. Pesticide exposure induces aggregation of ligand-functionalized AuNPs, leading to visible color changes detectable without advanced equipment. Recent studies demonstrate the successful detection of pesticides such as paraquat, glyphosate, and deltamethrin across diverse matrices. These biosensors offer advantages like visual output, simplicity, and fast response. Despite some limitations, functionalized AuNPs show strong potential for developing efficient biosensing systems, supporting safer agriculture and environmental monitoring.

Finally, in Chapter 7, "Nanomedicines and Its Applications Beyond," Popat Mohite, Roshan Dave, Shubham Munde, Anil Pawar, Sudarshan Singh, Bhupendra Prajapati, and Aarati Budhar explore the broader scope of nanomedicine. This chapter covers not only drug delivery and therapeutic applications but also diagnostic and regenerative medicine. The authors critically examine the challenges of biocompatibility, toxicity, and regulatory frameworks, while also highlighting the potential of nanomedicines to revolutionize healthcare across various disciplines. The chapter serves as an essential resource for understanding the multifaceted role of nanotechnology in medicine and its far-reaching implications for the future of healthcare.

Ultimately, the core theme of this volume is the critical role of research in translating complex scientific principles into practical solutions. As the chapters of this book demonstrate, while the applications of supramolecular systems and nanotechnology hold immense promise, it is through rigorous and ongoing research that we will address the persistent problems—whether in drug delivery, material science, or environmental sustainability—that stand in the way of their full potential.

This book is not just a collection of scientific knowledge; it is a call to action. To all researchers, students, and professionals in the field: we urge you to explore these concepts, to challenge existing boundaries, and to contribute to the collaborative effort of transforming these technologies into tangible solutions that will improve lives and address the most urgent global challenges.

Chapter 1
Molecular Docking and Its Applications in Nanotechnology

Popat Mohite
https://orcid.org/0000-0002-4536-4444
AET's St. John Institute of Pharmacy and Research, Palghar, India

Shubham Munde
AET's St. John Institute of Pharmacy and Research, Palghar, India

Sudarshan Singh
https://orcid.org/0000-0002-7929-3322
Faculty of Pharmacy, Chiang Mai University, Thailand

Bhupendra Prajapati
https://orcid.org/0000-0001-8242-4541
Shree S.K. Patel College of Pharmaceutical Education and Research, Ganpat University, India

ABSTRACT

This chapter explores the intersection of molecular docking and nanotechnology, highlighting how computational methods drive innovations at the nanoscale. The fundamental principles and algorithms of molecular docking are highlighted before moving on to its pivotal role in nanotechnology. The applications are diverse, from improving drug delivery systems through enhanced interactions between nanoparticles and drugs to guiding the precise design of nanomaterials through protein-nanomaterial interactions. The chapter includes case studies that provide real-world examples of molecular docking's applications. It also addresses challenges and offers a balanced perspective on integrating molecular docking in nanotechnology. The

DOI: 10.4018/979-8-3693-2774-6.ch001

chapter explores future directions and emerging trends, envisioning advancements in docking algorithms and their potential impact on future nanotechnology innovations. In conclusion, the chapter outlines implications for nanotechnology advancements and encourages the exploration of molecular docking in nanotechnology.

INTRODUCTION

In drug discovery and design, molecular docking is a computer tool commonly used to predict the binding affinities and modes of small compounds, or ligands, to target proteins or receptors, (Abdelsattar, Dawoud, & Helal, 2021). Molecular docking (MD) helps to find promising drug candidates with therapeutic efficacy by modelling the atomic-level interaction between ligands and receptors, (Abo-zeid et al., 2020). The most energetically advantageous binding pose is found using algorithms that search the conformational space of the ligand and the receptor. Molecular docking has developed into an effective drug development method, helping clarify structure-activity correlations and optimize lead molecules, (Agarwal & Mehrotra, 2016). MD is widely used in drug discovery as it is a cost-effective and simplified computational method. The MD is widely applied to find new molecules from normal to rare diseases, (Agu et al., 2023). Apart from drug discovery, it is also used to assess the toxicity of drug candidates. Various studies have been conducted during the pandemic to select and discover therapeutically effective vaccines or drug molecules for treating COVID-19, (Akbarzadeh et al., 2022). Apart from drug discovery, MD is widely used in formulation development. It is pivotal in formulation development because it predicts how drug candidates interact with target proteins or receptors. It expedites drug discovery through virtual screening of compound libraries, identifying potential candidates for further study. Docking simulations also optimize formulations by assessing drug-excipient interactions and predicting stability. Crucially, it elucidates molecular interactions, guiding the design of drugs with enhanced efficacy and reduced side effects, (Anselmo & Mitragotri, 2019). Additionally, docking aids in predicting ADMET properties early, influencing drug development strategy and resource allocation. Overall, molecular docking integrates computational tools with biochemical insights, significantly advancing the efficiency and success of pharmaceutical formulation development. Integrating molecular docking with nanotechnology represents a promising frontier in scientific research, offering novel approaches for drug delivery, diagnostics, and therapeutic interventions, (Ayesha et al., 2023). There several studies have reported on the application of MD in formulation development. Yousuf et al. performed the MD on *Piper nigrum* seed extract-loaded emugel to check its anti-ageing activity with targeted proteins, (Bansal et al., 2021). Integrating molecular docking with nanotechnology represents a promising frontier in scientific research, offering novel approaches for drug delivery, diagnostics, and therapeutic interventions, (Bertilsson et al., 2002). Nanotechnology involves the manipulation of particulate matter to the nanoscale, ranging from 1 to 1000 nanometers. Nanomaterials exhibit surface area-to-volume ratios, tuneable physicochemical properties, and an ability to interact with biological systems at molecular levels to generate functional structures with unique properties. Recently, the use of nanotechnology in molecular docking has revolutionized aspects from improving the accuracy of binding prediction to facilitating the targeted delivery of therapeutic agents, (Chaudhaury & Tyagi, 2024). Developing drug delivery systems based on nanoparticles is one of the main areas where molecular docking and nanotechnology converge. By using nanoparticles as carriers, these systems deliver medications to precise target areas inside the body, increasing their effectiveness and reducing systemic toxicity, (Chen & Riviere, 2017). Because it can predict how medicinal molecules will interact with the surface of nanoparticles, molecular docking is an essential tool in the rational design of nanoparticle formulations. By optimizing nanoparticle surface chemistry and functionalization, scientists can modify these particles' characteristics to enhance drug loading capacity, stability, and targeting specificity. Additionally, the problems with conventional drug delivery techniques, such as poor solubility, quick clearance, and off-target effects, can be creatively solved by nanotechnology. Hydrophobic medications can be encapsulated in nanoparticles to prevent degradation and enable controlled release at the intended site of action. Using molecular docking simulations, scientists may evaluate various nanoparticle formulations and identify the best chance of producing therapeutic effects, (Chibber & Ahmad, 2016). Furthermore, computational predictions can be verified and improved by combining experimental methods like surface plasmon resonance and atomic force microscopy with molecular docking algorithms, producing more effective nanomedicines. The combination of molecular docking and nanotechnology promises potential uses beyond drug delivery, such as the creation of biosensors and diagnostic tools. The sensitive and selective detection of biomolecules, pathogens, and disease indicators provided by nanomaterial-based biosensors makes early diagnosis and condition monitoring possible. By foretelling the binding interactions between biomolecular targets and receptor-functionalized nanomaterials, molecule docking facilitates the logical design of nano biosensors. Researchers can improve nano sensors' sensitivity, specificity, and reaction kinetics by fine-tuning their surface architecture and composition. This will facilitate the development of quick and precise diagnostic tests. The combination of molecular docking and nanotechnology has broad applications beyond biological domains, including materials research, environmental cleanup, and energy storage, (Cui et al., 2016). To solve global concerns connected to sustainability and environmental pollution, customized nanomaterials can be created for pollutant removal, sensing, and catalysis. By forecasting the adsorption energies and reaction kinetics of catalytic species on nanoparticle surfaces, molecular docking aids in the sensible design of nanocatalysts. Researchers can speed up the identification and optimization of nanomaterials for various applications by utilizing the predictive powers of computer modelling. This will promote innovation in nanotechnology-enabled solutions for various societal demands. Utilizing the combined strengths of molecular docking and nanotechnology, this approach leverages computational modelling and nanomaterials engineering to tackle intricate scientific and technological problems. This multidisciplinary field has enormous promise for improving human health, environmental sustainability, and technology innovation, from drug discovery and delivery to diagnostics and beyond. Future breakthroughs that revolutionize illness diagnosis, treatment, and prevention and open the door to long-term solutions to urgent global concerns are anticipated as scientists continue to investigate the synergies between molecular docking and nanotechnology, (Dar, 2017; Deng et al., 2022). The chapter, ever mindful of challenges, addresses limitations and considers computational and biological intricacies, providing a balanced perspective on integrating molecular docking in nanotechnology. Looking forward, the chapter explores the future directions and emerging trends, envisioning advancements in docking algorithms, their integration with experimental techniques, and the potential impact on future nanotechnology innovations. In conclusion, the chapter covers the key findings, outlines implications for nanotechnology advancements, and extends a compelling call to action for the continuous exploration of molecular docking in the nanotechnological landscape.

BASICS OF MOLECULAR DOCKING

A fundamental method in computational chemistry is molecular docking, which enables scientists to forecast the affinities and patterns of binding tiny molecules, usually ligands, to bigger biomolecular targets, which are frequently proteins or nucleic acids, (Dolai, Mandal, & Jana, 2021). Fundamentally, molecular docking aims to anticipate the most energetically favourable binding position by simulating the interaction between a ligand and a receptor. This technique entails examining the conformational space of the ligand and the receptor and assessing their complementarity to find possible binding configurations. Two primary processes are usually involved in the basic algorithmic approach to molecular docking: scoring function evaluation and ligand conformational sampling. Different computational methods, including Monte Carlo simulations or genetic algorithms, produce a heterogeneous ensemble of ligand conformations during the ligand conformational sampling step, (Edris et al., 2023). These conformations show various ligand orientations and conformations that could exist within the receptor's binding region. Each ligand shape's interaction energy or affinity with the receptor is then assessed using a scoring system. Based on variables like hydrogen bonding, desolvation effects, van der Waals interactions, and electrostatic interactions, the scoring function determines the fitness or acceptability of each ligand posture. Ultimately, the anticipated binding affinities of the ligand poses are used to rank them, and the most promising candidates are chosen for additional examination or experimental confirmation, (El-Sayed et al., 2024).

The phases of docking go as follows:

Step 1: Setting up the chemical database: Structural analogues were designed in ACD/ Chemsketch and saved in .cdx format.

Step 2: Software installation: PyRx and DSV were installed from their parent sites:

https://pyrx.sourceforge.io/ and https://discover.3ds.com/discovery - studio-visualizer-download.

Step 3: Protein retrieval and preparation:

The proteins were collected from the RCSB PDB site (https://www.rcsb.org/) and downloaded in PDB format. In DSV software, downloaded proteins were opened for their preparation. Raw proteins associated with solvent molecules, heteroatoms, cofactors, and ligands were removed. This was done by right-clicking each of these and then clicking on 'Cut' to eliminate water molecules, hetero atoms, and ligands. Both the proteins were single-chained and thus required no obliteration. Click on File > Save as > sdf file. Ergo, refined protein is set to dock.

Step 4: Downloading Co-crystallized ligand:

The ligand linked with the protein termed a co-crystallized ligand was downloaded in sdf format from PubChem. This ligand was thereupon applied for the blind docking process.

Step 5: Downloading Reference standard:

Reference standard for in-vitro anti-inflammatory activity evaluation was also downloaded to establish its binding to the target protein and compare it to the co-crystallized ligand with proposed compounds. Reference standards can be downloaded from PubChem in the SDF file.

Step 6: Conversion of ligands (chemical database) in sdf and pdbqt format:

The designed compound is known as the chemical database, along with co-crystallized ligands and

Reference standards were opened in PyRx as directed:

PyRx > Control Panel > Open Babel > Insert new item toolbar (+) to the right of the Control panel. Once loaded, right-clicking the compounds save as – sdf file format (except for Cocrystallized ligand and reference standard. Later, the compounds were right-clicked, and energy was minimized. Yet again, compounds were converted to. pdbqt format by right-clicking and converting all to AutoDock ligand (pdbqt format). Now, the ligands were ready to dock.

Step 7: Uploading refined protein in PyRx:

The upper left button above the navigator panel was clicked to load the macromolecule. After converting ligands and macromolecules to input. pdbqt files, ligands, and macromolecules were spotted under the Autodock tab.

Step 8: Blind Docking:

The start button in the Vina Wizard tab was clicked in the lower right corner. Then, in the select molecules of Vina Wizard, ligands and macromolecules (refined protein) from the AutoDock selection of the Navigator panel were selected. When the Forward button at the lower right corner was clicked, it took to the Run Vina window. The grid box with white spherical handles was shown on the Run Vina tab, and the entire macromolecule was positioned in the grid box using the maximize button beneath the Vina Search space. The co-crystallized ligand could freely interact with the macromolecule in the search space for blind docking. To begin Vina's calculations, the Forward button was clicked. After the computations were finished, ten findings with the best poses were identified, and their binding affinities (kcal/mol), mode, and RMSD (upper) and RMSD (lower) were displayed in the result. These results were stored in a specified folder as Comma Separated values (CSV) and SDF formats.

Step 9: Analysis of results:

To open docked ligand RLI out. For pdbqt files in DSV, click Files > Open > Select all files in the format in the open pop-up window. First protein. pdbqt in one window was opened, and ligand RLI_ was out. pdbqt files in another window. Right-click on the ligand > Copy > Paste into the protein window. Under the Receptor-ligand interactions tab, define the receptor by clicking on the protein and represent the ligand by clicking on the ligand. Then, interacting amino acids were noted, and the types of interactions were recorded.

Step 10: Docking of ligands:

Likewise, the ligands and the reference standard were selected subsequently and docked with the macromolecule. As a replacement for selecting whole protein for the Vina Calculations, only binding sites derived during blind docking were selected, and the docking continued. The binding affinities were saved, visualized, and analyzed in DSV similarly.

APPLICATIONS OF MOLECULAR DOCKING IN NANOTECHNOLOGY

Molecular docking, a computational technique rooted in molecular mechanics and quantum chemistry principles, stands as a cornerstone in the realm of nanotechnology. By simulating the interaction between small molecules and larger macromolecular targets such as proteins or nucleic acids, molecular docking elucidates binding configurations at the atomic level, guiding the rational design and optimization of nanomaterials with tailored properties and functionalities. In the burgeoning field of nanotechnology, molecular docking finds multifaceted applications across diverse domains, from drug delivery and biosensing to materials science and beyond. Through computational simulations, researchers can predict binding affinities, elucidate molecular recognition mechanisms, and accelerate the discovery and development of novel nanomaterials with applications ranging from targeted therapeutics to environmental sensing. This convergence of computational modelling and nanotechnology holds immense promise in addressing complex biomedical, environmental, and technological challenges, ushering in a new era of precision-engineered nanomaterials with transformative potential. Nanotechnology has wide applications in healthcare, including antimicrobial assessment, bioimaging, cancer treatments, etc., (Ghasemi, Abdolmaleki, & Shiri, 2017). Currently, the use of nanoparticles in clinical use is also approved by regulatory bodies such as the FDA, (Jakhar et al., 2020). Reportedly, MD is widely used to determine the interaction of nanoparticles with drug candidates or the human protein. Chubber et al. reported the interaction of human serum albumin protein and Cuo and TiO_2 nanoparticles, revealing the significant binding affinity for human serum proteins, (Kamal & Chakrabarti, 2023). Another study was reported by Abo-zeid et al. (2020) for FDA-approved iron oxide nanoparticles for the treatment and control of COVID-19 infection. The study reveals the efficient binding of iron oxide nanoparticles with the spike protein receptor domain of SARS-CoV-2, (Kaushik et al., 2018). Edris et al. (2023) performed MD studies on newly synthesized Letrozole analogue incorporated in gold nanoparticles for breast cancer treatment. The study reveals that a newly designed molecule has a significant binding affinity for aromatase protein, (Kaur et al., 2019).

MD is widely used in nanotechnology to predict their therapeutic role in drug delivery systems. Performed MD for Aloe vera and incorporated chitosan nanoparticles against H Pylori protein. The finding reveals that the major constituents of Aloe vera, chlorogenic acid and pyrocatechol, confirm the binding affinity with respective proteins, (Khembar et al., 2021). Deng et al. (2022) performed MD on curcumin nanoformulation to assess the activity against triple-negative breast cancer cell lines; the finding revealed a better binding affinity with selected proteins, (Kunjiappan et al., 2020). Ayesh et al. (2023) performed MD of Polyvinylpyrrolidone and chitosan magnetite nanoparticles for antimicrobial activity; the study shows significant binding affinity with DNA gyrase and dihydrofolate reductase, (Liang et al., 2023).

Figure 1. Binding of DHFR with magnetite in the pocket and with chitosan complex

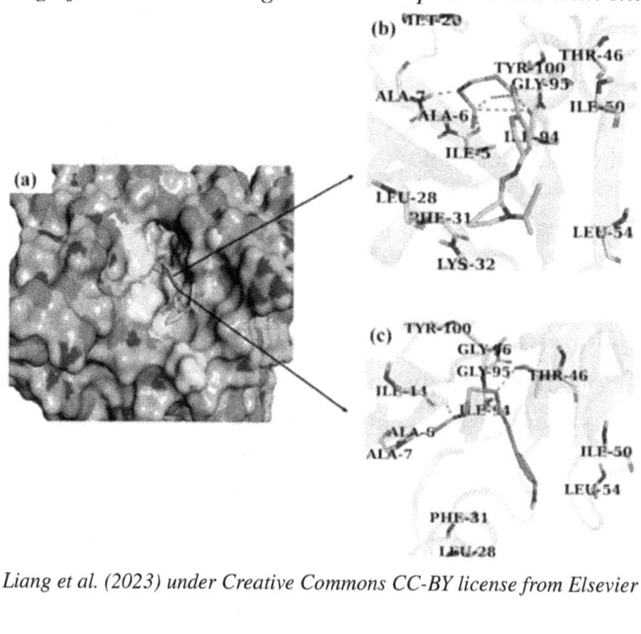

Adapted from Liang et al. (2023) under Creative Commons CC-BY license from Elsevier

Drug Delivery Nanocarriers Design

Molecular docking guided the development of multifunctional nanocarriers for targeted drug delivery and imaging in cancer theranostics. Iman Akbarzadeh and colleagues investigated the in-silico properties of Gingerol (Gin) and Letrozole (Let) as potential anticancer drugs, predicting significant ADME drawbacks. To address these issues, they developed mesoporous silica nanoparticles (MSNs) functionalized with zinc, amine, and graphene oxide (GO) (MZNG) for loading and delivering Gin and Let to breast cancer cells in vitro. Biophysical analyses revealed that Let and Gin-loaded MZNGs had a spherical structure with a mean diameter of approximately 210 nm, offering high entrapment efficiency and pH-sensitive sustained release. Cytotoxicity assays showed that MZNG significantly enhanced the inhibitory effects of the drugs on breast cancer cells due to improved cellular uptake. The treatment led to the downregulation of Bcl-2, MMP-2, MMP-9, CyclinD, and CyclinE genes and the upregulation of Bax, Caspase3, and Caspase9 genes, resulting in G0/G1 phase cell arrest, apoptosis, and inhibition of migration and invasion. The study concluded that MZNG is a promising nanosystem for co-delivery of Gin and Let, showing synergistic effects in treating breast and potentially other cancers, (Mirza, Ahmad, & Shah, 2020). Selvaraj Kunjiappan and colleagues (2020) explored a targeted drug delivery system using folic acid (FA)-conjugated bovine serum albumin (BSA) nanoparticles (NPs) to encapsulate myricetin (Myr) for treating folate receptor (FR)-positive breast cancer cells. The Myr-loaded BSA NPs were formulated using a modified desolvation cross-linking technique and further functionalized with FA to enhance targeting capabilities. Characterization through spectroscopy confirmed the successful synthesis and functional integrity of the FA-conjugated carrier. These biocompatible FA-Myr-BSA NPs exhibited appropriate morphology, size, stability, and drug release kinetics, with molecular docking studies indicating strong non-covalent binding to folate receptors, facilitating active endocytosis. The NPs released Myr rapidly in acidic conditions (pH 5.4) and demonstrated high biocompatibility in physiological conditions. The FA-Myr-BSA NPs significantly reduced the viability of MCF-7 cells, with an IC50 value of 72.45 μg ml−1, and enhanced Myr uptake, inducing apoptosis with characteristic nuclear condensation and membrane distortion. Additionally, these NPs targeted mitochondria, increasing reactive oxygen species release and disrupting mitochondrial membrane integrity. The study concluded that FA-Myr-BSA NPs effectively enhance Myr's anticancer activity, presenting a promising carrier for targeted chemotherapeutic delivery, (Mohammadjani et al., 2024).

NANOMATERIALS FOR BIOSENSING APPLICATIONS

Molecular docking guided the development of nanomaterial-based biosensors for rapid and specific detection of viral pathogens. Yingqi Liang and colleagues highlighted the critical need for novel diagnostic strategies to combat viral infections, which pose significant global health, social, and economic threats. Their review focused on quantum dots (QDs), fluorescent nanomaterials known for their high quantum yield, broad absorbance range, narrow and size-dependent emission, and stability. QDs-based nanotechnology has shown rapid response, easy operation, high sensitivity, and reasonable specificity, making it a promising tool for virus detection. The review systematically covered various QDs-based detection methods, including synthesis, biofunctionalization, viral nucleic acid detection strategies, and immunoassays.

Additionally, it discussed the challenges and future perspectives of applying QDs in virus detection, filling a gap in the literature by providing a comprehensive analysis of this emerging field, (Muhammed & Aki-Yalcin, 2024). Using the co-precipitation process, El-Sayed A.F. reported the synthesis of copper-doped ZnO nanoparticles, $Zn1-x(Cu)O$ ($x = 0.0, 0.03, 0.05$, and 0.07). The structural and chemical properties were analyzed using various techniques, revealing a hexagonal wurtzite crystal structure with successful incorporation of Cu ions, as confirmed by FT-IR spectra and P-XRD patterns. HR-TEM and FE-SEM/EDX analyses demonstrated changes in morphology, homogeneity, and the presence of Cu signals, while zeta potential measurements indicated increased nanoparticle stability with higher Cu doping. The Cu-doped ZnO nanoparticles exhibited significant in vitro antioxidant properties and antimicrobial efficacy against pathogenic microorganisms. Notably, the 7% Cu-doped sample showed the highest inhibitory effect on the α-amylase enzyme, suggesting potential anti-diabetic activity. However, no significant variations were observed for acetylcholinesterase and proteinase enzyme activities. Molecular docking provided insights into the interactions of Cu-doped ZnO with α-amylase, FabH of E. coli, and penicillin-binding proteins of S. aureus, suggesting these materials could inhibit enzymes and be effective against microbial infections and diabetes. Further studies are recommended to explore these nanoparticles' in vitro anti-diabetic potential, (Nowack et al., 2015).

Figure 2. 3D representation of interaction between amylase, FaBH of E coli, and penicillin

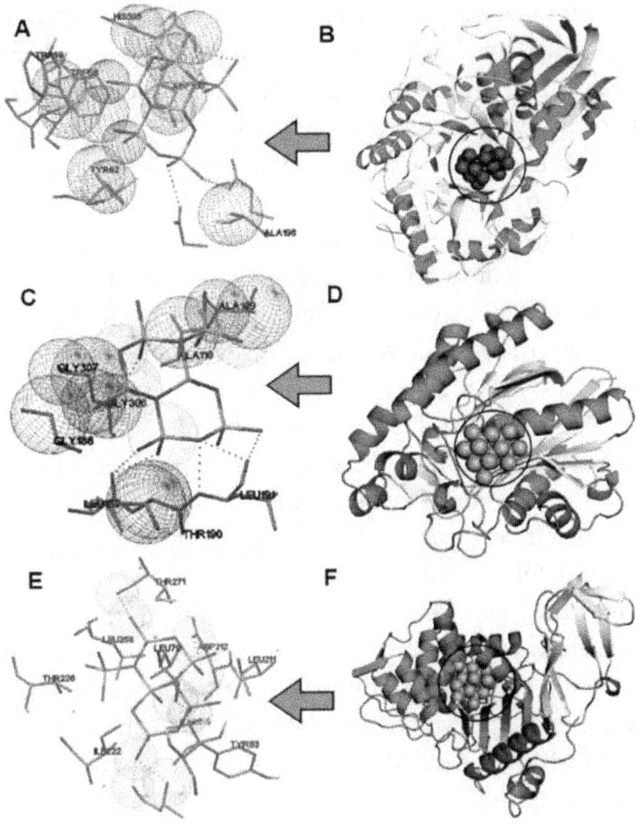

Adapted from Nowack et al. (2015) under the terms of the Creative Commons CC BY license from Springer Nature publisher

Design of Nanoparticle-Based Therapeutics

Molecular docking aided in developing nanoparticle-based therapeutics for neurodegenerative diseases such as Alzheimer's. Mal Yousfan and colleagues tackled the challenge of targeted drug delivery to the brain by developing predictive models based on a comprehensive data set of 403 data points, incorporating both numerical and categorical features. Utilizing machine learning techniques and extensive literature data analysis, they created models to predict nanoparticle (NP) delivery to the brain, focusing on the physicochemical properties of loaded drugs and NPs and analyzing pharmacodynamic parameters such as plasma area under the curve. The study highlighted the effectiveness of various linear models, particularly linear mixed-effect models (LMEMs), which showed exceptional accuracy in predicting NP biodistribution. The models were validated by preparing and administering two distinct NP formulations via intranasal and intravenous routes. Among the findings, LMEMs outperformed other models in capturing underlying patterns, revealing that factors like release rate and molecular weight negatively impact brain targeting. However, being a P-glycoprotein substrate slightly enhances it. This research offers insights into optimizing NP-based drug delivery systems for brain therapeutics, (Ozboyaci et al., 2016).

Molecular docking facilitated the development of targeted nanoparticle-based therapeutics for treating antibiotic-resistant bacterial infections. Navid Mohammadjani and colleagues (2024) investigated the interactions of metal oxide nanoparticles (MONPs) coated with a carbon layer (specifically Ag2O3, CdO, CuO, Fe2O3, FeO, MgO, MnO, and ZnO) with targets related to cancer and bacterial infections using molecular docking with AutoDock 4.2.6 software. They compared the toxicity (LD50) and molecular weight of these MONPs to standard drugs using the PRO TOX-II online tool. The study found that in several cases, MgO and Fe2O3 NPs demonstrated superior performance to standard medications based on their binding energies. Despite MONPs generally having a lower 50% lethal dose and higher molecular weight than standard drugs, they exhibited only minor differences in binding energy across various targets, likely due to their specific physicochemical and pharmacophoric properties. However, the toxicity of MONPs remains a significant challenge in drug development. Overall, MgO and Fe2O3 NPs emerged as the most efficient MONPs studied, (Pinzi & Rastelli, 2019).

Molecular docking aided in the development of nanoparticle-based vaccines for emerging viral diseases. Yasmin Abo-zeid et al. explored the potential of FDA-approved iron oxide nanoparticles (IONPs), specifically Fe2O3 and Fe3O4, as antiviral agents against COVID-19 through docking studies that examined their interactions with the SARS-CoV-2 spike protein receptor-binding domain (S1-RBD). The study found that Fe3O4 formed a more stable complex with the SARS-CoV-2 S1-RBD, while Fe2O3 showed a stronger affinity for hepatitis C virus (HCV) glycoproteins E1 and E2. These interactions are believed to induce conformational changes in the viral proteins, leading to viral inactivation. Based on these promising in silico results, the researchers recommend progressing to clinical trials to assess the efficacy and safety of IONPs in treating COVID-19, highlighting the potential of repurposing existing medications to address urgent global health challenges, (Rajput et al., 2023). Molecular docking facilitated the development of nanoparticle-based therapeutics for neurodegenerative disorders such as Parkinson's disease. Aman Chandra Kaushik and colleagues investigated the potential of cerium oxide (CeO2) nanoparticles (NPs) as inhibitors of α-synuclein, an essential protein involved in Parkinson's disease (PD), using a computational biology approach. Their study compared the efficacy of CeO2 NPs with biocompatible metal NPs, such as gold NPs and superparamagnetic iron oxide nanoparticles (SPIONs), in inhibiting α-synuclein activity. The results demonstrated that CeO2 NPs effectively fit into the active site of α-synuclein, showing significant interactions and contacts that surpassed those of conventional PD treatments like L-DOPA, used as a positive control. Consequently, CeO2 NPs were identified as promising inhibitors of α-synuclein, suggesting their potential application as nano-drugs for PD treatment, (Saikia & Bordoloi, 2019).

These case studies underscore the utility of molecular docking in guiding the rational design of nanoparticle-based therapeutics for various diseases, ranging from bacterial infections to viral diseases and neurodegenerative disorders. By leveraging computational techniques, researchers can expedite the development of targeted and efficacious nanoparticle-based therapies with enhanced therapeutic outcomes.

CHALLENGES AND CONSIDERATIONS

Molecular docking is a powerful computational technique widely used in nanotechnology to predict the interactions between molecules and nanomaterials. However, its application faces several significant challenges and limitations. One primary issue is the accuracy of docking predictions, primarily due to scoring functions that may not adequately capture the complexities of molecular interactions at the nanoscale. Additionally, the conformational flexibility of both ligands and targets poses a challenge, as accurately modelling this flexibility is computationally demanding and often leads to approximations, (Volgin et al., 2018).

The computational complexity of molecular docking is another significant hurdle. The high computational demand, especially when considering large molecules or nanoparticles, requires substantial resources and time. The vast number of possible ligand conformations and orientations increases the search space, making it challenging to find the global minimum energy configuration efficiently. This complexity is further exacerbated by the inadequacies of existing force fields, which are often parameterized for conventional biomolecules rather than nanomaterials, leading to inaccuracies in interaction energy calculations, (Yahya et al., 2022).

Solvent effects present additional challenges, with most docking studies relying on implicit solvent models that can overlook critical solvent-mediated interactions. While explicit solvent models are more accurate, they significantly increase computational demands. Moreover, the surface chemistry of nanoparticles, including coatings and functional groups, interacts with the solvent in complex ways that are difficult to model accurately, (Yan et al., 2013).

Experimental validation remains a critical bottleneck, as there can be significant discrepancies between docking predictions and experimental results. The limited availability of experimental binding data for many nanomaterial systems further complicates validation efforts, (Yousfan et al., 2024). Scalability is also a concern, with large nanomaterials requiring scalable algorithms to efficiently handle their complex surface interactions. Implementing high-throughput docking simulations for extensive compound libraries demands substantial computational infrastructure and optimized workflows.

Docking studies, which typically focus on binding affinities, often neglect thermodynamic and kinetic considerations. Accurate modelling of binding kinetics and thermodynamic stability at the nanoscale is complex and essential for understanding the dynamics of interactions, (Yousuf et al., 2022). Nanomaterial-specific challenges, such as surface heterogeneity and anisotropy, add further complexity to the docking process, (Zhou, Jiang, & Chen, 2022).

Lastly, integrating molecular docking with broader multiscale models and ensuring effective interdisciplinary collaboration between computational chemists, materials scientists, and biologists is crucial yet challenging. Addressing these challenges requires advancements in computational methods, improved force fields, and better integration with experimental data to enhance the reliability and applicability of molecular docking in nanotechnology.

FUTURE DIRECTIONS AND EMERGING TRENDS

The future of molecular docking in nanotechnology is poised to witness several exciting advancements and emerging trends. One significant direction is the integration of artificial intelligence (AI) and machine learning (ML) algorithms to enhance the accuracy and efficiency of docking simulations. These technologies can help refine scoring functions, predict binding affinities more accurately, and reduce computational time by learning from vast datasets of molecular interactions. Advancements in high-performance computing (HPC) and quantum computing are also set to revolutionize molecular docking. HPC enables handling more complex systems and larger datasets, while quantum computing promises to solve currently intractable problems in molecular simulation by leveraging quantum mechanical principles. Multiscale modelling approaches are gaining traction, combining molecular docking with larger-scale simulations to provide a more comprehensive understanding of interactions at the nanoscale. These models can capture the dynamic behaviour of molecules and nanomaterials in more realistic environments, accounting for factors like solvent effects and thermodynamic stability.

The development of improved force fields specifically tailored for nanomaterials will enhance the accuracy of docking predictions. These specialized force fields can better account for the unique physicochemical properties of nanomaterials, leading to more reliable simulations. Additionally, there is a growing trend towards integrating molecular docking with experimental techniques such as cryo-electron and atomic force microscopy. This integration allows for the validation of docking predictions and provides a feedback loop to improve computational models. Overall, the future of molecular docking in nanotechnology looks promising, with AI, advanced computing, multiscale modelling, specialized force fields, and experimental integration driving significant improvements in accuracy and applicability, (Zhou, Jiang, & Chen, 2022).

CONCLUSION

Molecular docking is a pivotal tool in advancing nanotechnology, offering invaluable insights into the interactions between molecules and nanomaterials. This chapter has provided a comprehensive overview of molecular docking principles and algorithms, highlighting its significance and the critical components in predicting binding affinities. Through detailed discussions on applications, we've seen how molecular docking can revolutionize drug delivery systems by optimizing nanoparticle-drug interactions and enhancing targeted drug delivery. Additionally, its role in nanomaterial design is underscored by the ability to tailor nanomaterials for specific applications and understand protein-nanomaterial interactions. Exploring bioconjugation strategies reveals how molecular docking facilitates protein-ligand conjugation, which is essential for developing nanoscale bioconjugates with diverse applications. Case studies illustrate real-world applications and outcomes, offering practical insights and lessons learned from integrating molecular docking in nanotechnology projects. Despite its promising potential, molecular docking faces challenges such as computational complexity, accuracy of predictions, and experimental validation. These obstacles necessitate ongoing advancements in computational methods, force field development, and interdisciplinary collaboration.

REFERENCES

Abdelsattar, A. S., Dawoud, A., & Helal, M. A. (2021). Interaction of nanoparticles with biological macromolecules: A review of molecular docking studies. *Nanotoxicology, 15*(1), 66–95. DOI: 10.1080/17435390.2020.1842537 PMID: 33283572

Abo-zeid, Y., Ismail, N. S. M., McLean, G. R., & Hamdy, N. M. (2020). A molecular docking study repurposes FDA-approved iron oxide nanoparticles to treat and control COVID-19 infection. *European Journal of Pharmaceutical Sciences, 153*, 105465. DOI: 10.1016/j.ejps.2020.105465 PMID: 32668312

Agarwal, S., & Mehrotra, R. (2016). An overview of molecular docking. *JSM Chemistry, 4*(2), 1024–1028.

Agu, P. C., Afiukwa, C. A., Orji, O. U., Ezeh, E. M., Ofoke, I. H., Ogbu, C. O., Ugwuja, E. I., & Aja, P. M. (2023). Molecular docking as a tool for the discovery of molecular targets of nutraceuticals in diseases management. *Scientific Reports, 13*(1), 13398. DOI: 10.1038/s41598-023-40160-2 PMID: 37592012

Akbarzadeh, I., Saremi Poor, A., Khodarahmi, M., Abdihaji, M., Moammeri, A., Jafari, S., Salehi Moghaddam, Z., Seif, M., Moghtaderi, M., Lalami, Z. A., Heydari, M., Adelnia, H., & Farasati Far, B. (2022). Gingerol/letrozole-loaded mesoporous silica nanoparticles for breast cancer therapy: In-silico and in-vitro studies. *Microporous and Mesoporous Materials, 337*, 111919. DOI: 10.1016/j.micromeso.2022.111919

Anselmo, A. C., & Mitragotri, S. (2019). Nanoparticles in the clinic: An update. *Bioengineering & Translational Medicine, 4*(3), e10143. DOI: 10.1002/btm2.10143 PMID: 31572799

Ayesha, I., Imran, M., Haider, A., Shahzadi, I., Moeen, S., Ul-Hamid, A., Nabgan, W., Shahzadi, A., Alshahrani, T., & Ikram, M. (2023). Polyvinylpyrrolidone and chitosan-coated magnetite (Fe3O4) nanoparticles for catalytic and antimicrobial activity with molecular docking analysis. *Journal of Environmental Chemical Engineering, 11*(3), 110088. DOI: 10.1016/j.jece.2023.110088

Bansal, P., Kumar, R., Singh, J., & Dhanda, S. (2021). In silico molecular docking of SARS-CoV-2 surface proteins with microbial non-ribosomal peptides: Identification of potential drugs. *Journal of Proteins and Proteomics, 12*(3), 177–184. DOI: 10.1007/s42485-021-00072-z PMID: 34456530

Bertilsson, L., Dahl, M.-L., Dalén, P., & Al-Shurbaji, A. (2002). Molecular genetics of CYP2D6: Clinical relevance with focus on psychotropic drugs. *British Journal of Clinical Pharmacology, 53*(2), 111–122. DOI: 10.1046/j.0306-5251.2001.01548.x PMID: 11851634

Chaudhary, M., & Tyagi, K. (2024). A review on molecular docking and its application. *International Journal of Advanced Research, 12*(3), 1141–1153. DOI: 10.21474/IJAR01/18505

Chen, R., & Riviere, J. E. (2017). Biological and environmental surface interactions of nanomaterials: Characterization, modeling, and prediction. *Wiley Interdisciplinary Reviews. Nanomedicine and Nanobiotechnology, 9*(3), e1440. DOI: 10.1002/wnan.1440 PMID: 27863136

Chibber, S., & Ahmad, I. (2016). Molecular docking, a tool to determine the interaction of CuO and TiO2 nanoparticles with human serum albumin. *Biochemistry and Biophysics Reports, 6*, 63–67. DOI: 10.1016/j.bbrep.2016.03.004 PMID: 28955863

Cui, Q., Hernandez, R., Mason, S. E., Frauenheim, T., Pedersen, J. A., & Geiger, F. (2016). Sustainable nanotechnology: Opportunities and challenges for theoretical/computational studies. *The Journal of Physical Chemistry B, 120*(31), 7297–7306. DOI: 10.1021/acs.jpcb.6b03976 PMID: 27388532

Dar, A., & Mir, S. (2017). Molecular docking: Approaches, types, applications, and basic challenges. *Journal of Analytical & Bioanalytical Techniques, 8*(2), 356. DOI: 10.4172/2155-9872.1000356

Deng, Z., Chen, G., Shi, Y., Lin, Y., Ou, J., Zhu, H., Wu, J., Li, G., & Lv, L. (2022). Curcumin and its nanoformulations: Defining triple-negative breast cancer targets through network pharmacology, molecular docking, and experimental verification. *Frontiers in Pharmacology, 13*, 1001046. DOI: 10.3389/fphar.2022.920514 PMID: 36003508

Dolai, J., Mandal, K., & Jana, N. R. (2021). Nanoparticle size effects in biomedical applications. *ACS Applied Nano Materials, 4*(7), 6471–6496. DOI: 10.1021/acsanm.1c00987

Edris, A., Abdelrahman, M., Osman, W., Sherif, A. E., Ashour, A., Garelnabi, E. A. E., Ibrahim, S. R. M., Bafail, R., Samman, W. A., Ghazawi, K. F., Mohamed, G. A., & Alzain, A. A. (2023). Design of novel letrozole analogues targeting aromatase for breast cancer: Molecular docking, molecular dynamics, and theoretical studies on gold nanoparticles. *Metabolites, 13*(5), 583. DOI: 10.3390/metabo13050583 PMID: 37233624

El-Sayed, A. F., Aboulthana, W. M., Sherief, M. A., El-Bassyouni, G. T., & Mousa, S. M. (2024). Synthesis, structural, molecular docking, and in vitro biological activities of Cu-doped ZnO nanomaterials. *Scientific Reports, 14*(1), 9027. DOI: 10.1038/s41598-024-59088-2 PMID: 38641640

Ghasemi, J. B., Abdolmaleki, A., & Shiri, F. (2017). Molecular docking challenges and limitations. In *Pharmaceutical sciences: Breakthroughs in research and practice* (pp. 770–794). IGI Global. DOI: 10.4018/978-1-5225-1762-7.ch030

Jakhar, R., Dangi, M., Khichi, A., & Chhillar, A. K. (2020). Relevance of molecular docking studies in drug designing. *Current Bioinformatics*, *15*(4), 270–278. DOI: 10.2174/1574893615666191219094216

Kamal, I. M., & Chakrabarti, S. (2023). MetaDOCK: A combinatorial molecular docking approach. *ACS Omega*, *8*(6), 5850–5860. DOI: 10.1021/acsomega.2c07619 PMID: 36816658

Kaur, T., Madgulkar, A., Bhalekar, M., & Asgaonkar, K. (2019). Molecular docking in formulation and development. *Current Drug Discovery Technologies*, *16*(1), 30–39. DOI: 10.2174/1570163815666180219112421 PMID: 29468973

Kaushik, A. C., Bharadwaj, S., Kumar, S., & Wei, D. Q. (2018). Nanoparticle mediated inhibition of Parkinson's disease using computational biology approach. *Scientific Reports*, *8*(1), 9169. DOI: 10.1038/s41598-018-27580-1 PMID: 29907754

Khemnar, M., Galave, V., Kulkarni, V., Chandrakant, M., & Otari, K. (2021). A review on molecular docking. *International Research Journal of Pure and Applied Chemistry*, *22*(3), 60–68. DOI: 10.9734/irjpac/2021/v22i330396

Kunjiappan, S., Govindaraj, S., Parasuraman, P., Sankaranarayanan, M., Arunachalam, S., Palanisamy, P., Mohan, U. P., Babkiewicz, E., Maszczyk, P., Vellaisamy, S., & Panneerselvam, T. (2020). Design, in silico modeling, and functionality theory of folate-receptor-targeted myricetin-loaded bovine serum albumin nanoparticle formulation for cancer treatment. *Nanotechnology*, *31*(15), 155102. DOI: 10.1088/1361-6528/ab5c56 PMID: 31775133

Liang, Y., Mao, G., Dai, J., & Ma, Y. (2023). Biofunctionalized semiconductor quantum dots for virus detection. *Journal of Semiconductors*, *44*(2), 023101. DOI: 10.1088/1674-4926/44/2/023101

Mirza, S., Ahmad, M. S., & Shah, M. I. A. (2020). Magnetic nanoparticles: Drug delivery and bioimaging applications. In Shah, M. R., Imran, M., & Ullah, S. (Eds.), *Metal nanoparticles for drug delivery and diagnostic applications* (pp. 189–213). Elsevier. DOI: 10.1016/B978-0-12-816960-5.00011-2

Mohammadjani, N., Karimi, S., Moetasam Zorab, M., Ashengroph, M., & Alavi, M. (2024). Comparative molecular docking and toxicity between carbon-capped metal oxide nanoparticles and standard drugs in cancer and bacterial infections. *BioImpacts*, *14*(3), 27778. DOI: 10.34172/bi.2023.27778 PMID: 38505671

Muhammed, M. T., & Aki-Yalcin, E. (2024). Molecular docking: Principles, advances, and its applications in drug discovery. *Letters in Drug Design & Discovery*, *21*(3), 480–495. DOI: 10.2174/1570180819666220922103109

Nowack, B., Baalousha, M., Bornhöft, N., Chaudhry, Q., Cornelis, G., Cotterill, J., Gondikas, A., Hassellöv, M., Lead, J., Mitrano, D. M., von der Kammer, F., & Wontner-Smith, T. (2015). Progress towards the validation of modelled environmental concentrations of engineered nanomaterials by analytical measurements. *Environmental Science. Nano*, 2(5), 421–428. DOI: 10.1039/C5EN00100E

Ozboyaci, M., Kokh, D. B., Corni, S., & Wade, R. C. (2016). Modeling and simulation of protein–surface interactions: Achievements and challenges. *Quarterly Reviews of Biophysics*, *49*, e4. DOI: 10.1017/S0033583515000256 PMID: 26821792

Pinzi, L., & Rastelli, G. (2019). Molecular docking: Shifting paradigms in drug discovery. *International Journal of Molecular Sciences*, *20*(18), 4331. DOI: 10.3390/ijms20184331 PMID: 31487867

Rajput, A., Shevalkar, G., Pardeshi, K., & Pingale, P. (2023). Computational nanoscience and technology. *OpenNano*, *12*, 100147. DOI: 10.1016/j.onano.2023.100147

Saikia, S., & Bordoloi, M. (2019). Molecular docking: Challenges, advances and its use in drug discovery perspective. *Current Drug Targets*, *20*(5), 501–521. DOI: 10.2174/1389450119666181022153016 PMID: 30360733

Volgin, I. V., Larin, S. V., Lyulin, A. V., & Lyulin, S. V. (2018). Coarse-grained molecular-dynamics simulations of nanoparticle diffusion in polymer nanocomposites. *Polymer*, *145*, 80–87. DOI: 10.1016/j.polymer.2018.04.058

Yahya, R., Al-Rajhi, A. M. H., Alzaid, S. Z., Al Abboud, M. A., Almuhayawi, M. S., Al Jaouni, S. K., Selim, S., Ismail, K. S., & Abdelghany, T. M. (2022). Molecular docking and efficacy of Aloe vera gel based on chitosan nanoparticles against Helicobacter pylori and its antioxidant and anti-inflammatory activities. *Polymers*, *14*(15), 2994. DOI: 10.3390/polym14152994 PMID: 35893958

Yan, Z., Zheng, X., Wang, E., & Wang, J. (2013). Thermodynamic and kinetic specificities of ligand binding. *Chemical Science (Cambridge)*, *4*(6), 2387–2395. DOI: 10.1039/c3sc50478f

Yousfan, A., Al Rahwanji, M. J., Hanano, A., & Al-Obaidi, H. (2024). A comprehensive study on nanoparticle drug delivery to the brain: Application of machine learning techniques. *Molecular Pharmaceutics*, *21*(2), 333–345. DOI: 10.1021/acs.molpharmaceut.3c00880 PMID: 38060692

Yousuf, M., Khan, H. M., Rasool, F., Khan, K. U., Usman, F., Ghalloo, B. A., Umair, M., Babalghith, A. O., Kamran, M., Aadil, R. M., Al Jaouni, S. K., Selim, S., Korma, S. A., & Conte-Junior, C. A. (2022). Chemical profiling, formulation development, in vitro evaluation, and molecular docking of Piper nigrum seed extract loaded with emulgel for anti-ageing. *Molecules (Basel, Switzerland)*, *27*(18), 5990. DOI: 10.3390/molecules27185990 PMID: 36144719

Zhou, Y., Jiang, Y., & Chen, S. J. (2022). RNA–ligand molecular docking: Advances and challenges. *Wiley Interdisciplinary Reviews. Computational Molecular Science*, *12*(3), e1571. DOI: 10.1002/wcms.1571 PMID: 37293430

Chapter 2
Applications in Nanomedicine and Beyond

P. Selvakumar
https://orcid.org/0000-0002-3650-4548
Department of Science and Humanities, Nehru Institute of Technology, Coimbatore, India

S. Seenivasan
https://orcid.org/0000-0002-2155-9781
Rathinam Technical Campus, India

Sumanta Bhattacharya
https://orcid.org/0000-0003-2563-2787
Maulana Abul Kalam Azad University of Technology, India

D. Sudha
https://orcid.org/0000-0003-0354-9261
KPR Institute of Engineering and Technology, India

A. Akila
Avinashilingam Institute for Home Science and Higher Education for Women, India

Manjunath T. C.
https://orcid.org/0000-0003-2545-9160
Rajarajeswari College of Engineering, India

ABSTRACT

Materials that are biological, nonbiological, biomimetic, or hybrid are employed in a variety of nanomedicine research applications, such as high-throughput screening platforms, wearable devices, antibacterial, antiviral, and diagnostic instruments. The systematic, biomimetic, and rational process of creating the optimal therapeutic combinations has included NPs-based drug delivery. Numerous noteworthy developments in cellular, preclinical, and clinical nanomedicine research have been made, encompassing both fundamental and applied aspects. It is anticipated that in the years to come this excessive emphasis will be rectified, with research efforts shifting to other medical problems including antibiotic resistance and artificial organs. Because of the diversity of the design of the nanomaterial platform, nano-

DOI: 10.4018/979-8-3693-2774-6.ch002

medicine is well-positioned to be advantageous in many domains, whether through multimodal beneficial approach or exceedingly specialized multifunctional design for pertinent natural application.

INTRODUCTION

The engineering, science and exercise of diagnosing, treating, curing, tracking, anticipating, and avoiding illnesses is known as medicine. But there is a wider spectrum of research and development in this sector. In order to work in all the ways listed below, materials and technologies with nanoscale length scales must be developed and applied. This is how nanomedicine differs from previous forms of medicine:

The interdisciplinary field of nanomedicine involves the interaction of nanoscience and nanotechnology with the biological sciences (Altınoğlu et al., 2008). We anticipate that nanomedicine will eventually touch on every facet of medicine due to its vast scope. Furthermore, nanomedicine has the same potential to enter clinics as traditional medicine as long as all prerequisites related to translation-such as safety, ethics, and regulations-are met. The goal of nanomedicine is to improve patient quality of life by facilitating the creation of more precise, effective, and personalized tools, medications, and other application for early diagnosis and behaviour of a variety of diseases. This Nano Focus aims to give a summary of current development in the field of nanomedicine, rather than to define it. Materials and technologies included are the focus of this article (Baker & Baker, 2010).

One of the main ingredients of nanomedicine are nanoparticles (NPs), of which there are many different kinds at the moment. Numerous nanomaterials can mimic a few of the roles played by spherical biological macromolecules. Numerous researchers have observed that even minor modifications to NPs' size and form can have a big impact on their characteristics. Since one must be aware of and comprehend what is being injected into the body, thorough description of NP samples used in medicinal applications is also essential. Following synthesis, a sample of NPs may exhibit heterogeneity with discrete subpopulations. The usual method is microscopic imaging, but this approach might not be enough since it can only image a minute number of NPs, which might or might not be representative of the entire sample (Baughman et al., 2002). Therefore, information regarding surface functionalization, composition, and other factors that determine properties may not be sufficiently provided by microscopic imaging (Yu et al., 2005). Additional methods that are beginning to be included in the characterization plan of nanoparticles before they are used on humans. But before being used in people, NPs should not meet any set standards for characterisation, and here is where nanomedicine applications need to concentrate.

The primary cause is that multiple NPs will distribute differently in a heterogeneous sample and may potentially show toxicity or undesirable consequences due to the size- and surface-dependent nature of NPs' biodistribution and interactions with proteins (Torchilin, 2007). Apart from characterisation, novel and enhanced techniques for NP separation and purification must be devised to generate ideal samples for nanomedical uses and to investigate NP behavior within the body (a crucial aspect in creating medically effective NP formulations). Notwithstanding the requirement to harmonize characterisation techniques, NPs are anticipated to enhance illness identification and diagnosis. Initially, following insertion into the body, Secondly, nanoparticles are essential parts of numerous high-throughput diagnostic devices that are capable of analyzing extracted materials (blood, tissue, etc.) from outside the body in order to quickly identify molecular changes and biological markers (Chow & Ho, 2013). Simultaneous analysis of many biomarkers may increase the accuracy of diagnosis. Furthermore, theranostic or multifunctional NPs are being developed that have the ability to treat, diagnose, and even track the effectiveness of treatments all at once.

Additionally, nanoparticles (NPs) are being produced for medical applications; they are employed as bioactive materials, drug delivery vehicles, or key components of implants. When it comes to distribution, NP-based carrier systems are special since they can get through biological barriers (Abhilasha et al., 2022). As a result, NPs can enter tumors through their specific, leaky and are then trapped there by the tumor microenvironment's inadequate lymphatic drainage. It's critical to determine the safety and effectiveness of NPs in biological systems in addition to applying nanomedicine for illness diagnosis and treatment. What happens to the carrier particle once the NP has entered the body and performed as intended, either for the delivery of the medicine or for imaging the tissue. The particles may be eliminated through hepatobiliary or renal clearance (Colombo et al., 2012). The NPs' long-term destiny is unclear if they are not cleared. Because they can pass past the kidney's filtration slits, these particles may either disintegrate and be removed by the kidneys, or they may accumulate in other organs and interact with cells that are not their intended targets. Understanding nanobiokinetics (nanopharmacokinetics and pharmacodynamics) is important because it may be related to the distinctive and intriguing toxicological responses of NPs. The in vivo fate of NPs may be a dynamic process (Nurunnabi et al., 2014). The evaluation of colloidal materials and technology for in vivo applications is not the exclusive domain of nanomedicine. The concept of the "magic particle bullet" is not the extent of advances in nanomedicine. While some of these ideas have already been put into clinical practice, many are still in the early phases of development.

APPLICATIONS OF MATERIALS

In Vivo Diagnosis ("Smart Imaging")

Using nanoparticles as contrast agents for functional and anatomical imaging is a major area of interest in nanomedicine. Physicians can see inside body structures, distinguish healthy from diseased tissues, and prescribe the right course of action by using nanomaterials as contrast agents (Melamed et al., 2016). It is possible to manufacture nanoparticles with various contrast qualities. These include computed tomography (CT), magnetic resonance imaging (MRI), fluorescence imaging, photoacoustic imaging, and imaging of radioactivity using techniques like positron emission tomography (SPECT) and positron emission tomography (PET). Material development is essential for all these approaches because the NPs act as contrast agents, making biological tissues visible. NPs can be designed with the ability to localize in particular tissues and possibly generate strong contrast for this purpose.

Computed Tomography

The workhorse of clinical diagnostics is computed tomography imaging because of its ease of use, low infrastructure requirements, quick picture generation, and affordable cost of a standard examination. In the United States, around 250 out of every 1000 individuals had CT imaging performed in the recent past (Selvakumar et al., 2024). Because of their higher relative electron density than soft tissue, bones therefore produce more contrast than the latter. Contrast agents that contain elements with a high electron density, such iodine and barium, are commonly used to see blood arteries in soft tissues, malignancies, and the gastrointestinal system (Matsumura & Maeda, 1986). Because of the high signal-to-noise ratio made possible by the massive photon flux, CT performs better in terms of spatial resolution than other 3D imaging methods. However, the weak tissue-X-ray beam interaction leads to comparatively low CT sensitivity, which may account for the near-complete lack of contrast methods designed to target specific areas of the body.

Magnetic Resonance Imaging

The physiological information provided by MRI-responsive contrast chemicals is noteworthy since it enhances standard anatomical imaging. Unmatched soft tissue contrast, limitless depth of penetration, and no requirement for ionizing radiation are features of magnetic resonance imaging (MRI). MRI requires a lot of contrast chemicals and takes longer to get results than nuclear and optical methods because to its lower sensitivity. The sensitivity of the target nuclei has increased thanks to the introduction of hyperpolarized MRI, which allows for real-time metabolic profile using steady isotope precursors and quantitative in vivo imaging (Lobatto et al., 2011). Hyperpolarized MRI is defined as nuclear magnetic moment polarization that is well beyond thermal equilibrium conditions. The use of modern contrast agents, such as SPIONs or paramagnetic agents, is crucial for improving dissimilarity in T1 or T2 images, which increase MRI sympathy and accuracy while imaging live people. When using either kind of agent on humans or in veterinary medicine, it is important to consider their potential for hazardous effects.

Figure 1. Multifunctional asymmetrical polymer vesicles for efficient cancer-targeted medication administration and ultrasensitive T1 MRI

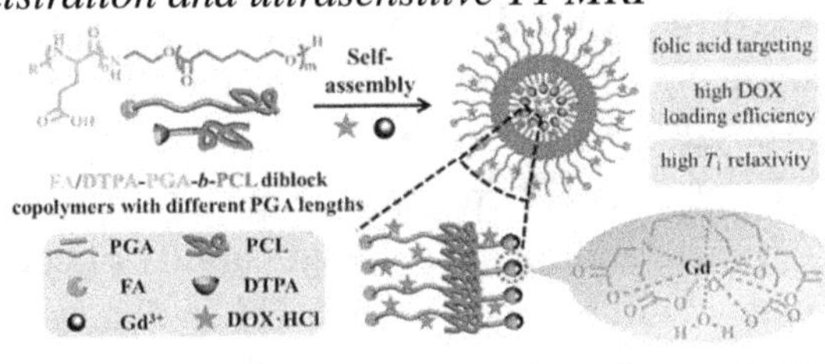

Abnormal redox reactions have been linked to a number of illnesses, including liver damage and HIV. Redox reactions are essential to biological functions. Redox processes in the tumor microenvironment are used by a third class of "smart" NP probes for MRI to cleave atoms from the NP structure. Redox activatable NPs were developed by Kikuchi and colleagues to visualize decreasing conditions. Disulfide linkers anchored Gd^{3+} complexes to the surface of NPs containing ^{19}F. The Gd^{3+} complexes are separated from the NP surface when the disulfide connections that allow Gd^{3+} to bind are broken. Subsequently, the encapsulated ^{19}F compounds exhibit an increase in T_2 NMR/MRI signal, signifying the NPs' existence in a redox active environment.

Figure 2. Design of redox activatable NPs

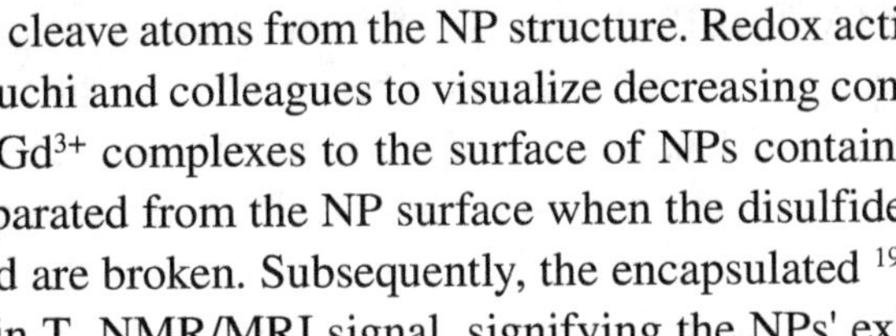

Imaging Radiolabels

For this reason, a combination of amphiphiles including ligands for targeting, chelators containing radiolabels. Micelle encapsulation was introduced for this purpose. It is possible to successfully use the same procedure for several NP types, including core/shell UCNPs, plasmonic NPs, and SPIONs. Since the use of radiolabels yields excellent signals, an adequate image quality can be obtained with a low dose to facilitate diagnostic interpretation. The ability to read out radiolabels with various γ-emission energies in simultaneously via multiplexing is an additional benefit (Lee et al., 2007). In recent times, this has been employed to independently document the biodistribution of many NP compounds. The NPs may disintegrate after being administered in vivo if different portions of the NPs are labeled with radiolabels of varying energies. This would cause the organic surface capping to separate from the purposeful particle core. Future research of this kind will make it possible to image the various NP components and determine what will happen to them individually. Understanding the biodegradability of each component of an NP and how to eliminate it will be aided by this. This will have a significant impact on how NPs are designed for in vivo delivery applications.

Fluorescence Imaging

Fluorophores that are suitable for particular uses are widely available. Aggregation-caused quenching (ACQ) affects a large number of conventional organic fluorophores, which restricts design strategies where particular interactions encourage fluorophore localization. Due of this, researchers are now compelled to employ diluted fluorophores, frequently at the nanomole level. The degree of contrast that can be achieved is limited because of how easily photobleached they are due to their tiny amount of fluorophore (Mulvaney P. et.al. 2016). Aggregation-induced emission (AIE) is the process where higher fluorescent signal is produced by increased aggregation, in contrast to ACQ. For bioimaging applications, this makes AIE fluorogens (AIEgens) excellent choices. An increased fluorescence signal is produced when AIEgens gather in areas of interest and form nanoaggregates, or AIE dots. High-quality bioimaging across a broad temporal window is made possible by the AIE dots' resistance to photobleaching. Stem-cell therapy, drug delivery and release, and tumor-metastasis monitoring are among the long-term theranostic processes that they are appropriate for tracking. With the exception of cytotoxicity, organic AIE dots have all the benefits of inorganic QDs without any of their drawbacks. To enable active targeting, functional groups that have a specific binding affinity for certain receptors can also be added to AIE molecules and nanoaggregates.

Nevertheless, because semiconductor QDs produce metal ions that can be harmful to cells, many researchers construct QDs with this in mind when doing long-term imaging investigations. For example, to prevent potentially hazardous semiconductor core materials from dissolving, ZnS polymer shells or SiO2 coatings can be applied to CdSe QDs. The specifics of the possible toxicity's mechanism are still up for debate. For the transportation of intracellular sensing moieties, organic polymer NPs show a number of benefits. There are numerous harmless polymer matrices that can be customized for the best cell penetration by adjusting their hydrophilic/hydrophobic characteristics. Moreover, the diverse techniques accessible for modifying polymer nanoparticles facilitate the functionalization of fluorophores, cell-targeting ligands, and recognition ligands, while also reducing nonspecific adsorption. In order to build polymer-based NPs for intracellular sensing applications, two general techniques are shown in Figure 3A fluorophore is incorporated into the NP matrix in one type of sensing NPs, panel I, where it functions as an optical transducer and binding ligand.

Figure 3. Polymer NPs with chemical modifications for intracellular sensing: (I) NPs act as ligands that are responsive to light. (II)

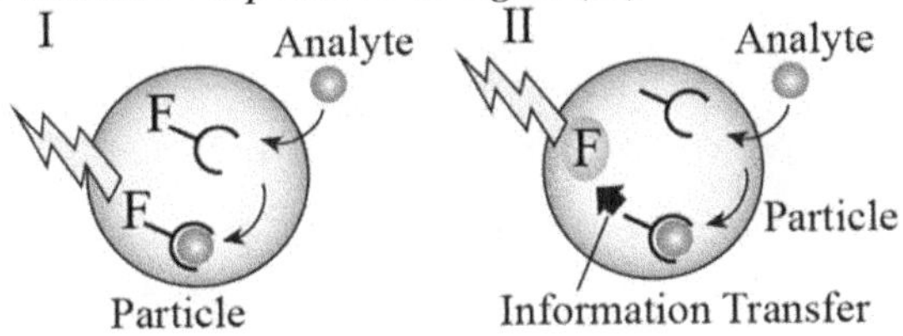

In Vivo Treatment ("Particle-Based Delivery")

Although other diseases have been targeted by nanoparticle design, cancer has received the greatest attention. Several NP formulations for cancer handling have already been licensed by authoritarian bodies and have reach hospitals, demonstrating significantly less side effects than bare medicines (Langer & Weissleder, 2015). The literature is full of big sets of in vivo data collected from nature. But typically, they have little to no improvement in their therapeutic efficacies. In order to create systems with high therapeutic efficacies, efforts are still being made. The secret to these advancements is to fine-tune the nanodelivery system's architecture by carefully regulating the materials' attributes and functions.

Nanoparticles as Anticancer Drugs

The CAPIR cascade, which comprises five crucial steps for cancer treatment delivery to a solid tumor. Therefore, each step's effectiveness determines a nanomedicine's overall therapeutic efficacy (Figure 4). A high beneficial index is desirable for a nanomedicine that effectively completes the entire CAPIR cascade. In accordance with this, a nanomedicine of this kind need to possess 2R2SP qualities, which stand for "tumor penetration (P)", "surface stealthy vs sticky (2S)," and "drug retention vs release (2R)." The two letters (2R) stand for the necessary characteristics of the loaded drug in the nanomedicine (Madgule et al., 2023). In addition to its 2R2S characteristics, the nanomedicine ought to have the capacity to infiltrate tumor tissues and reach distant tumor cells from blood arteries (Kowalczyk et al., 2011). In order to effectively complete the CAPIR cascade, the current focus on creating nanomedicines with a high beneficial index is on modifying the fundamental physiochemical property of NPs, or their "size, surface properties, and stability (3S nanoproperties)".

Figure 4. (A) CAPIR cascade in five steps for targeted delivery of cancer drugs

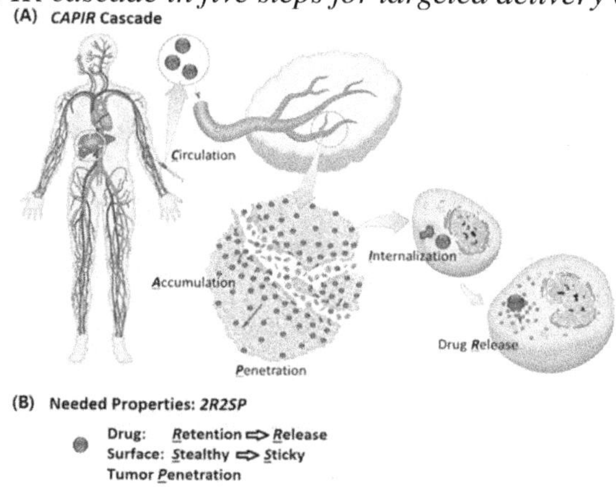

Nanoparticulate Delivery Vehicles

Certain cancer treatments already use liposome-based methods to encapsulate medications. Nevertheless, liposomes have a few noteworthy disadvantages: They are produced using organic solvents, have a limited ability to encapsulate lipophilic medicines, are frequently leaking, and are unstable in aqueous solutions and biological fluids (Wolf, 2012). Among the most promising carbon nanomaterials for theranostic, imaging, and drug delivery are nanodiamonds. The optical, mechanical, and thermal capabilities of ~5 nm nanodiamonds are paired with a broad accessible surface and customizable surface chemistry. The intentional surface modification of NDs facilitates improved drug adsorption and chemical binding for either triggered or sustained release of the medication (Kotov N. A. 2010). The surface chemistry can regulate the strength of adsorption as well as the amount of drug adsorbed. Anticancer chemotherapeutics can be made more potent by attaching medicines to the surface of NDs and obtaining regulated release. In a similar vein, the creation of NDs for the targeted delivery of antibiotics and other medications may help to decrease systemic toxicity and restore drug efficiency (Singh & Lillard, 2009). Nucleic acids, insulin, and anticancer chemotherapy are now being studied for their potential to be delivered and released over time using NDs. It has also been shown that nanodiamonds can covalently link to proteins, occasionally increasing their activity.

Delivery of Macromolecular Biopharmaceuticals across Biological Barriers

Infectious, neurological, and cardiovascular disorders are among the other conditions for which nanomedicines are being explored. Although most anticancer medications are tiny organic compounds, macromolecular biopharmaceuticals—such as polypeptides and polynucleotides—are becoming far more prevalent and significant (Selvakumar et al., 2024). Large size and complexity, along with poor metabolic stability and rapid elimination, pose serious obstacles to their utilization as safe and effective medications. Additionally, this particular class of chemicals is infamous for having poor permeability through biological barriers, such epithelial tissues or cell membranes (Kotov, 2010). However, their medical application presents the potential to treat or even prevent diseases, either directly addressing hormonal imbalances (like insulin in diabetes) or by reprogramming, suppressing, or repairing genetically aberrant cells. Thus, another significant potential use of nanomedicines is the delivery of macromolecular biopharmaceuticals, preferably using noninvasive ("needle free") methods. This methodology capitalizes on the ability of nanoparticles (NPs) to traverse epithelial tissue barriers, such as those found in the skin, lungs, or gastrointestinal (GI) tract, and simultaneously maintain and protect the therapeutic cargo.

Antibacterial Coatings

The development of biofilms and the spread of bacteria resistant to antibiotics through contaminated surfaces and indirect transmission are growing threats to worldwide public health (Ramakrishnan et al., 2022). Surfaces made of steel and aluminum, which are frequently found in medical settings as knobs, buttons, frames, taps, etc., have very little or no antibacterial action. Realizing persistent and efficient antimicrobial surfaces by nanostructured coatings is a potential strategy for reducing indirect transmission. Analogously, infections brought on by medical equipment have grown to be a serious problem (Kim et al., 2010). The first 24 hours after an implant are thought to be critical for the adherence of bacteria to the device surfaces, hence designing the best antiadhesion of microbes at this time is crucial to preventing implant-associated illnesses. Designing dynamic, multifunctional antibacterial surfaces with multilayer coating by LbL self-assembly is a sound technique.

Nanoparticle-Labeled Stem Cells for Targeted Imaging and Therapy

The use of nanotechnology in stem-cell research and development is known as "stem cell nanotechnology," a recently-emerging multidisciplinary area. Numerous clinical trials involving autologous or allogeneic MSCs are presently in progress, with the aim of guaranteeing the safety and effectiveness of these advanced therapeutic medications. Even though the field of stem cells has made and continues to make great strides, a number of challenges need to be resolved before stem cells can be used therapeutically (Guo, 2010). These actions involve the creation of cutting-edge tools for tracking and directing transplanted stem cells in addition to the comprehension and manipulation of the roles of signals from the microenvironment.

CONCLUSION

Nanomedicine is constantly developing novel nanoparticles (NPs) for applications, especially in drug delivery, imaging, diagnostics, antibiotic resistance, and cancer treatments. Recent study suggests that a single nanoplatform functionalized with multiple beneficial agents or multiple nanomaterials (NDs and proteins, for example) may be utilized to scan and treat tumors more efficiently. NPs-based medicine delivery has been considered in the methodical, biomimetic, and logical formulation of optimal therapy combinations. A great deal of significant progress has been made in both fundamental and translational areas of nanomedicine. However, a large portion of these developments have been made in the identification and management of cancer. Future research foci are anticipated to remedy this disproportionate attention by addressing other medical concerns including antibiotic resistance and artificial organs. Scientific obstacles have also surfaced, mostly because of the paucity of information regarding the behavior of nanomaterials inside living bodies, despite the fact that nanomedicine has aroused high expectations for many medical problems.

Nevertheless, because of the fundamental research conducted on these problems, we are now getting closer to finding solutions and achieving “real” nanomedicine-based medical treatments.

REFERENCES

Abhilasha, A., Sreenivasulu, A., & Manimozhi, T. (2022). The Model of Smart Sensing Device For Sensitive Nanoclusters Modification in Sensing Properties. 2022 2nd International Conference on Advance Computing and Innovative Technologies in Engineering (ICACITE), 1043-1047, DOI: 10.1109/ICACITE53722.2022.9823846

Altınoğlu, E. I., Russin, T. J., Kaiser, J. M., Barth, B. M., Eklund, B. C., Kester, M., & Adair, J. H. (2008). Near-Infrared Emitting Fluorophore-Doped Calcium Phosphate Nanoparticles for In Vivo Imaging of Human Breast Cancer. *ACS Nano*, *2*(10), 2075–2084. DOI: 10.1021/nn800448r PMID: 19206454

Baker, S. N., & Baker, G. A. (2010). Luminescent Carbon Nanodots: Emergent Nanolights. *Angewandte Chemie International Edition*, *49*(38), 6726–6744. DOI: 10.1002/anie.200906623 PMID: 20687055

Baughman, R. H., Zakhidov, A. A., & de Heer, W. A. (2002). Carbon Nanotubes—The Route toward. *Applied Sciences (Basel, Switzerland)*, *297*(5582), 787–792. DOI: 10.1126/science.1060928 PMID: 12161643

Chow, E. K., & Ho, D. (2013). Cancer Nanomedicine: From Drug Delivery to Imaging. *Science Translational Medicine*, *5*(216), 216rv4. Advance online publication. DOI: 10.1126/scitranslmed.3005872 PMID: 24353161

Colombo, M., Carregal-Romero, S., Casula, M. F., Gutiérrez, L., Morales, M. P., Böhm, I. B., Heverhagen, J. T., Prosperi, D., & Parak, W. J. (2012). Biological Applications of Magnetic Nanoparticles. *Chemical Society Reviews*, *41*(11), 4306–4334. DOI: 10.1039/c2cs15337h PMID: 22481569

Guo, P. (2010). The Emerging Field of RNA Nanotechnology. *Nature Nanotechnology*, *5*(12), 833–842. DOI: 10.1038/nnano.2010.231 PMID: 21102465

Kim, B. Y. S., Rutka, J. T., & Chan, W. C. W. (2010). Nanomedicine. *The New England Journal of Medicine*, *363*(25), 2434–2443. DOI: 10.1056/NEJMra0912273 PMID: 21158659

Kotov, N. A. (2010). Inorganic Nanoparticles as Protein Mimics. *Science*, *330*(6001), 188–189. DOI: 10.1126/science.1190094 PMID: 20929766

Kowalczyk, B., Lagzi, I., & Grzybowski, B. A. (2011). Nanoseparations: Strategies for Size and/or Shape-Selective Purification of Nanoparticles. *Current Opinion in Colloid & Interface Science*, *16*(2), 135–148. DOI: 10.1016/j.cocis.2011.01.004

Langer, R., & Weissleder, R. (2015). Nanotechnology. *Journal of the American Medical Association, 313*(2), 135–136. DOI: 10.1001/jama.2014.16315 PMID: 25585320

Lee, Y., Fukushima, S., Bae, Y., Hiki, S., Ishii, T., & Kataoka, K. A. (2007). Protein Nanocarrier from Charge-Conversion Polymer in Response to Endosomal pH. *Journal of the American Chemical Society, 129*(17), 5362–5363. DOI: 10.1021/ja071090b PMID: 17408272

Lobatto, M. E., Fuster, V., Fayad, Z. A., & Mulder, W. J. M. (2011). Perspectives and Opportunities for Nanomedicine in the Management of Atherosclerosis. *Nature Reviews. Cancer, 10*(12), 963–963. DOI: 10.1038/nrd3614 PMID: 22015921

Mahadev Madgule, N. (2023). Mechanical properties and microstructure of activated TIG welded similar joints of Inconel alloys by desirability approaches. *Materials Today: Proceedings, 77*, 528–533. DOI: 10.1016/j.matpr.2022.12.250

Matsumura, Y., & Maeda, H. (1986). A New Concept for Macromolecular Therapeutics in Cancer Chemotherapy: Mechanism of Tumoritropic Accumulation of Proteins and the Antitumor Agent Smancs. *Cancer Research, 46*, 6387–6392. PMID: 2946403

Melamed, J. R., Riley, R. S., Valcourt, D. M., & Day, E. S. (2016). Using Gold Nanoparticles To DIsrupt the Tumor Microenvironment: An Emerging Therapeutic Strategy. *ACS Nano, 10*(12), 10631–10635. DOI: 10.1021/acsnano.6b07673 PMID: 28024339

Mulvaney, P., Parak, W. J., Caruso, F., & Weiss, P. S. (2016). Standardizing Nanomaterials. *ACS Nano, 10*(11), 9763–9764. DOI: 10.1021/acsnano.6b07629 PMID: 27934085

Nurunnabi, M., Khatun, Z., Reeck, G. R., Lee, D. Y., & Lee, Y. (2014). Photoluminescent Graphene Nanoparticles for Cancer Phototherapy and Imaging. *ACS Applied Materials & Interfaces, 6*(15), 12413–12421. DOI: 10.1021/am504071z PMID: 25054687

Ramakrishnan, T., Mohan Gift, M. D., Chitradevi, S., Jegan, R., & Subha Hency Jose, P. (2022). Study of Numerous Resins Used in Polymer Matrix Composite Materials. Advances in Materials Science and Engineering.DOI: 10.1155/2022/1088926

Selvakumar, P. (2024). Prediction of Earthquakes, Volcanic Eruptions, Tornadoes, Wildfires, and Droughts Through Artificial Intelligence. In Utilizing AI and Machine Learning for Natural Disaster Management (pp. 133–145). IGI Global.DOI: 10.4018/979-8-3693-3362-4.ch008

Selvakumar, P. (2024). AI-Based Cameras to Detect a Leak of Chemicals or Gas in Real Time. In Industry Applications of Thrust Manufacturing: Convergence with Real-Time Data and AI (pp. 221–236). IGI Global. DOI: 10.4018/979-8-3693-4276-3.ch009

Singh, R., & Lillard, J. W.Jr. (2009). Nanoparticle-Based Targeted Drug Delivery. *Experimental and Molecular Pathology*, *86*(3), 215–223. DOI: 10.1016/j.yexmp.2008.12.004 PMID: 19186176

Torchilin, V. P. (2007). Micellar Nanocarriers: Pharmaceutical Perspectives. *Pharmaceutical Research*, *24*(1), 1–16. DOI: 10.1007/s11095-006-9132-0 PMID: 17109211

Wolf, L. K. (2012). Sweating the Small Stuff. *Chemical and Engineering News*, *90*(22), 48–50. DOI: 10.1021/cen-09022-scitech1

Yu, S. J., Kang, M. W., Chang, H. C., Chen, K. M., & Yu, Y. C. (2005). Bright Fluorescent Nanodiamonds: No Photobleaching and Low Cytotoxicity. *Journal of the American Chemical Society*, *127*(50), 17604–17605. DOI: 10.1021/ja0567081 PMID: 16351080

Chapter 3
Supramolecules as Efficient Tools for Modern Targeted Drug Delivery Approaches

Nikita Patel
https://orcid.org/0009-0005-1883-5158
Krishna School of Science, KPGU, India

Miral Vora
Krishna School of Science, KPGU, India

Asha Joshi
Krishna School of Science, KPGU, India

Bhupesh Bhatt
Sardar Patel University, India

ABSTRACT

The field of supramolecular chemistry includes dynamic covalent chemistry as well as the investigation of all forms of intermolecular non-covalent bond formation in engineered molecular systems. Reversible non-covalent interactions, a key component of supramolecular chemistry, include hydrogen-bonding, π-π stacking, hydrophobic, van der Waals, and electrostatic interactions. Advances in supramolecular chemistry in recent times have largely aided biomedical applications. For an understanding of how pharmaceuticals interact with binding sites in biological systems, supramolecular chemistry is crucial. Therapeutic compounds should be momentarily contained within the carrier vehicles and released at a predetermined location or time in order

DOI: 10.4018/979-8-3693-2774-6.ch003

to create effective drug delivery systems (DDS). This chapter summarizes the current trends and notable developments in supramolecular systems with specific functions, along with their many biomedical uses in drug delivery, gene transfection, protein delivery, molecular recognition, and biomimetic chemistry.

INTRODUCTION

Supramolecular chemistry has garnered a lot of interest lately for its potential applications in material design such as polymer chemistry and general chemistry. When a large number of constituents spontaneously self-assemble in a particular phase, a polymolecular assembly known as a supramolecular polymer is created. Depending on the phase's nature, these assemblies can take the form of a film, layer, membrane, vesicle, mesomorphic phase, crystal, etc., and have highly specific microscopic organization and macroscopic properties.(Krieg et al., 2016) The building blocks for a relatively novel method of creating intelligent drug carriers and other functional medicinal materials are these molecular characteristics. Although systemic therapy benefits from the use of drug carriers, regulated release of pharmaceuticals from localized depots has been a significant area of research in drug delivery, (Wang et al., 2018).

Once more, the fact that this early drug delivery system was clinically successful suggests that employing an engineering approach to enhance treatment outcomes is a promising approach. In the realm of medication administration, approaches that can achieve any or all of the following are required: enhanced therapeutic efficacy, decreased off-site drug exposure or toxicity, decreased dosage or frequency of administration, enhanced therapeutic adherence, and enhanced patient-specific therapeutic function.(Chen et al., 2018) The best way to address issue of a medicine being safely transported through body to its intended location is through drug delivery. Targeted drug delivery will address a number of issues, including preventing medication deterioration and unintended interactions with biological molecules, enhancing drug absorption by tumor cells in a targeted manner, tracking pharmacokinetics, and boosting drug bioavailability inside tumor cells.

Therapeutic compounds should be momentarily contained within carrier vehicles and released at a predetermined location or time in order to create effective drug delivery systems (DDS).(Mura, Nicolas, & Couvreur, 2013) Numerous supramolecular systems, including polymersomes, micelles, vesicles, and physical gels, are presently being studied for their potential as drug delivery methods.(Yoon & Jang, 2010) These materials may be a good fit for the demands of nanomedicine because of their complex chemical design.(Xing & Zhao, 2016) Nanotechnology is an interdisciplinary field of study that has the potential to significantly transform scientific and industrial technologies. It finds application in a wide range of fields, including biomolecular diagnostics and detection, enzyme immobilization, nanoparticles, therapeutics, gene therapy, microelectronics, DNA sequencing, optics, optoelectronics, photonics, pharmaceuticals, and sensors (Figure 1). Nanoparticles are of particular relevance among the different types of nanomaterials employed in biosensing because of their intrinsic characteristics, such as their size, high signal tag loading, high surface area, and dynamic nature for signal amplification. (Gontero et al., 2017) Particularly, the field of nano-biotechnology, which is the use of molecular and biological systems to the creation of nanoscale materials is presently yielding exciting new possibilities for medicines and diagnostics.(Toy, Bauer, Hoimes, Ghaghada, & Karathanasis, 2014) Stimulus-responsive nanocarriers that are sensitive to pH, temperature, redox potential, glucose, and enzymes have been developed for drug carriers in order to effectively target tumor tissue. Due to their safety and efficacy in cancer therapy, stimuli-responsive nanocarriers for drug delivery have attracted a lot of attention lately.(A. Zhang, Jung, Li, Liu, & Boyer, 2019)

Figure 1. Represent the various application of supramolecular drug delivery system (SDDs)

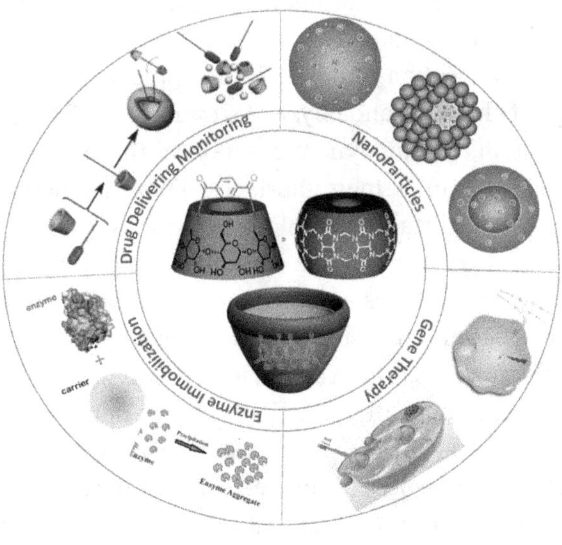

Supramolecular Affinity Between a Macrocyclic Host and a Pharmacological Guest

A cyclic "host" interacting with an insertable guest that gets supramolecularly bound inside might be seen as the host-guest complexation motif. To generate stable host-guest complexes, the macrocyclic host molecules offer the appropriate hydrophobic cavities for the guest molecules to be effectively incorporated and bound. (Miyamae, Nakahata, Takashima, & Harada, 2015) The physical and chemical characteristics of the macrocyclic host molecules, as well as their interactions with different guest molecules, varied greatly because different types of macrocyclic host molecules have distinct chemical constitutions, cavity shapes, and cavity dimensions. Cyclodextrins, calix[n]arenes, crown ethers, cucurbit[n]urils, pillar[n]arenes, and other significant macrocyclic host molecules have drawn a lot of attention in host-guest chemistry research.(Braegelman & Webber, 2019) To given their notable biofriendliness and biocompatibility, the macrocyclic host molecules CD and CBn homologues have garnered the greatest attention in supramolecular chemistry research, particularly with respect to host-guest systems potential applications in various biomedical contexts. These host molecules are all equipped with macrocyclic cavities that allow for partial or complete encapsulation of appropriate guest molecules, which shown in Figure 2.

This facilitates easy creation of host-guest complexes, which are frequently responsive to a range of chemical and physical stimuli.(Yan, Wang, Zheng, & Huang, 2012) Biological systems exhibit dynamic responses that are significantly impacted by arrange of external stimuli at their interfaces. Thus, the incorporation of diverse host-guest chemistry into surface technology represents a significant advancement towards the creation of multifunctional bio-interfaces with effective stimulus responsiveness. These biointerfaces exhibit strong bio-compatibility to replicate the intricate molecular recognition systems found in nature, and are highly adaptable to environmental fluctuations.(Hui Yang, Yuan, Zhang, & Scherman, 2014) Interfacial molecular recognition is often the first step in a number of biological system activities, including bio-transformation, matter transport, and energy transduction.

Several innovative approaches are covered in particular biological applications of the host-guest systems, including bioimaging, drug/gene delivery, and photodynamic treatment (PDT). (Dong et al., 2015) These days, host-guest supramolecular polymeric systems, hydrogen-bonded supramolecular polymeric systems, and supramolecular drug complexes are most common types of supramolecular drug delivery systems (SDDs).(K. Yang, Pei, Wen, & Pei, 2016) One of the most useful delivery methods for drugs and non-viral genes is the use of supramolecular nanoparticles, or SNPs. Functional medication delivery with controlled release and effective gene administration through transfection have benefited greatly from the adaptable self-assembly processes and appropriate association strengths of SNPs.(Wen, Zhang, & Li, 2014) The synthesis of cationic supramolecular polymers with dynamic structures and programmable performance, which have potential to be employed as effective gene delivery vectors is made possible by supramolecular chemistry.

Figure 2. Represent the biomedical implementations of supramolecular systems driven by host–guest interactions

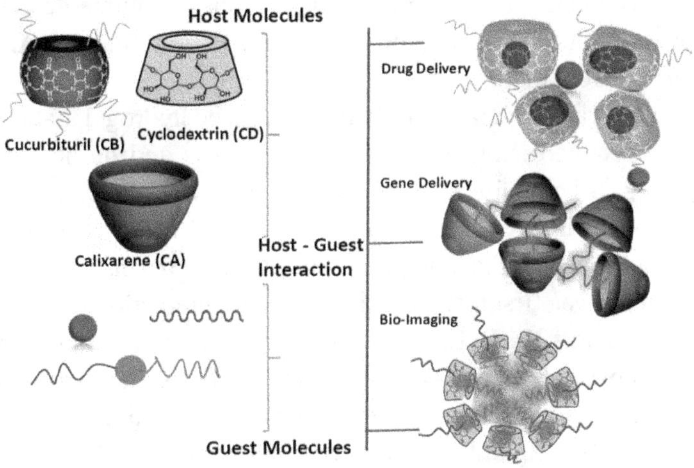

(Ma & Zhao, 2015)

Supramolecular Hydrogels Induced by Host–Guest Interaction

Hydrogels are three-dimensional networks of polymers created by physically or chemically cross-linking chains of polymers with hydrophilic qualities.(Liu et al., 2019) Hydrogels can be made cross-linked through a variety of interactions, including ionic, covalent, hydrogen, and hydrophobic interactions.(Yoon & Jang, 2010) A noncovalent interaction that has been extensively studied as a method for creating supramolecular hydrogels is macrocyclic host-guest complexation. The following classes of compounds have been studied: crown ethers, cyclophanes, catenanes, cavitands, porphyrins, and cryptophanes. Cavitands, like cyclodextrin and cucurbit[n]urils, are commonly employed in formation of supramolecular hydrogels.(Harada, Takashima, & Nakahata, 2014) When creating physical hydrogels, the host-guest interaction is a crucial noncovalent bonding that results from the interplay of several dynamic interactions between corresponding molecules, such as hydrophobic interactions, hydrogen bonding, π-π stacking, and so on.

Host monomers exhibit greater complexation strength and more precise identification with guest monomers in contrast to certain other physical interactions. Strong binding results from the sequestration of a hydrophobic guest into the hydrophobic inner cavity of the host, which causes host-guest interactions. These interactions are shown in water in cyclodextrins (α, β, and γ - CD) and cucurbit[n]urils (CB[n], n = 5 -- 8 and 10). CB and CD hosts are excellent options for biomedical and pharmaceutical applications due to their biocompatible and nontoxic nature.(Amin, Ahmad, Pandey, Abeer, & Mohamad, 2015)

Here, a novel supramolecular SF (silk fibroin) based hydrogel that is self-healing and injectable was created using a straightforward process, and the fabrication strategy shown in Figure 3. SF contains aromatic amino acids like tyrosine, phenylalanine, and tryptophan, which have been reported to function as functional groups generating host–guest interactions with host polymers (like Cyclodextrin, Calixerene, etc), the hydrogels precursor was first formed via host–guest interaction between host polymer and SF.(Raia et al., 2017) The host-guest interaction between aromatic amino acids of SF and host polymers has allowed for successful fabrication of a unique supramolecular SF-based hydrogel with self-healing and injectability.(Chouhan et al., 2018) The mechanical qualities, long-term stability, self-healing behavior, injectability, and biocompatibility of this SF/host polymers hydrogel were all enhanced. When used as wound dressings, these SF-based hydrogels demonstrated improved wound healing performance in terms of wound area, anti-inflammatory, epidermis, granulation tissue, and collagen deposition.

Figure 3. Represent the self- healing host-guest interaction of hydrogel between SF/Host polymer

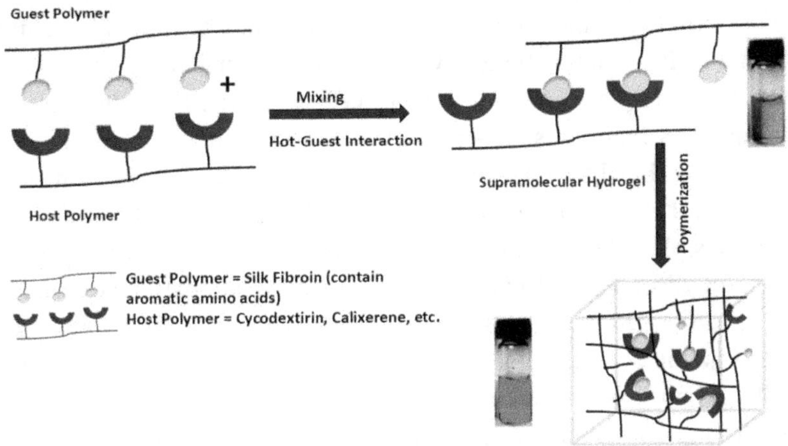

(Yu, Yang, He, Li, & Guo, 2021)

Hydrogels of several kinds have been created to be used as bioinjectable materials. Changes in the swelling characteristics in reaction to external chemical and physical stimuli, such as temperature, pH, light, pressure, and an electric or magnetic field, regulate the release of the medicine. (Forero-Doria et al., 2020) The release of pharmaceuticals is mostly dependent on the erosion or degradation of the gelling components, despite the fact that hydrogels have a remarkably high drug loading capacity. Given that the injected gels come into constant touch with nearby tissues, the biodegradability and biocompatibility of the gelling materials are crucial for the safe delivery of drugs. Useful supramolecular Hydrogels and their applications in the biomedical field, including as tissue engineering, medication transport, gene transfection, protein delivery, bioimaging and diagnostics, and biomimetic chemistry.

Supramolecular Vesicles Induced by Host–Guest Interaction

The recently proposed supramolecular vesicles with numerous medicinal activities heavily rely on host-guest interactions and supramolecular macrocycles. Artificial vesicles have long been made using supramolecular chemistry techniques, such as the polar/nonpolar interaction offered by amphiphilic surfactants, to create non-covalent contacts.(Ravoo, 2012) Now a days, a range of amphiphilic molecules have been employed in the vesicle-building process; additionally, due to their exceptional dynamic reversibility, which has resulted in the formation of supramolecular vesicles, amphiphilic host molecules hold great promise for vesicle creation.(T. Xiao et al., 2019)

Hydrophobic or hydrophilic medicines can be easily inserted into the centre of produced vesicles to facilitate medication distribution. The rationale behind this is that supramolecular vesicles typically have an appropriate/adjustable size for the enhanced permeability and Retention (EPR) effect, and their structural resemblance to cell membranes facilitates easier internalization.(Soussan, Cassel, Blanzat, & Rico-Lattes, 2009) Facilitating the creation of supramolecular amphiphiles, serving as the catalyst for vesicle formation, and enabling simple surface alterations are among the roles played by the host-guest inclusion in vesicle formation.

As for the hydrophobic drugs that can be simultaneously loaded into vesicle membranes, the aqueous cavities of vesicles aid in the efficient loading of hydrophilic medications and drugs. It is easy to develop multifunctional theranostic platforms based on supramolecular vesicles through the careful chemical design of organic building components (Figure 4). External stimuli such as voltage, pH, UV irradiation, oxidation/reduction, enzymes, and even magnetism can readily disrupt weak non-covalent connections.(Hu, Zhang, & Liu, 2012) Novel drug carriers may be developed by creating vesicles with the ability to alter their aggregation state in response to pH. (G. Yu, Jie, & Huang, 2015) An extracellular microenvironment with a pH of 6.6–7.0 that is slightly acidic is an interesting stimulation in tumor tissue.(Gillies, Verduzco, & Gatenby, 2012)

Figure 4. Demonstration of supramolecular vesicles-based stimulus-responsive carriers

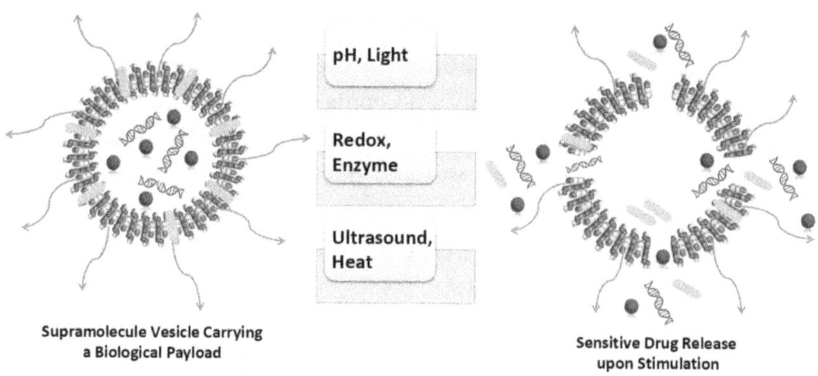

The biomedical payload that has been demonstrated consists of DNA Helix (grayish blue helices), hydrophobic drugs (yellow rods), hydrophilic drugs (red spheres), and targeting ligands (blue arrows).

Drugs that are hydrophilic or hydrophobic can be put into the aqueous inner void or membrane layer of vesicles, which have dual membrane layers. Because of the inner cavity's enormous volume, vesicle-mediated DDS has advantages for protein administration and cocktail therapy. Multi- stimuli responsive vesicles, which respond to several stimuli, were employed as delivery systems for drug vehicles. The vesicles have the potential to effectively encapsulate an anticancer medication (DOX) with a loading efficiency of 86% (Figure 5). When a competitive supramolecular host (α-CD) was added or the vesicles disassembled, the loaded anticancer medication was liberated from them. It is scientifically required to create responsive drug delivery vesicles because controlled drug delivery can significantly reduce side effects caused by the drug prerelease before reaching the target cells.

Figure 5. The proposed mechanism of cyclodextrin vesicle formation and its stimulus responsiveness

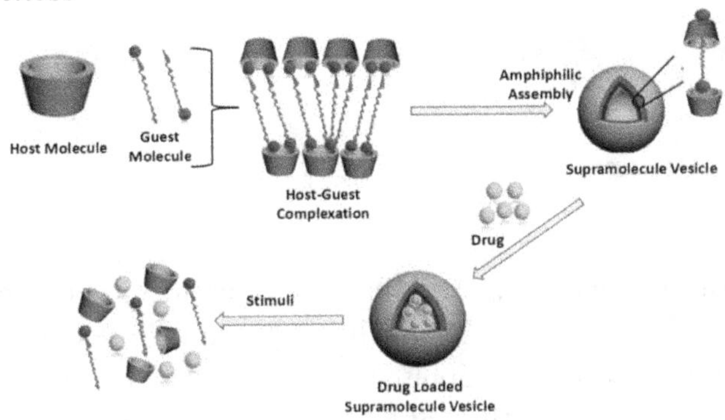

(Anis Abdul Karim, Qingqing Dou, Zibiao Li, & Xian Jun Loh, 2016)

Supramolecular Micelles Induced by Host–Guest Interaction

Functional groups have been used to create a variety of supramolecular micelles that are used to carry medicines and genes. It is usually generated by an amphiphilic surfactant with hydrophobic "heads" facing the outside world and hydrophobic tails pointing in the direction of the micelle core.(Q. Zhang, Re Ko, & Kwon Oh, 2012) Since the hydrophilic surface of micelles provides solubility in an aqueous media and its hydrophobic compartments can serve as storehouses, they offer an efficient platform for delivery.(A. A. Karim, Q. Dou, Z. Li, & X. J. Loh, 2016) Research is being done on supramolecular block copolymers to carry out drug delivery activities, as micelles are particularly interesting in the delivery of hydrophobic medications or cargo. Stimulus-responsive block connections, also known as stimuli-responsive blocks, appear to be the preferred method, particularly when considering the triggered release of drug cargo.(Schmidt & Barner-Kowollik, 2016)

Figure 6. The host-guest interaction via DNA supra-amphiphiles D18-CB[7] and their external stimuli

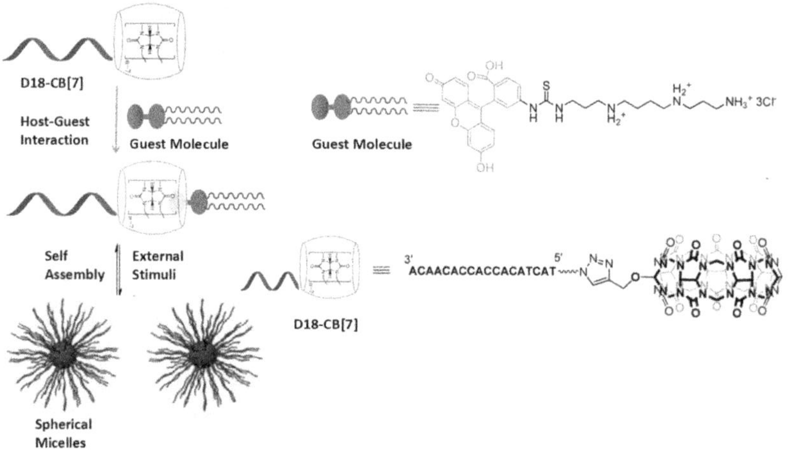

(Yuan, Ma, Zhao, & Liu, 2020)

A CB[7]-scaffold, which was covered in many sugars to bind proteins having a high degree of selectivity. This glycosylated CB interacted with several guests to create stable complexes. The spermine–fluorescein isothiocyanate (FITC) combination was employed as a guest for the scaffold of β-Gal-CB[7]. Using the host-guest interaction between cucurbit[7]uril (CB[7]) and the guest molecule spermine–fluorescein isothiocyanate (FITC), create two DNA supra-amphiphiles here by noncovalent recognition. There has been extensive investigation in supramolecular chemistry about the host-guest interaction between CB[7] and the guest spermine–fluorescein isothiocyanate (FITC). The supramolecular DNA amphiphiles may be modularly built by host-guest interaction and then go through their self-assembly process and formed spherical micelles, which shown in Figure 6. Wei Yuan et.al., reported DNA-based supra-amphiphiles (D18-CB[7]/G1 and D18-CB[7]/G2) through host–guest interaction and their self-assembly behavior. The DNA strands at the corona of the micelles could further anchor other functional molecules via DNA hybridization. (Yuan et al., 2020)

METALLOPOLYMER-BASED DDSS

In current years, scientists' consideration as a DDSs to polymers-based has grown significantly due to their flexibility, the diversity of polymers in terms of composition and properties, their easy functionalization, and the possibility of smart performance in response to environmental incentives.(Dzhardimalieva, Rabinskiy, Kydralieva, & Uflyand, 2019) Owing to the various orders of magnitude among the size of cancer cells and polymer materials, it is preferably right as *in vivo* DDSs. Three various protocols for generating polymer-based DDSs have been established: (1) physical encapsulation, (2) In polymer chain, complexation in between drugs and donor atoms chains, and (3) conjugation of drugs with a polymeric transporter via polymeric prodrug. The certain advantage of the approach is the delay in the release and even the activate of the issue when the DDS reaches its target.

In the 20th century, discovered the main paths of polymers was MPs. In polymer backbone, the transition and lanthanides series metal comparative have been defined uses Wolf as Type I, Type II or Type III (Figure 7). Metallopolymers are linear and more branched structure it containing covalent bonds between the metals and polymer, which principals to mainly irreversible noncovalent, which could consent possibly reversible 'dynamic' binding.(Homberg, 2018) Metallopolymers as individual stage of chemical natural, distinguishing features, the presence of the polymeric chain are consenting reliably classify such as inorganic compounds, organic biological nature and metal cores.(Dragutan, Dragutan, Simionescu, Demonceau, & Fischer, 2015)

Figure 7. In metallopolymer structural illustrations of Wolf Type I, II and III

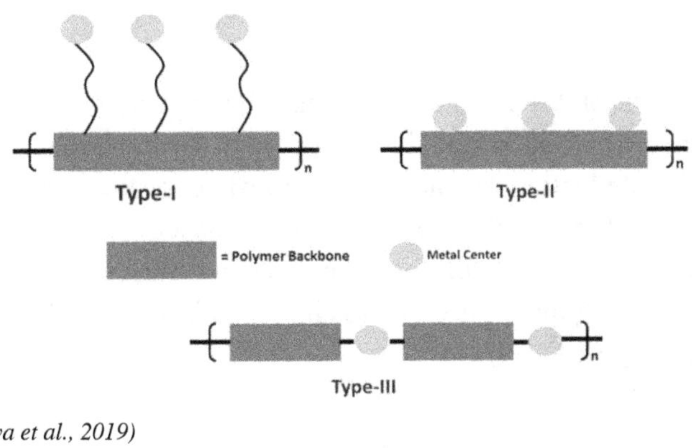

(Dzhardimalieva et al., 2019)

Metallopolymers are good applications including functional materials, catalysis and biocatalysis, electroactive, photoactive and magnetic materials, environmental science. An important discovery was made in the development of new synthesis method such as ring-opening polymerization, polycondensation, living ionic polymerization, electro polymerization polymerization with radical controlled and delayed access to characterization techniques like XPS, MALDI-TOF, ESI-MS, electron microscopy, spatially determined optical spectroscopy. Synthesis of metallopolymer (MPs) with Type I and II were recognized by developing new synthetic method that are well-matched with inorganic metal centres (Figure 8). To date, around 20,000 metal–organic frameworks have been produced for a range of applications with various combinations of organic compounds with metal nodes to forms the complex polymer compounds. The key approaches, synthetic approaches and post-synthetic changes of complex polymers.(Dzhardimalieva et al., 2019)

Figure 8. Arrangement of the methods for preparation of metallopolymers of Type I and II

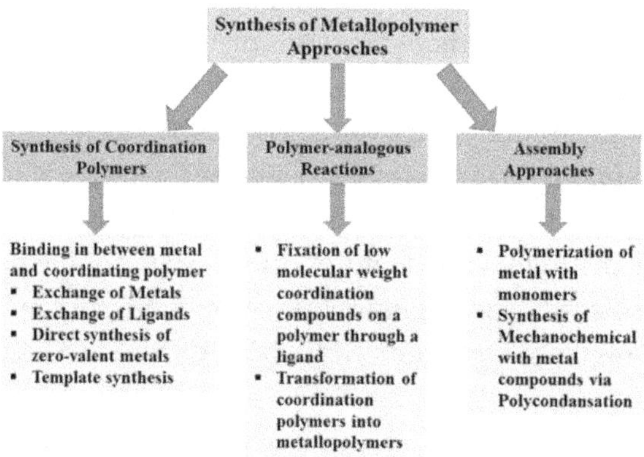

(Dzhardimalieva et al., 2019)

The development of MPs is a new strategy for importantly more effectiveness of biomedical materials, in particular, to address current scenario of traditional approaches of prophylactics and action. Synthesis coordination compounds having different donor atoms enclosing ligands binding with various metals for these compounds have due potential for use in an extensive variety of DDSs. The branch of metal containing polymers-based DDSs has used in various therapy and as multimodal systems (Figure 9).

Figure 9. The key stages of metal containing polymers-based DDSs with use MPs as incentives- and multistimuli-responsive DDSs, for combination therapy and as multimodal systems

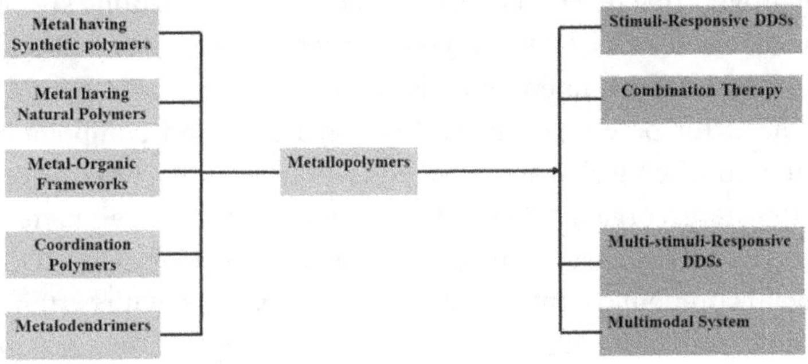

(Dzhardimalieva et al., 2019)

Metal with Natural Polymer Compounds

Proteins and peptides as a part of natural biomacromolecules and it show great potential as transporters for small metal involving drug compound owing to their intrinsic qualities of biodegradability, biocompatibility, minimal toxicity and elevated water solubility. Furthermore, it is crucial that biological carriers for metallodrugs can discern tumor markers overexpressed by cancer cells, facilitating their utilization in targeted treatment approaches. As a examples, the PKKKRKV peptide derived from the antigen (SV40 large T), recognized as the nuclear localization sequence (NLS) peptide, was employed as a vehicle for Pt(II) drug delivery. (Wlodarczyk et al., 2018) The hybrid exhibited significant biological efficacy in ovarian cancer cell lines, surmounting resistance to platinum, possibly linked to an increased Pt(II) content in resistant cells. (Trokowski, Akine, & Nabeshima, 2008)Quantitative assessment of Pt(II) within the nucleus of two isogenic cell lines, A2780 (Pt sensitive) and CP70 (Pt resistant), revealed that in former cells, one Pt(II) was bound to every 20th turn of DNA, whereas in the latter cells, it occurred at every 10th turn. (Dzhardimalieva et al., 2019)

Natural polymers employed in DDSs encompass polysaccharides such as agarose, hyaluronic acid (HA), alginate, carrageenan, etc. and protein-based polymers like collagen, albumin, gelatine, etc. Specifically, HA stands out as a polyanionic polysaccharide that serves dual roles, functioning both as a DDSs and as a target motif for drug delivery. Hence, the HA-Pt(IV) conjugate was derived by coupling Pt(IV) prodrugs to ethylenediamine (EDA)-modified HA through a succinimide bond (HA-EDA-Pt(IV)). (Ling et al., 2015)

A hyaluronan-nanogel with targeting capability, templated by cisplatin, has been created. Similarly, nanogels carrying cisplatin-loaded chondroitin sulphate were synthesized and subsequently integrated into bioresorbable PEG-poly(b-aminoester urethane) hydrogels that respond to pH and temperature, enabling specific cisplatin drug passing to cancer cells. DDSs relying on inorganic metal compound containing ordinary polymers offer numerous benefits, providing control over drug release, eliminating need for potentially hazardous organic liquid compounds in sample manufacturing, enhancing pharmacokinetics, and augmenting biodistribution of therapeutic agents in target organs. These advantages contribute to heightened efficiency, encompassing a non-invasive approach to drug delivery. However, DDSs relying on inorganic metal compound contain natural polymers come with certain drawbacks. These include hazard of bacterial contamination during production stemming from their natural sources, potential for heavy metal contamination often linked to plant polymers, variability from batch to batch due to differences in resources and their geographical origins, a relatively slow production process, and more. These deficiencies are contingent on environmental factors and various other considerations, and including potential impurities may trigger undesired immune reactions.

Metal with Synthetic Polymers

DDSs often rely on polymer carriers to transport bioactive molecules. Synthetic polymers can be branched or linear and serve as either bioactive component carrier for covalently linked drugs. Biodegradable polymers are preferred for DDSs are crucial due to their capability to break down in body over time and such examples include recombinant protein-based polymers, polyesters, and polypeptides. Metallodrugs are therapeutic agents containing metal ions (e.g., cisplatin and oxaliplatin). Various approaches for metallodrug binding with polymers: (1) Stable Ligand Approach, (2) Leaving Group Approach, polymers act as stable ligands, coordinating with metal ions. This approach enhances stability and controlled release of metallodrugs. Polymers serve as leaving groups for metallodrugs. Upon drug release, polymer remains as bioactive component carrier.(Chakraborty & Newkome, 2018)

Metal-polymer interactions must be wisely synthesised with optimize drug delivery. This interaction conjugates with a ruthenium metal center and a cyclopentadienyl ligand into a polymer structure. The polymer backbone, often composed of polylactide chains, provides stability and allows for controlled drug release. Ruthenium containing compound have illustrated cytotoxic effects against cancer cells. By incorporating these complexes into conjugates based polymeric material, scientist aim to enhance their therapeutic potential. A critical requirement for DDSs is stability in aqueous environments. The polymeric backbone ensures that the conjugates remain stable and functional in biological fluids. These conjugates can serve as drug carriers, delivering ruthenium-based cytotoxic agent to tumor sites. The combination of cytotoxicity and stability makes them capable applicants designed for personalized cancer therapy. (Côrte-Real et al., 2015) In another compelling instance, a poly(L-glutamic acid)-graft-methoxy poly(ethylene glycol) (PGA-g-mPEG) has been suggested as a carrier for incorporation of cisplatin. The resultant metal compound exhibited extended blood circulation duration and remarkable accumulation in tumour. Specifically, its plasma area-under-curve (AUC) was 31-fold greater compared to that of free cisplatin, and its AUC within Lewis lung carcinoma tumour was elevated by a factor of

Metalloproteins (MPs) exhibited notably elevated in vivo micellar stability, diminished nephrotoxicity, heightened accumulation in neoplastic tissues, and achieved an efficacy level comparable to that of cisplatin. (Ahn et al., 2015)

Palladium catalysts were employed to facilitate intracellular reactions, releasing diverse model prodrugs, such as doxorubicin (DOX) prodrug (pro-DOX), leading to efficient in vivo tumour treatment. Cyclic peptides capable of self-assembly were linked to a biocompatible polymer, poly(2-hydroxypropyl methacrylamide), and modified with exceptionally potent organoiridium anticancer compounds. The anti-cancer study of three substances like incorporated into the polymer, the conjugate and free drug was assessed in relation to human ovarian fibroblasts (HOF), serving as a model for healthy, non-cancerous cells, and was juxtaposed with the study against A2780 ovarian cancer cells. Nanotubes loaded with the organic and inorganic drug molecule showed similar or increased toxicity against human ovarian cancer cells compared to free drug. Interestingly, the drug-conjugated nanotubes did not increase iridium uptake; instead, they exhibited a more effective mode of action. (Lu, Chen, Noy, Lu, & Stenzel, 2017)

Researchers have developed a dual-DDSs for battered therapy by combining Pt(IV) prodrug-based micelles and dichloroacetate within poly(vinyl alcohol) nanofibers. Notably, the metallodrug component features a self-cross-linked nanostructure derived from Pt(IV) prodrug, resulting in exceptionally high cisplatin loading efficiency. Advantages of using inorganic metal compounds containing on synthetic polymers in DDSs and it has included consistent properties across batches and simplified large-scale synthesis. These complexes provide numerous opportunities for after synthesis variations to achieve desired properties. Their main focus is on improving drug solubility and stability, fine-tuning dosage, enhancing bioavailability, optimizing pharmacokinetics, ensuring controlled drug release, minimizing tissue toxicity, and enabling targeted drug delivery.

Coordination Polymers

In coordination species, formation of ligand-metal via coordination bonds, multiple discrete coordination complexes can also assemble into supramolecular structures through weaker intermolecular forces.(Xie et al., 2016) These forces include hydrophobic interactions, π–π interactions, and electrostatic effects. Recently, Liu and colleagues showcased a triad supramolecular array with promising applications in drug deliver.(M. Yu, Chen, Zhang, & Liu, 2010) The triad supramolecular array consists of two distinct coordination complexes: Zn[4-amino-N-(2-methylquinolin-8-yl)benzenesulfonamidemodified permethyl-cyclodextrin (MQAS-PMCD)]2 (referred to as Zn12) and Zn[meso-tetrakis-(4-sulfonatophenyl)-porphyrin (TSPP)] (referred to as ZnTSPP). In an aqueous environment, these complexes form a linear triad Zn12–ZnTSPP supramolecular array through strong interactions between the cyclodextrin cavity and anionic porphyrin shows in Figure 10. These supramolecular interactions not only prevent unfavourable porphyrin-porphyrin aggregation but also ensure the high stability of such systems in biological environments.(Z. Ma & Moulton, 2011)

Figure 10. Graphical representation of complex molecules reaction in linear triad Zn12–ZnTSPP complex

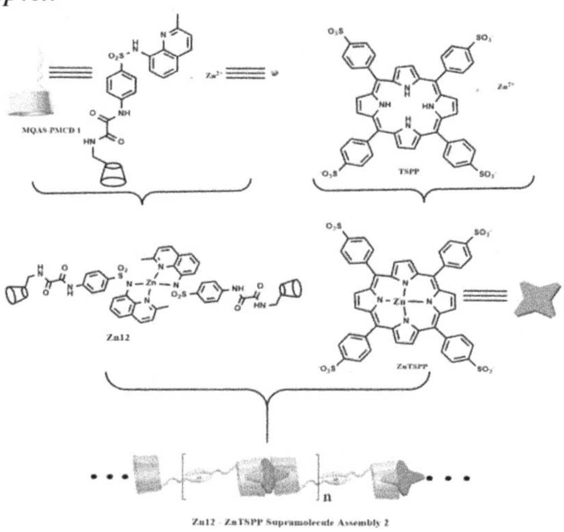

(Ma & Moulton, 2011)

Numerous research studies focus on utilizing coordination polymers as DDSs. One noteworthy example is the utilising of long-circulating core-shell CPs, which efficiently carry various drugs with diverse mechanisms of action, enhancing synergistic therapeutic outcomes. (He, Poon, Chan, Yamada, & Lin, 2016) Coordination polymers (CPs), modified with PEG (polyethylene glycol) by incorporating a poly-histidine (pHis)-PEG copolymer, demonstrate effective passive accumulation within tumors. In this context, the reduced pH (approximately 6.5) triggers charge changes and size expansion, thereby enhancing tumor retention and cell internalization (Figure 11). Upon cellular uptake, CPs accumulate in cell endo-/lysosomes. Resulting in release of drug with decreases pH and decomposition of CPs. In animal tumor models, chemotherapy utilizing CPs has demonstrated remarkable efficacy even at minimal amount the drug molecule doses. In zinc with CPs, organic moiety contain ibuprofen act as a ligand molecule, boast several advantageous properties. These include high drug content, excellent stability, low cytotoxicity, good biodegradability, and high biocompatibility. For the transfer of small-molecule chemotherapeutic agents, such as 5-fluorouracil (5-FU), a one-dimensional (1D) zinc CP based on benzene-1,3,5-tricarboxylic acid (trimesic acid) was employed. The 5-FU-loaded CP demonstrated time-dependent anti-proliferation cytotoxicity against MCF-7 and 4T1 cell line. Intratumorally treatment with 5-FU-loaded CP exhibited favourable antitumor efficacy by effectively inhibiting tumor growth.(Lago, Pino-Cuevas, Carballo, & Vázquez-López, 2016)

Molecular scale CPs deserve special attention for their ability to overcome existing drawbacks in DDSs. These CPs exhibit varying compositions, sizes, shapes, along with drug entering, and fundamental biodegradable nature. As an alternative to other nanocarriers, CPs offer several advantages due to their inherent adaptability and diversity. Moreover, CPs can combine the strengths of both organic and inorganic polymer nanoparticles like drug entering, compositional variability, biodegradability, well-defined particle morphologies and affluence of surface functionalization resulting in a new drug transport stage.(Dzhardimalieva et al., 2019)

Figure 11. Schematics representation of one pot reaction of PEGylated nanoscale coordination polymers and their use for tumor microenvironment (TME)

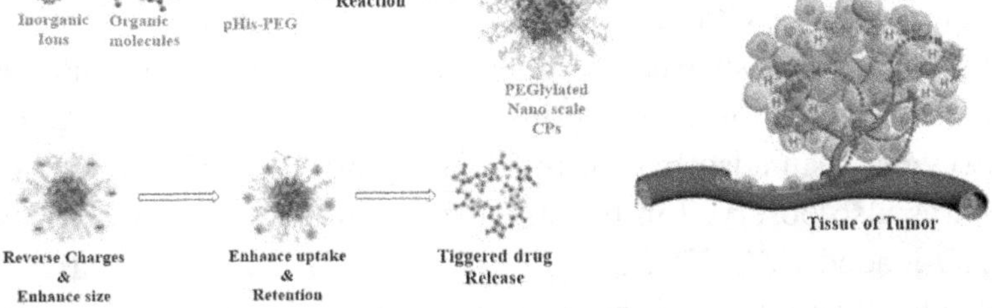

(Y. Yang et al., 2018)

Bulk CP-based Drug Transport Platforms

Coordination polymers, often referred to as hybrid inorganic–organic materials, are composed of transition metal ions/clusters and multidentate organic ligands. In recent years, researchers have explored the application of CPs in drug transport. The significant potential of using Cps as DDSs is attributed to several factors: (a) high surface area and permeability, (b) low toxicities of certain transition metals, i.e Mn and Cu, (c) good physical and chemical strengths and (d) well-defined chemical structures. (Perles, Snejko, Iglesias, & Monge, 2009)

In the realm of biodegradable coordination polymers (CPs), Serre and colleagues pioneered a drug transport approach. They created biodegradable therapeutic CPs by coordinating the bioactive molecule vitamin B_3 (niacin) to non-toxic Fe (iron) ions. (S. R. Miller et al., 2010) An advantageous feature is that the bioactive compounds itself serves as the ligand, eliminating the need for additional drug loading steps. The Fe-vitamin B_3 coordination polymer contains 71.5% w/w vitamin B_3 based on its formula. In drug release experiments conducted at 37°C in PBS (pH = 7.4), rapid release of vitamin B3 ensued as a result of the breakdown of the coordination polymer. Notably, full drug liberation was accomplished within a 1 hour.

Nitric oxide plays crucial roles in the nervous, immune systems and cardiovascular system. Solid NO carry out an important biomedical applications such as anti-bacterial, anti-thrombic, and wound healing properties. (M. Miller & Megson, 2007) In this context, we delve into the delivery of NO using CPs in a dedicated section. Unlike regular organic and inorganic drug compounds in solid and liquid states, NO occurs as a gas at normal temperature and pressure. (Tsai et al., 2005) The distinct molecular weight and dimensions of NO compared to small molecule drugs lead to variations in the delivery mechanism and release kinetics within coordination polymers. Researchers have explored a range of chemical materials, including polymers, nanoparticles, and zeolites, as potential NO delivery platforms (DeRosa et al., 2007).

In a more recent study by Morris and co-workers (McKinlay et al., 2008), two coordination polymers, CPO-27–Co and CPO-27–Ni (Dietzel, Panella, Hirscher, Blom, & Fjellvåg, 2006), showed exceptional performance over the whole adsorption storage-transport cycle for nitric oxide (Figure 12). Both activated CPO-27 coordination polymers can adsorb ~7.0 mmol NO/g of solids. Considerable isotherm hysteresis was also observed in the study. Xiao author and co-author explain the method, NO-loaded CPO-27–Co and NO-loaded CPO-27–Ni released ~7 mmol NO/g of solids, illustrating that the entire amount of adsorbed NO was recoverable. (B. Xiao et al., 2007) CPO-27–Co/Ni can deliver ~7 mmol NO/g of solids upon exposure to moisture, which is 3 orders of magnitude greater than that by HKUST-1 and 7 times greater than that by the best zeolites. It's understandable that metals might influence the release kinetics because of variations in the interactions between metal and NO binding. For instance, CPO-27–Ni released NO faster than CPO-27–Co due to the relatively weaker interactions between NO and Ni. NO release rate also depends on the release medium.(McKinlay et al., 2008) The half-life of NO delivery from CPO-27 coordination polymers was reduced from several hours in flowing wet gas to approximately 10 min in PBS buffer.

A different method to transport NO by post-synthetic changes of CPs has also been explored. Rosseinsky and co-workers informed a two-step post-synthetic transformation of HKUST-1 to generate diazeniumdiolates inside of the coordination polymer pores. The diazeniumdiolate functional group formed by strong binding of NO and secondary amine sites is regarded as a "masked" version of NO because NO release from diazeniumdiolate can be proficiently activated by water (Figure 13).(Ingleson, Heck, Gould, & Rosseinsky, 2009)

Figure 12. Diagrammatic image of determines the success of a gas-storage material (Cobalt and Nickel-MOF from activation cycle, loading & storage, and delivery)

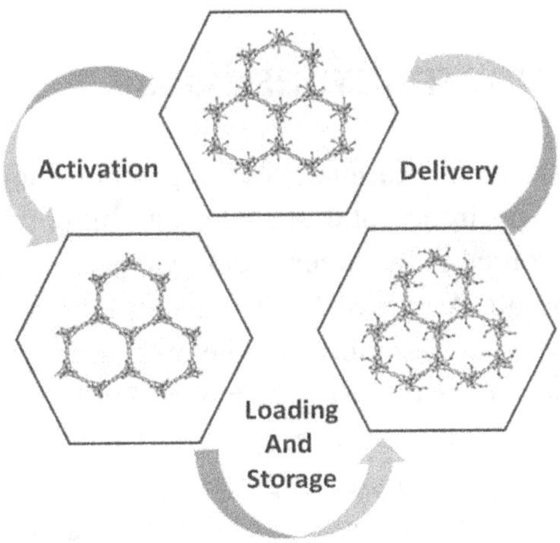

(McKinlay et al., 2008)

Figure 13. The NO group removal mechanism with the formation of diazeniumdiolate group

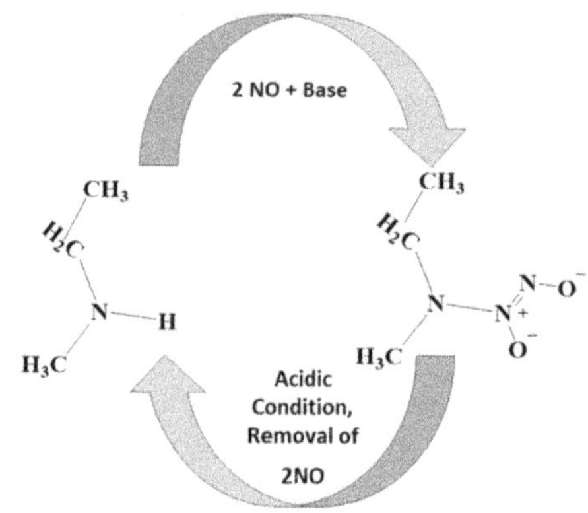

(McKinlay et al., 2008)

CORE-SHELL NANOSTRUCTURES DDSS

In nanotechnology science continually development in biomedical, engineering, medical technologies, environmental science, agricultural chemistry, materials technology, and electronics as an interdisciplinary stream compassing science. (Ahmadivand, Gerislioglu, Ahuja, & Mishra, 2020) Despite the advancements in therapies and technologies, communication between nanoparticles and biological systems remains unclear in many aspects. For example, basic nanoparticles like Au, Ag, Pd, Pt, and Fe exhibit specific properties. To utilize nanoparticles in drug delivery, biomedical study, and tissue engineering, nanostructures must be transferred into water and exhibit improved colloidal stability and mono dispersity. (Han, Ghosh, De, & Rotello, 2007) However, several parameters, including the chemical composition of nanoparticles and the surface interaction strength of the hydrophilic ligand, influence the ligand exchange process, which should be stronger than that of the hydrophobic ligand.(L. Zhang et al., 2008)

The method for synthesizing core-shell nanomaterials have been well-established for few years, resulting in homogeneous nanomaterials by precise control over size and shape using selective stabilizers.(Dutt, Kumar, & Siril, 2015) These stabilizers exhibit various stages of interactions with various facets of nanoparticles. The collective characteristics of core-shell nanoparticles render them captivating for biological and advance innovative applications. (Newman, Brown, Siril, Lee, &

Wilson, 2006) For example, the design of core-shell nanoparticles, such as magnetic nanomaterials coated with Au allows drug loading through thiol derivative compounds of chemistry with gold. Moreover, these magnetic nanoparticles can be guided to deliver drugs at specific sites of action within the human body using an external magnetic field.(Xavier-Jr, Gueutin, Chacun, Vauthier, & Egito, 2019) While bare nanomaterials can be toxic and potentially harmful to host tissues, core-shell nanoparticles offer several advantages. These include reduced cytotoxicity, high dispersibility, improved biocompatibility, enhanced conjugation with biomolecules and drugs of their properties like modified surface properties, and increased chemical and thermal stability.(Cho et al., 2011) By coating core materials with biocompatible shells, nanoparticles become less toxic and more suitable for various applications. Moreover, shell materials can influence or enhance the characteristics of the core nanomaterials beyond just reducing toxicity.(Xiong, Yang, Yang, Xu, & Li, 2010)

In biological applications, the importance of water affinity of nanomaterials is well-known. Hence, water-repellent nanoparticles treated by hydrophilic compound can be appropriate for medicinal study, and functionalization with improved dispersibility and Core-shell nanoparticles.(Fratoddi, 2017) Functionalization of core-shell nanoparticles, like polymer, silane, dendrimers, and Au shown in Figure 14(B). This is attributed to the predominant core material over the shell composition. The attachment of biomolecules on the nanoparticle surface is pivotal for the successful development, delivery, and efficacy of therapeutic applications, particularly in biomedical and drug delivery fields. The different methods of inorganic metal nanoparticles surface engineering have representation in Figure 14(C), but in many cases, Conjugating biomolecules onto the surface of target materials presents challenges. These materials can be employed as coatings on appropriate core materials, serving as stimuli-responsive nanocarriers.(R. Kumar, Mondal, et al., 2020)

A structure of Layer-by-layer nanoparticles are various stages in Figure 14(A). The broadly used core materials are such as magnetic nanoparticles, biomolecules (drug/gene/protein/amino acid) nanoparticles, silica nanoparticles, metal nanoparticles and shell materials such as polymers, carbohydrate, proteins, organic, metals, and metal oxides.(R. Kumar, Aadil, Ranjan, & Kumar, 2020) Core-shell nanoparticles exhibit diverse physicochemical properties and can be categorized into various types. These include core-shell metallic nanoparticles, core-shell magnetic nanoparticles, core-shell polymer nanoparticles, core-shell silica nanoparticles, core-shell up conversion nanoparticles, and core-shell nanoparticles based on carbon nanomaterials. (V. B. Kumar, Kumar, Gedanken, & Shefi, 2019)

Figure 14. (A) Different kinds of core-shell nanomaterials used for DDSs. (B) Various functional compounds of Core-shell nanoparticles. (C) The plans of surface engineering of metal nanoparticles. (i) covalent ligand conjugation, (ii) amphiphilic polymer assembly, (iii) electrostatic layer-by-layer assembly, (iv) in situ ligand coating during synthesis, (v) Host-guest supramolecular ligand self-assembly, (vi) lipid shell coating

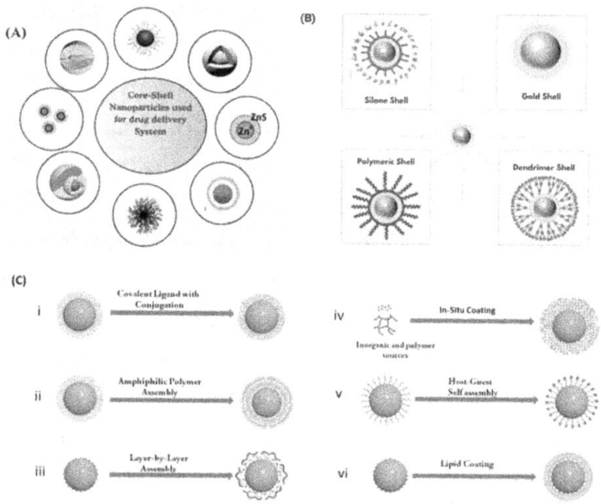

(Shen, Zhang, Qi, Mao, & Shen, 2018)

Researchers have access to a wide variety of polymers, allowing them to select the most suitable option based on specific requirements. However, biopolymers stand out as promising candidates for drug transport study due to their distinctive physicochemical and biological properties. Polymer coating on magnetic nanoparticles is one of the broadly utilised techniques to advance the stability of nanoparticles. The polyethylene glycol (PEG) is also among one of the best candidates that exhibits desired biocompatibility. PEG coating induces required properties such as water solubility, lowered toxicity, enzymatic degradation, and enhances drug half-life in-vivo.(Molineux, 2002) Polymers such as starch, dextran, and chitosan are commonly employed as coating materials for magnetic nanoparticles. Their hydrophilic properties reduce dipole-dipole attractions between the nanoparticles, resulting in well-dispersed aqueous suspensions. (R. Kumar, Soni, & Siril, 2019) Chitosan coated magnetic (Fe_2O_3) core shell nanoparticles exhibited promising results in gene and drug delivery, examined by Kumar et al. (A. Kumar et al., 2010) In targeted photodynamic therapy, a analogous type of nanoparticles (chitosan-coated magnetic NPs) is also used. (Hong Yang et al., 2017) Using further advantages are offered in conjugation with ligands with drugs and protein.

BIOMEDICAL APPLICATIONS OF SUPRAMOLECULAR HYDROGELS

The host-guest supramolecular hydrogels possess multiple characteristics that make them desirable for a range of biomedical applications. Hydrogel have garnered significant interest because of versatile properties, including biocompatibility, stimuli-responsiveness, and tunable mechanical properties. The network structure of supramolecular hydrogels allows for the encapsulation of drugs or beneficial substances for prolonged sustained release. Supramolecular hydrogels are ideal biomaterials for applications in a range of biomedical fields since they show separate physicochemical properties and specific functions, as well as the ability to undergo reversible swelling and gel–sol transition according to different environmental stimuli. In this section, we mainly focus on the application of supramolecular hydrogels such as biodetection, drug delivery, wound healing and tissue engineering (Figure 15).

Figure 15. Schematic illustrations of supramolecular hydrogels as potential versatile vehicles for drug delivery, wound healing, 3D-bioprinting and tissue engineering

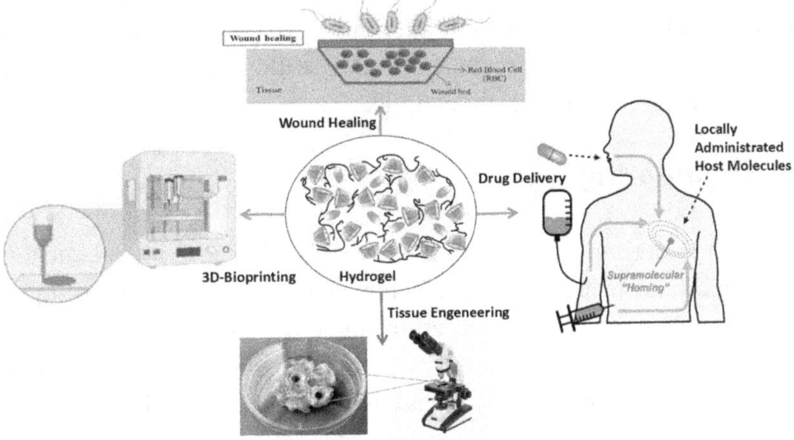

Biodetection

One of the most significant challenges is the research and development of biodetection techniques for diagnosing illnesses and biological agent analysis on the environment. Recently, an extensive variety of optical and mechanical supramolecular hydrogel-based sensors have been actively developed for the purpose of identifying and sensing a wide range of bio-relevant molecules or species, including polyanions, polyamines, glycosidases, alkaline phosphatase, glycoconjugates, proteins, and cancer cells. For example, Hamachi recently succeeded in fabricating a supramolecular hydrogel capsule by co-assembly of a prostate specific antigen (PSA)-cleavable additive with a glycolipid-based stiff and stable hydrogel matrix. Supramolecular hydrogels as nanostructured functional biomaterials for applications such as intelligent controlled release and cell culture matrices and actuators. However, to detect ultralow concentrations of analytes for use in biomedical, environmental, and national security applications, it is becoming more crucial to take benefit of simple, low-cost, portable, and user-friendly naked eye biosensing devices. The combination of a supramolecular hydrogel with bola-amphiphilic glycolipid-based supramolecular hydrogels that exhibit a quick color change in response to glycosidases was recently reported. This hybridization could be used for constructing a colorimetric sensor array chip that would allow visual detection of glycosidases. The superaptamer hydrogels displayed a significant decrease in volume, which was detectable even at femtomolar protein concentrations. A challenging interaction between protein-aptamer supramolecular cross-links and the hydrogel's progressive reduction in excluded volume may be essential for this remarkable macromolecular amplification. Therefore, the hydrogel response's visibility in this biomarker identification technique provided an alternative to standard analytical techniques, which call for expensive equipment and highly skilled employees, and makes them perfect for point-of-care or portable applications.

Tissue Engineering

In tissue engineering, sick or defective tissue is allowed to regenerate through the use of cells, biomaterials, and growth factors. The objective of tissue engineering is to generate live tissues that are structurally and biochemically comparable to native tissues so that the generated tissue works as intended *in vivo*. This means that in order to support the development of functional tissues, it is essential to better control the gelation conditions for cell encapsulation and *in vivo* creation as well as to modify the hydrogels chemistry, macroscopic characteristics, and degradation. The expected therapeutic effects are decreased by the low survival of transplanted cells; tissue engineering has been utilized as a second generation of cell therapy to solve these issues. The fundamental concept of tissue engineering is the concept that extracellular matrix (ECM) can be substituted with prefabricated porous three-dimensional (3D) scaffolds (such as collagen gel, fibrin gel, and aminotic membranes), into that living cells are seeded. Patients may receive these bioengineered tissues as transplant to repair or increase the original tissue function. Various bioengineered tissues have already been successfully used in clinical applications in regenerative medicine.

Recent research has shown that bioactive supramolecular nanostructures constructed from peptide-based building blocks provide a lot of potential as dynamic biomaterials for application in tissue engineering. Studies on the self-assembly of peptide amphiphiles were pioneered by Stupp and Zhang.(Guler & Stupp, 2007) Tissue engineering was effectively conducted using the resultant peptide-based supramolecular nanofibers, which varied in terms of form, surface chemistry, and bioactivity. After the hydrogen-bonded supramolecular polymers self-assembled into nanoscale fibers, bioactivity was introduced to these supramolecular nanofibers with supramolecular intercalation of peptides produced by the extracellular matrix that had been modified by UPy. This procedure allowed for the electrospinning of the bioactive nanofibers to form supramolecular biomembranes.

Drug Delivery

The term "drug delivery" describes procedures, systems, formulations, and techniques for safely delivering pharmaceutical compounds into human tumor tissues with the goal to achieve the desired therapeutic effect. Hydrogels are considered to have great potential for medication delivery and wound dressing because of their high water content and porous structure, which are characteristics of biological tissues. This can be linked to their flexible polymer networks. Hydrogels with pore sizes that can be controlled by adding cellulose derivatives can likewise be used as "on-off" release mechanisms. Cellulose based hydrogels have a wide range of applications in the field of drug release *in vivo* when combined with the benefits of hydrogels and biomaterials. Sultana and colleagues (Sultana, Van Hai, Park, Lee, & Lee, 2020) reported thermosensitive hydrogel barrier by combining mitomycin C (MMC) with modified tempo oxidized nanocellulose (cTOCN) via hydrogen bonds followed by integration with methyl cellulose (MC). The effective combination of MMC and cTOCN allowed for the controlled release of MMC from hydrogels during a duration of 14 days. Together with the well-managed administration of MMC, this study demonstrated the fast thermoreversible phenomenon (30 s) of tailored hydrogels because of excellent loading, non-cytotoxicity, good injectability, degradation, and mechanical stability.

The low cytotoxicity and weak immunogenicity of CDs and their derivatives have made them useful in the development of functional supramolecular materials for the administration of drugs, as well as for the construction of multiple host-guest supramolecular nanostructures with multiple dimensions and morphologies. In the pharmaceutical fields, CDs have been effectively used to form inclusion complexes with hydrophobic drugs through host-guest interactions. These interactions have been shown to reduce local and systemic toxicity, increase drug solubility and stability, enhance drug absorption, control drug-release pathways, and improve drug permeability across biological barriers, among other benefits. Li's group (Li et al., 2006) recently published a wide range of CD-based supramolecular drug delivery systems, such as supramolecular hydrogels, CD-drug complexes, and supramolecular nanoparticles. Using the host-guest complexation of α-CD and a biodegradable PEG-poly[(R)-3-hydroxybutyrate]-PEG triblock copolymer, they created a novel type of supramolecular hydrogels which have a thixotropic and reversible nature. As a consequence, they could be employed in injectable formulations that deliver macromolecular drugs over a relatively long period of time. Other supramolecular hosts, such as pillar[n]arene, cucurbit[n]uril, and calix[n]-arene, may be utilized as well to create promising supramolecular drug carriers for drug delivery in addition to CDs.

Wound Healing

Hydrogels have demonstrated significant potential in several applications within the domains of tissue engineering and wound treatments. Hydrogels have the potential to be biocompatible and biodegradable, and they can also have favorable characteristics like elasticity comparable to that of native tissue, humidity provision, the capacity to absorb wound discharge, a porous structure necessary for gas exchange, a pivotal function in preventing bacterial infections, and the provision of an environment that improves cellular activities like migration and proliferation. There is substantial overlap among the three main stages of wound healing, which are proliferation, remodelling, and inflammation. The first tissue response to a wound to stop bleeding involves clot formation. The physical, chemical, and mechanical qualities can be improved by modification, complexation, or cross-linking with other polymers and/or cross-linking agents, in addition to its use as wound dressings. Stimuli-responsive hydrogels can release drugs and bioactive molecules on-demand in reaction to the wounds acidic environment and high glucose levels, they are promising biomaterials for diabetic wound healing. An innovative approach to wound care could be the development of smart hydrogel-based wound dressings with the ability to recognize organisms and release antibacterial drugs at the area of the wound. Other drugs, such as anti-inflammatory ones, antioxidants, amino acids, vitamins, and minerals, can also be delivered via hydrogels. These medications may decrease inflammatory responses, promote wound healing, and support healthy wound tissue.

Tissue regeneration and wound healing have recently drawn attention to dressings based on hydrogels and nanofibers. Nahideh Asadi et al.,(Asadi et al., 2021) reported the challenges of individual nanofibers and hydrogels are taken into consideration as first concentrate on various methods for fabricating nanofiber-hydrogel based scaffolds. These methods primarily include bilayers, nanofiber reinforcement dressings, and self-assembled scaffolds as potential structures for wound healing applications.

CONCLUSION

This review summarizes recent developments in supramolecular host-guest self-assemblies with regard to various material types by employing individually tunable molecules as host and guest polymers (primarily four active domains: hydrogels, vesicles, nanoparticles, and micelles) and their diverse morphologies transition materials responsive to external stimuli. Thus, it would be desirable to add new recognition motifs with superior qualities based on these macrocyclic hosts to collection of supramolecular amphiphiles. Metallopolymers are good applications including functional materials, catalysis and biocatalysis, electroactive, photoactive and magnetic materials, environmental science. In NPs, HA stands out as a polyanionic polysaccharide that serves dual roles, functioning both as a DDSs and as a target motif for drug delivery. Ruthenium complexes have illustrated cytotoxic effects against cancer cells. By incorporating these complexes into conjugates based polymeric material, scientist aim to enhance their therapeutic potential. Coordination polymers is the utilising of long-circulating core-shell CPs, which efficiently carry various drugs with diverse mechanisms of action, enhancing synergistic therapeutic outcomes. The significant potential of using Cps as DDSs is attributed to several factors such as low toxicities of transition metals, well-defined chemical structures, high surface area and permeability. The collective characteristics of core-shell nanoparticles render them captivating for biological and advance innovative applications.

ACKNOWLEDGMENT

Dr. Nikita Patel is very grateful to Dr. Miral Vora and Dr. Asha Joshi for their enormous contributing, supports, and encouragement. The authors thank to Dr. Bhupesh Bhatt (Department of chemistry, Sardar Patel University) for creative input and fruitful scientific discussions. A valuable thanks to Dr. Keyur Bhatt for providing me this opportunity.

REFERENCES

Ahmadivand, A., Gerislioglu, B., Ahuja, R., & Mishra, Y. K. (2020). Terahertz plasmonics: The rise of toroidal metadevices towards immunobiosensings. *Materials Today*, *32*, 108–130. DOI: 10.1016/j.mattod.2019.08.002

Ahn, J., Miura, Y., Yamada, N., Chida, T., Liu, X., Kim, A., & Yasunaga, M. (2015). Antibody fragment-conjugated polymeric micelles incorporating platinum drugs for targeted therapy of pancreatic cancer. *Biomaterials*, *39*, 23–30. DOI: 10.1016/j.biomaterials.2014.10.069 PMID: 25477168

Amin, M. C. I. M., Ahmad, N., Pandey, M., Abeer, M. M., & Mohamad, N. (2015). Recent advances in the role of supramolecular hydrogels in drug delivery. *Expert Opinion on Drug Delivery*, *12*(7), 1149–1161. DOI: 10.1517/17425247.2015.997707 PMID: 25547588

Asadi, N., Pazoki-Toroudi, H., Del Bakhshayesh, A. R., Akbarzadeh, A., Davaran, S., & Annabi, N. (2021). Multifunctional hydrogels for wound healing: Special focus on biomacromolecular based hydrogels. *International Journal of Biological Macromolecules*, *170*, 728–750. DOI: 10.1016/j.ijbiomac.2020.12.202 PMID: 33387543

Braegelman, A. S., & Webber, M. J. (2019). Integrating Stimuli-Responsive Properties in Host-Guest Supramolecular Drug Delivery Systems. *Theranostics*, *9*(11), 3017–3040. DOI: 10.7150/thno.31913 PMID: 31244940

Chakraborty, S., & Newkome, G. R. (2018). Terpyridine-based metallosupramolecular constructs: Tailored monomers to precise 2D-motifs and 3D-metallocages. *Chemical Society Reviews*, *47*(11), 3991–4016. DOI: 10.1039/C8CS00030A PMID: 29594272

Chen, H., Chen, Y., Wu, H., Xu, J.-F., Sun, Z., & Zhang, X. (2018). Supramolecular polymeric chemotherapy based on cucurbit[7]uril-PEG copolymer. *Biomaterials*, *178*, 697–705. DOI: 10.1016/j.biomaterials.2018.02.051 PMID: 29545011

Cho, N., Cheong, T., Min, J., Wu, J., Lee, S., & Kim, D. (2011). JS yang, S. Kim, YK Kim and SY Seong. *Nature Nanotechnology*, *6*(10), 675–682. DOI: 10.1038/nnano.2011.149 PMID: 21909083

Chouhan, D., Thatikonda, N., Nilebäck, L., Widhe, M., Hedhammar, M., & Mandal, B. B. (2018). Recombinant Spider Silk Functionalized Silkworm Silk Matrices as Potential Bioactive Wound Dressings and Skin Grafts. *ACS Applied Materials & Interfaces*, *10*(28), 23560–23572. DOI: 10.1021/acsami.8b05853 PMID: 29940099

Côrte-Real, L., Robalo, M. P., Marques, F., Nogueira, G., Avecilla, F., Silva, T. J., & Valente, A. (2015). The key role of coligands in novel ruthenium (II)-cyclopentadienyl bipyridine derivatives: Ranging from non-cytotoxic to highly cytotoxic compounds. *Journal of Inorganic Biochemistry*, *150*, 148–159. DOI: 10.1016/j.jinorgbio.2015.06.015 PMID: 26150132

DeRosa, F., Kibbe, M. R., Najjar, S. F., Citro, M. L., Keefer, L. K., & Hrabie, J. A. (2007). Nitric oxide-releasing fabrics and other acrylonitrile-based diazeniumdiolates. *Journal of the American Chemical Society*, *129*(13), 3786–3787. DOI: 10.1021/ja0686864 PMID: 17343382

Dietzel, P. D., Panella, B., Hirscher, M., Blom, R., & Fjellvåg, H. (2006). Hydrogen adsorption in a nickel based coordination polymer with open metal sites in the cylindrical cavities of the desolvated framework. *Chemical Communications*, (9), 959–961. DOI: 10.1039/b515434k PMID: 16491175

Dong, R., Zhou, Y., Huang, X., Zhu, X., Lu, Y., & Shen, J. (2015). Functional supramolecular polymers for biomedical applications. *Advanced Materials*, *27*(3), 498–526. DOI: 10.1002/adma.201402975 PMID: 25393728

Dragutan, I., Dragutan, V., Simionescu, B. C., Demonceau, A., & Fischer, H. (2015). Recent advances in metathesis-derived polymers containing transition metals in the side chain. *Beilstein Journal of Organic Chemistry*, *11*(1), 2747–2762. DOI: 10.3762/bjoc.11.296 PMID: 26877797

Dutt, S., Kumar, R., & Siril, P. F. (2015). Green synthesis of a palladium–polyaniline nanocomposite for green Suzuki–Miyaura coupling reactions. *RSC Advances*, *5*(43), 33786–33791. DOI: 10.1039/C5RA05007C

Dzhardimalieva, G. I., Rabinskiy, L. N., Kydralieva, K. A., & Uflyand, I. E. (2019). Recent advances in metallopolymer-based drug delivery systems. *RSC Advances*, *9*(63), 37009–37051. DOI: 10.1039/C9RA06678K PMID: 35539076

Forero-Doria, O., Polo, E., Marican, A., Guzmán, L., Venegas, B., Vijayakumar, S., Wehinger, S., Guerrero, M., Gallego, J., & Durán-Lara, E. F. (2020). Supramolecular hydrogels based on cellulose for sustained release of therapeutic substances with antimicrobial and wound healing properties. *Carbohydrate Polymers, 242*, 116383. DOI: 10.1016/j.carbpol.2020.116383 PMID: 32564841

Fratoddi, I. (2017). Hydrophobic and hydrophilic Au and Ag nanoparticles. Breakthroughs and perspectives. *Nanomaterials (Basel, Switzerland), 8*(1), 11. DOI: 10.3390/nano8010011 PMID: 29280980

Gillies, R. J., Verduzco, D., & Gatenby, R. A. (2012). Evolutionary dynamics of carcinogenesis and why targeted therapy does not work. *Nature Reviews. Cancer, 12*(7), 487–493. DOI: 10.1038/nrc3298 PMID: 22695393

Gontero, D., Lessard-Viger, M., Brouard, D., Bracamonte, A. G., Boudreau, D., & Veglia, A. V. (2017). Smart multifunctional nanoparticles design as sensors and drug delivery systems based on supramolecular chemistry. *Microchemical Journal, 130*, 316–328. DOI: 10.1016/j.microc.2016.10.007

Guler, M. O., & Stupp, S. I. (2007). A self-assembled nanofiber catalyst for ester hydrolysis. *Journal of the American Chemical Society, 129*(40), 12082–12083. DOI: 10.1021/ja075044n PMID: 17854188

Han, G., Ghosh, P., De, M., & Rotello, V. M. (2007). Drug and gene delivery using gold nanoparticles. *NanoBiotechnology, 3*(1), 40–45. DOI: 10.1007/s12030-007-0005-3

Harada, A., Takashima, Y., & Nakahata, M. (2014). Supramolecular polymeric materials via cyclodextrin-guest interactions. *Accounts of Chemical Research, 47*(7), 2128–2140. DOI: 10.1021/ar500109h PMID: 24911321

He, C., Poon, C., Chan, C., Yamada, S. D., & Lin, W. (2016). Nanoscale coordination polymers codeliver chemotherapeutics and siRNAs to eradicate tumors of cisplatin-resistant ovarian cancer. *Journal of the American Chemical Society, 138*(18), 6010–6019. DOI: 10.1021/jacs.6b02486 PMID: 27088560

Homberg, A. (2018). Synthesis and Applications of Chiral Polyether Macrocycles. éditeur non identifié, Hu, J., Zhang, G., & Liu, S. (2012). Enzyme-responsive polymeric assemblies, nanoparticles and hydrogels. *Chemical Society Reviews, 41*(18), 5933–5949. DOI: 10.1039/c2cs35103j

Ingleson, M. J., Heck, R., Gould, J. A., & Rosseinsky, M. J. (2009). Nitric oxide chemisorption in a postsynthetically modified metal– organic framework. *Inorganic Chemistry, 48*(21), 9986–9988. DOI: 10.1021/ic9015977 PMID: 19795833

Karim, A. A., Dou, Q., Li, Z., & Loh, X. J. (2016). Emerging Supramolecular Therapeutic Carriers Based on Host-Guest Interactions. *Chemistry, an Asian Journal, 11*(9), 1300–1321. DOI: 10.1002/asia.201501434 PMID: 26833861

Karim, A. A., Dou, Q., Li, Z., & Loh, X. J. (2016). Emerging supramolecular therapeutic carriers based on host–guest interactions. *Chemistry, an Asian Journal, 11*(9), 1300–1321. DOI: 10.1002/asia.201501434 PMID: 26833861

Krieg, E., Bastings, M. M. C., Besenius, P., & Rybtchinski, B. (2016). Supramolecular Polymers in Aqueous Media. *Chemical Reviews, 116*(4), 2414–2477. DOI: 10.1021/acs.chemrev.5b00369 PMID: 26727633

Kumar, A., Jena, P. K., Behera, S., Lockey, R. F., Mohapatra, S., & Mohapatra, S. (2010). Multifunctional magnetic nanoparticles for targeted delivery. *Nanomedicine (London), 6*(1), 64–69. DOI: 10.1016/j.nano.2009.04.002 PMID: 19446653

Kumar, R., Aadil, K. R., Ranjan, S., & Kumar, V. B. (2020). Advances in nanotechnology and nanomaterials based strategies for neural tissue engineering. *Journal of Drug Delivery Science and Technology, 57*, 101617. DOI: 10.1016/j.jddst.2020.101617

Kumar, R., Mondal, K., Panda, P. K., Kaushik, A., Abolhassani, R., Ahuja, R., Rubahn, H.-G., & Mishra, Y. K. (2020). Core–shell nanostructures: Perspectives towards drug delivery applications. *Journal of Materials Chemistry. B, 8*(39), 8992–9027. DOI: 10.1039/D0TB01559H PMID: 32902559

Kumar, R., Soni, P., & Siril, P. F. (2019). Engineering the morphology and particle size of high energetic compounds using drop-by-drop and drop-to-drop solvent–antisolvent interaction methods. *ACS Omega, 4*(3), 5424–5433. DOI: 10.1021/acsomega.8b03214 PMID: 31459707

Kumar, V. B., Kumar, R., Gedanken, A., & Shefi, O. (2019). Fluorescent metal-doped carbon dots for neuronal manipulations. *Ultrasonics Sonochemistry, 52*, 205–213. DOI: 10.1016/j.ultsonch.2018.11.017 PMID: 30522849

Lago, A. B., Pino-Cuevas, A., Carballo, R., & Vázquez-López, E. M. (2016). A new metal–organic polymeric system capable of stimuli-responsive controlled release of the drug ibuprofen. *Dalton Transactions (Cambridge, England), 45*(4), 1614–1621. DOI: 10.1039/C5DT04031K PMID: 26692254

Li, J., Li, X., Ni, X., Wang, X., Li, H., & Leong, K. W. (2006). Self-assembled supramolecular hydrogels formed by biodegradable PEO–PHB–PEO triblock copolymers and α-cyclodextrin for controlled drug delivery. *Biomaterials, 27*(22), 4132–4140. DOI: 10.1016/j.biomaterials.2006.03.025 PMID: 16584769

Ling, X., Shen, Y., Sun, R., Zhang, M., Li, C., Mao, J., Xing, J., Sun, C., & Tu, J. (2015). Tumor-targeting delivery of hyaluronic acid–platinum (iv) nanoconjugate to reduce toxicity and improve survival. *Polymer Chemistry*, *6*(9), 1541–1552. DOI: 10.1039/C4PY01592D

Liu, Y., Liu, J., Chen, S., Lei, T., Kim, Y., Niu, S., & Tok, J. B.-H. (2019). Soft and elastic hydrogel-based microelectronics for localized low-voltage neuromodulation. *Nature Biomedical Engineering*, *3*(1), 58–68. DOI: 10.1038/s41551-018-0335-6 PMID: 30932073

Lu, M., Chen, F., Noy, J. M., Lu, H., & Stenzel, M. H. (2017). Enhanced Anti-metastatic Activity of the Ruthenium Anticancer Drug RAPTA-C Delivered in Fructose-Coated Micelles. *Macromolecular Bioscience*, *17*(10), 1600513. DOI: 10.1002/mabi.201600513 PMID: 28234423

Ma, X., & Zhao, Y. (2015). Biomedical Applications of Supramolecular Systems Based on Host-Guest Interactions. *Chemical Reviews*, *115*(15), 7794–7839. DOI: 10.1021/cr500392w PMID: 25415447

Ma, Z., & Moulton, B. (2011). Recent advances of discrete coordination complexes and coordination polymers in drug delivery. *Coordination Chemistry Reviews*, *255*(15-16), 1623–1641. DOI: 10.1016/j.ccr.2011.01.031

McKinlay, A. C., Xiao, B., Wragg, D. S., Wheatley, P. S., Megson, I. L., & Morris, R. E. (2008). Exceptional behavior over the whole adsorption– storage– delivery cycle for NO in porous metal organic frameworks. *Journal of the American Chemical Society*, *130*(31), 10440–10444. DOI: 10.1021/ja801997r PMID: 18627150

Miller, M., & Megson, I. (2007). Recent developments in nitric oxide donor drugs. *British Journal of Pharmacology*, *151*(3), 305–321. DOI: 10.1038/sj.bjp.0707224 PMID: 17401442

Miller, S. R., Heurtaux, D., Baati, T., Horcajada, P., Grenèche, J.-M., & Serre, C. (2010). Biodegradable therapeutic MOFs for the delivery of bioactive molecules. *Chemical Communications*, *46*(25), 4526–4528. DOI: 10.1039/c001181a PMID: 20467672

Miyamae, K., Nakahata, M., Takashima, Y., & Harada, A. (2015). Self-Healing, Expansion-Contraction, and Shape-Memory Properties of a Preorganized Supramolecular Hydrogel through Host-Guest Interactions. *Angewandte Chemie International Edition in English*, *54*(31), 8984–8987. DOI: 10.1002/anie.201502957 PMID: 26080301

Molineux, G. (2002). Pegylation: Engineering improved pharmaceuticals for enhanced therapy. *Cancer Treatment Reviews*, *28*, 13–16. DOI: 10.1016/S0305-7372(02)80004-4 PMID: 12173407

Mura, S., Nicolas, J., & Couvreur, P. (2013). Stimuli-responsive nanocarriers for drug delivery. *Nature Materials*, *12*(11), 991–1003. DOI: 10.1038/nmat3776 PMID: 24150417

Newman, A. D., Brown, D. R., Siril, P., Lee, A. F., & Wilson, K. (2006). Structural studies of high dispersion H 3 PW 12 O 40/SiO 2 solid acid catalysts. *Physical Chemistry Chemical Physics*, *8*(24), 2893–2902. DOI: 10.1039/b603979k PMID: 16775645

Perles, J., Snejko, N., Iglesias, M., & Monge, M. Á. (2009). 3D scandium and yttrium arenedisulfonate MOF materials as highly thermally stable bifunctional heterogeneous catalysts. *Journal of Materials Chemistry*, *19*(36), 6504–6511. DOI: 10.1039/b902954k

Raia, N. R., Partlow, B. P., McGill, M., Kimmerling, E. P., Ghezzi, C. E., & Kaplan, D. L. (2017). Enzymatically crosslinked silk-hyaluronic acid hydrogels. *Biomaterials*, *131*, 58–67. DOI: 10.1016/j.biomaterials.2017.03.046 PMID: 28376366

Ravoo, B. J. (2012). Vesicles in Supramolecular Chemistry. In *Supramolecular Chemistry*. DOI: 10.1002/9780470661345.smc032

Schmidt, B. V. K. J., & Barner-Kowollik, C. (2016). A Supramolecular Approach to Macromolecular Self-Assembly: Cyclodextrin Host/Guest Complexes. In *Macromolecular Self-assembly* (pp. 1-32). Academic Press.

Shen, J., Zhang, W., Qi, R., Mao, Z.-W., & Shen, H. (2018). Engineering functional inorganic–organic hybrid systems: Advances in siRNA therapeutics. *Chemical Society Reviews*, *47*(6), 1969–1995. DOI: 10.1039/C7CS00479F PMID: 29417968

Soussan, E., Cassel, S., Blanzat, M., & Rico-Lattes, I. (2009). Drug delivery by soft matter: Matrix and vesicular carriers. *Angewandte Chemie International Edition in English*, *48*(2), 274–288. DOI: 10.1002/anie.200802453 PMID: 19072808

Sultana, T., Van Hai, H., Park, M., Lee, S.-Y., & Lee, B.-T. (2020). Controlled release of Mitomycin C from modified cellulose based thermo-gel prevents post-operative de novo peritoneal adhesion. *Carbohydrate Polymers*, *229*, 115552. DOI: 10.1016/j.carbpol.2019.115552 PMID: 31826495

Toy, R., Bauer, L., Hoimes, C., Ghaghada, K. B., & Karathanasis, E. (2014). Targeted nanotechnology for cancer imaging. *Advanced Drug Delivery Reviews*, *76*, 79–97. DOI: 10.1016/j.addr.2014.08.002 PMID: 25116445

Trokowski, R., Akine, S., & Nabeshima, T. (2008). Synthesis, characterization and molecular recognition of a bis-platinum terpyridine dimer. *Chemical Communications*, (7), 889–890. DOI: 10.1039/b716459a PMID: 18253538

Tsai, F.-T., Chiou, S.-J., Tsai, M.-C., Tsai, M.-L., Huang, H.-W., Chiang, M.-H., & Liaw, W.-F. (2005). Dinitrosyl iron complexes (DNICs)[L2Fe (NO) 2]-(L= Thiolate): Interconversion among {Fe (NO) 2} 9 DNICs,{Fe (NO) 2} 10 DNICs, and [2Fe-2S] clusters, and the critical role of the thiolate ligands in regulating NO release of DNICs. *Inorganic Chemistry*, *44*(16), 5872–5881. DOI: 10.1021/ic0505044 PMID: 16060642

Wang, J., Liu, W., Luo, G., Li, Z., Zhao, C., Zhang, H., Zhu, M., Xu, Q., Wang, X., Zhao, C., Qu, Y., Yang, Z., Yao, T., Li, Y., Lin, Y., Wu, Y., & Li, Y. (2018). Synergistic effect of well-defined dual sites boosting the oxygen reduction reaction. *Energy & Environmental Science*, *11*(12), 3375–3379. DOI: 10.1039/C8EE02656D

Wen, Y., Zhang, Z., & Li, J. (2014). Highly Efficient Multifunctional Supramolecular Gene Carrier System Self-Assembled from Redox-Sensitive and Zwitterionic Polymer Blocks. *Advanced Functional Materials*, *24*(25), 3874–3884. DOI: 10.1002/adfm.201303687

Wlodarczyk, M. T., Dragulska, S. A., Camacho-Vanegas, O., Dottino, P. R., Jarzęcki, A. A., Martignetti, J. A., & Mieszawska, A. J. (2018). Platinum (II) complex-nuclear localization sequence peptide hybrid for overcoming platinum resistance in cancer therapy. *ACS Biomaterials Science & Engineering*, *4*(2), 463–467. DOI: 10.1021/acsbiomaterials.7b00921 PMID: 32042890

Xavier-Jr, F., Gueutin, C., Chacun, H., Vauthier, C., & Egito, E. (2019). Mucoadhesive paclitaxel-loaded chitosan-poly (isobutyl cyanoacrylate) core-shell nanocapsules containing copaiba oil designed for oral drug delivery. *Journal of Drug Delivery Science and Technology*, *53*, 101194. DOI: 10.1016/j.jddst.2019.101194

Xiao, B., Wheatley, P. S., Zhao, X., Fletcher, A. J., Fox, S., Rossi, A. G., & Thomas, K. M. (2007). High-capacity hydrogen and nitric oxide adsorption and storage in a metal– organic framework. *Journal of the American Chemical Society*, *129*(5), 1203–1209. DOI: 10.1021/ja066098k PMID: 17263402

Xiao, T., Zhong, W., Xu, L., Sun, X. Q., Hu, X. Y., & Wang, L. (2019). Supramolecular vesicles based on pillar[n]arenes: Design, construction, and applications. *Organic & Biomolecular Chemistry*, *17*(6), 1336–1350. DOI: 10.1039/C8OB03095B PMID: 30638249

Xie, T.-Z., Guo, K., Li, J.-Y., Zhang, B., Zheng, K., Moorefield, C. N., & Wesdemiotis, C. (2016). Coordination-Driven, Self-Assembly of a Polycyclic, Terpyridine-Based Nanobelt. *Journal of Inorganic and Organometallic Polymers and Materials*, *26*(4), 907–913. DOI: 10.1007/s10904-016-0393-8

Xing, P., & Zhao, Y. (2016). Multifunctional Nanoparticles Self-Assembled from Small Organic Building Blocks for Biomedicine. *Advanced Materials*, *28*(34), 7304–7339. DOI: 10.1002/adma.201600906 PMID: 27273862

Xiong, L., Yang, T., Yang, Y., Xu, C., & Li, F. (2010). Long-term in vivo biodistribution imaging and toxicity of polyacrylic acid-coated upconversion nanophosphors. *Biomaterials*, *31*(27), 7078–7085. DOI: 10.1016/j.biomaterials.2010.05.065 PMID: 20619791

Yan, X., Wang, F., Zheng, B., & Huang, F. (2012). Stimuli-responsive supramolecular polymeric materials. *Chemical Society Reviews*, *41*(18), 6042–6065. DOI: 10.1039/c2cs35091b PMID: 22618080

Yang, H., Chen, Y., Chen, Z., Geng, Y., Xie, X., Shen, X., Li, T., Li, S., Wu, C., & Liu, Y. (2017). Chemo-photodynamic combined gene therapy and dual-modal cancer imaging achieved by pH-responsive alginate/chitosan multilayer-modified magnetic mesoporous silica nanocomposites. *Biomaterials Science*, *5*(5), 1001–1013. DOI: 10.1039/C7BM00043J PMID: 28327716

Yang, H., Yuan, B., Zhang, X., & Scherman, O. A. (2014). Supramolecular Chemistry at Interfaces: Host–Guest Interactions for Fabricating Multifunctional Biointerfaces. *Accounts of Chemical Research*, *47*(7), 2106–2115. DOI: 10.1021/ar500105t PMID: 24766328

Yang, K., Pei, Y., Wen, J., & Pei, Z. (2016). Recent advances in pillar [n] arenes: Synthesis and applications based on host–guest interactions. *Chemical Communications*, *52*(60), 9316–9326. DOI: 10.1039/C6CC03641D PMID: 27332040

Yang, Y., Xu, L., Zhu, W., Feng, L., Liu, J., Chen, Q., Dong, Z., Zhao, J., Liu, Z., & Chen, M. (2018). One-pot synthesis of pH-responsive charge-switchable PEGylated nanoscale coordination polymers for improved cancer therapy. *Biomaterials*, *156*, 121–133. DOI: 10.1016/j.biomaterials.2017.11.038 PMID: 29195181

Yoon, H.-J., & Jang, W.-D. (2010). Polymeric supramolecular systems for drug delivery. *Journal of Materials Chemistry*, *20*(2), 211–222. DOI: 10.1039/B910948J

Yu, G., Jie, K., & Huang, F. (2015). Supramolecular Amphiphiles Based on Host–Guest Molecular Recognition Motifs. *Chemical Reviews*, *115*(15), 7240–7303. DOI: 10.1021/cr5005315 PMID: 25716119

Yu, M., Chen, Y., Zhang, N., & Liu, Y. (2010). Construction and transmembrane dissociation behavior of supramolecular assembly of quinolinocyclodextrin with porphyrin. *Organic & Biomolecular Chemistry*, *8*(18), 4148–4154. DOI: 10.1039/c0ob00080a PMID: 20657925

Yu, R., Yang, Y., He, J., Li, M., & Guo, B. (2021). Novel supramolecular self-healing silk fibroin-based hydrogel via host–guest interaction as wound dressing to enhance wound healing. *Chemical Engineering Journal*, *417*, 128278. DOI: 10.1016/j.cej.2020.128278

Yuan, W., Ma, J., Zhao, Z., & Liu, S. (2020). Self-Assembly of Supramolecular DNA Amphiphiles through Host–Guest Interaction and Their Stimuli-Responsiveness. *Macromolecular Rapid Communications*, *41*(9), 2000022. DOI: 10.1002/marc.202000022 PMID: 32196823

Zhang, A., Jung, K., Li, A., Liu, J., & Boyer, C. (2019). Recent advances in stimuli-responsive polymer systems for remotely controlled drug release. *Progress in Polymer Science*, *99*, 101164. DOI: 10.1016/j.progpolymsci.2019.101164

Zhang, L., Gu, F., Chan, J., Wang, A., Langer, R., & Farokhzad, O. (2008). Nanoparticles in medicine: Therapeutic applications and developments. *Clinical Pharmacology and Therapeutics*, *83*(5), 761–769. DOI: 10.1038/sj.clpt.6100400 PMID: 17957183

Zhang, Q., Re Ko, N., & Kwon Oh, J. (2012). Modulated morphologies and tunable thiol-responsive shedding of aqueous block copolymer aggregates. *RSC Advances*, *2*(21), 8079–8086. DOI: 10.1039/c2ra21209a

Chapter 4
Supramolecular Approaches to Green Drug Delivery

Bancha Yingngam
https://orcid.org/0000-0001-7215-9123
Ubon Ratchathani University, Thailand

ABSTRACT

In the pharmaceutical sciences, developing efficient, sustainable, and biocompatible drug delivery systems is crucial. Traditional methods face issues such as poor solubility and low bioavailability. Supramolecular chemistry, which involves the formation of complex structures through noncovalent interactions, offers solutions to these challenges. This chapter discusses the use of supramolecular systems for green drug delivery, aiming to enhance efficacy, specificity, and sustainability. The authors examined how these structures can improve drug solubility, target specificity, and controlled release, reducing environmental impact. This chapter covers fundamental supramolecular chemistry principles; the various structures, such as cyclodextrins, dendrimers, and micelles and their drug delivery applications. This chapter concludes with the future prospects of supramolecular approaches in green drug delivery, highlighting ongoing research, challenges, and directions such as integrating nanotechnology and personalized medicine for better drug delivery systems.

INTRODUCTION

The field of drug delivery systems (DDSs) has evolved dramatically over the past few decades, markedly changing the administration of therapeutic agents (Hu et al., 2024). The primary objective of DDSs is to ensure that the active pharmaceutical

DOI: 10.4018/979-8-3693-2774-6.ch004

ingredient (API), the component of a drug responsible for its therapeutic effects, reaches the target site in the body in an efficient and controlled manner (Pandey et al., 2024). APIs are distinct from excipients or additives; they are included in drug formulations to aid in delivery, stability, or other aspects but do not have therapeutic effects themselves (Puhlmann et al., 2024). By focusing on efficient API delivery, DDSs minimize systemic drug exposure, thereby reducing side effects and improving therapeutic efficacy (Gao et al., 2023). The evolution of drug delivery can be categorized into several generations, each marked by technological advancements and increased complexity (Fuhrmann, 2023). The first generation focused on immediate and sustained release formulations, aiming to improve patient compliance and maintain steady drug levels (Ezike et al., 2023). Subsequent generations introduced more sophisticated systems, such as targeted delivery, directing the drug to specific cell types or tissues, and responsive delivery systems that release the drug in response to specific physiological triggers (Joshi et al., 2023). These innovations have been crucial in treating various diseases, especially those requiring precise drug targeting, such as cancer or localized infections (Sun et al., 2023).

Despite significant advancements, traditional drug delivery systems still face several challenges. One of the primary issues is the inability to precisely control drug release, which often leads to suboptimal therapeutic levels and potential toxicity (Lin et al., 2024). For instance, drugs that are not specifically targeted at diseased cells can inadvertently affect healthy cells, resulting in adverse side effects. Encapusuating drugs in specialized drug delivery systems can significantly mitigate this issue. These systems reduce the impact on health by targeting the drug directly at diseased cells (Zhou et al., 2024). Thus, side effects should be minimized, and therapeutic efficacy should be increased (Saha et al., 2023). Additionally, many drugs suffer from poor solubility and bioavailability, meaning that a significant portion of the drug is wasted and fails to reach the target site in an active form (Wu et al., 2024). Encapsulation can improve the stability and solubility of a drug, ensuring that it remains effective until it reaches the target site (Buddhadev et al., 2024). Furthermore, such systems can provide controlled release of the drug, maintaining optimal therapeutic levels in the body for an extended period (Z. Shi et al., 2024). This controlled release not only enhances the effectiveness of the drug but also improves patient compliance, as it can reduce the frequency of dosing (Jindal et al., 2023). Another major challenge is the development of drug resistance, particularly in the treatment of chronic diseases and infections (Gautam & Kaur, 2024; Kirla et al., 2023). Over time, pathogens or cancer cells may become resistant to certain drugs, necessitating higher doses or the development of new medications, which further complicates treatment (Yong Joon Kim et al., 2023). Furthermore, patient compliance remains a significant hurdle, especially in treatments that require frequent or complex dosing schedules (Z. Wang et al., 2024). These challenges underscore the need for more sophisticated and patient-friendly drug delivery systems capable of addressing these multifaceted issues effectively (Parveen et al., 2023).

Supramolecular chemistry, which studies the assembly of molecules into larger complex structures through noncovalent interactions, offers promising solutions to challenges in traditional drug delivery systems (Attia et al., 2024). These interactions, such as hydrogen bonding, hydrophobic effects, electrostatic forces, van der Waals forces, and π-π interactions, facilitate the creation of highly organized structures capable of encapsulating and transporting drugs (R. Hu et al., 2024; Rajaram et al., 2023). Supramolecular drug delivery systems boast several advantages, including improved drug solubility and stability, targeted delivery to specific body sites, and controlled drug release (Xiang et al., 2024). These systems are designed to respond to specific physiological stimuli—such as pH changes, temperature shifts, or the presence of certain enzymes—enabling precise drug release at the target site (Alam et al., 2023). This level of control is crucial for reducing side effects and enhancing treatment efficacy (H.-Y. Zheng et al., 2024). Furthermore, the dynamic nature of supramolecular assemblies allows them to be adaptable to a variety of drug molecules and therapeutic contexts (Yu et al., 2024). The integration of supramolecular approaches into drug delivery represents a rapidly evolving area of research. This approach has demonstrated significant potential in overcoming many limitations of conventional systems, paving the way for more efficient and safer therapeutic treatments (Q. Wang et al., 2024).

The aims of this chapter are as follows:

(1) To elucidate the fundamental concepts of supramolecular chemistry and their application in drug delivery systems. This includes an in-depth examination of molecular interactions and self-assembly processes central to supramolecular structures, providing a solid foundation for understanding their role in drug delivery.
(2) To explore the advancement of environmental sustainability in drug delivery through supramolecular approaches. This involves assessing methods to decrease the ecological impact of pharmaceuticals, with a focus on reducing harmful solvents and utilizing biocompatible and biodegradable materials.
(3) To present and analyze current research and specific case studies demonstrating the application of supramolecular techniques in developing greener drug delivery systems. These examples will showcase successful applications and discuss the practical challenges encountered.
(4) To forecast future directions and innovative developments in supramolecular drug delivery. The aim here is to identify emerging trends and potential breakthroughs that could more profoundly integrate green chemistry principles with

supramolecular drug delivery, potentially leading to more environmentally friendly and efficient therapeutic solutions.

Through these aims, this chapter intends to provide a comprehensive and forward-looking perspective on the role of supramolecular chemistry in enhancing the eco-friendliness of drug delivery systems.

The rest of this chapter is organized as follows: Section 2 explores the intricacies of supramolecular structures in drug delivery, examining their unique properties and roles. This review includes a detailed exploration of cyclodextrins, emphasizing their diverse applications in pharmaceutical formulations. This section further investigates dendrimers, focusing on their utility in targeted delivery and adaptability for specific therapeutic targets. Additionally, the discussion on micelles centers on their role in enhancing the solubility of poorly soluble drugs, a crucial aspect of drug formulation. This chapter then examines macrocycles, discussing their versatile applications in drug delivery, followed by an exploration of supramolecular gels and their distinctive properties suitable for various drug delivery applications. Section 3 presents practical case studies and applications of these systems, offering insights into their utility and effectiveness. This section also addresses the environmental impact of drug delivery systems, highlighting the ecological implications of pharmaceuticals and the role of supramolecular systems in promoting eco-friendly practices. In Section 4, the focus shifts to the challenges and future directions in the field. This section emphasizes the current limitations of supramolecular drug delivery and forecasts future research trends and potential breakthroughs. In Section 5, the conclusion and perspectives section summarize the key findings of the chapter. This chapter explores the future scope of this field, offering a comprehensive overview of the potential and challenges of supramolecular approaches in green drug delivery.

SUPRAMOLECULAR STRUCTURES IN DRUG DELIVERY

Within the scope of drug delivery systems, supramolecular structures are intricate entities formed by the self-assembly of individual molecules through noncovalent interactions, including hydrogen bonding, van der Waals forces, hydrophobic effects, and electrostatic interactions (Zhuo et al., 2024). These reversible and dynamic interactions facilitate the formation of highly organized yet adaptable structures (T. Hu et al., 2024). Particularly suited for delivering BCS Class II and IV drugs, which are characterized by low solubility (and low permeability in the case of Class IV), supramolecular structures such as cyclodextrins, micelles, and dendrimers can effectively enhance the solubility and dissolution rates of these drugs. Cyclodextrins, for example, form inclusion complexes with hydrophobic drug molecules to increase aqueous solubility, while micelles and dendrimers encapsulate drugs within their hydrophobic core, improving solubility and protecting drugs from degradation (Rubim et al., 2024; Sanphui et al., 2024). Moreover, these structures can be engineered for targeted drug delivery, releasing drugs in a controlled manner at specific sites, thus maximizing therapeutic efficacy and minimizing systemic side effects. This is particularly vital for BCS Class IV drugs, for which poor permeability could otherwise hinder their effectiveness. Additionally, supramolecular carriers facilitate enhanced permeation through biological membranes for BCS Class IV drugs, often by leveraging endocytosis pathways or modulating drug interactions with cell membrane lipid bilayers (Jha et al., 2022).

Supramolecular systems for drug delivery, illustrated in Figure 1, offer significant advantages by improving the solubility and stability of therapeutic agents through encapsulation. These structures not only enhance the effectiveness and shelf life of drugs but also enable targeted delivery (Zhou et al., 2024). Targeting is achieved through surface modifications with molecules such as ligands or antibodies, which allow for selective binding to specific cells or tissues, thereby minimizing side effects. Additionally, the design of these systems facilitates controlled drug release, which is responsive to physiological triggers such as pH, temperature, or enzymatic activity, ensuring timely and location-specific drug release (Z. Zhang et al., 2024). The versatility and modularity of supramolecular systems are particularly noteworthy; they can be tailored to transport various therapeutic agents, from small molecules to proteins and nucleic acids, with customization extending to their size, shape, and functional properties (Y.-J. Zhang et al., 2024).

Table 1 provides an overview of the various supramolecular structures employed in drug delivery, highlighting their unique properties and applications. Cyclodextrins, composed of cyclic oligosaccharides with a hydrophobic cavity and hydrophilic exterior, are utilized for solubility enhancement, stability improvement, and taste masking in drug formulations (Kali et al., 2024). Dendrimers, characterized by their highly branched, tree-like structure and precisely defined molecular weight with multiple terminal functional groups, have applications in targeted drug delivery, gene therapy, and diagnostic imaging (Khatik et al., 2024). As amphiphilic materials with a hydrophobic core and hydrophilic shell, micelles self-assemble in aqueous solutions and are effective at enhancing the solubility of hydrophobic drugs, controlling drug release, and reducing systemic toxicity (Xue et al., 2024). Macrocycles, large ring-shaped molecules known for their ability to encapsulate guest molecules and chemical versatility, are used to enhance drug solubility and stability, in targeted delivery systems, and for the protection of sensitive drugs (Lyu et al., 2023). Finally, supramolecular gels, which form a 3D network of self-assembled fibers that are responsive to external stimuli and biocompatible and biodegradable, are applied in sustained and controlled drug release, topical and transdermal delivery, and tissue engineering applications (Shang et al., 2022). Each of these supramolecular structures, as detailed by various researchers, plays a vital role in advancing drug delivery technologies.

In addition, Table 2 presents a comparison of drug loading efficiencies in various supramolecular systems. Cyclodextrins generally exhibit moderate to high efficiency and are especially effective for small, hydrophobic molecules, but their efficacy depends on the cyclodextrin size and nature of the guest molecule (Safari et al., 2023; Saha et al., 2023). Dendrimers, characterized by numerous attachment points, offer high drug loading, influenced by dendrimer generation and drug type (Macedo et al., 2023; Pandey et al., 2023). Moreover, these micelles exhibit moderate efficiency and are particularly suitable for hydrophobic drugs due to their hydrophobic core, the efficiency of which is contingent on the core-shell properties (Cai et al., 2022; Shafiq et al., 2023). Macrocycles have variable efficiencies and heavily rely on their size and chemical compatibility with drugs, and selective binding can enhance their efficiency (He et al., 2024). Supramolecular gels range from moderate to high efficiency, vary in terms of their network structure and physicochemical properties, and can provide controlled release (Cheng et al., 2022). The efficiency of each system is influenced by specific structural and chemical factors, as detailed in various studies (Zhuo et al., 2024). The biocompatibility and biodegradability of these materials further ensure safe integration within the body, consistent with biological processes (W. Shi et al., 2024). This section explores various types of supramolecular systems, including cyclodextrin inclusion complexes, dendrimers, micelles, macrocycles, and supramolecular gels, each of which significantly contributes to enhancing the efficacy, specificity, and safety of drug delivery.

Figure 1. Illustration of supramolecular systems for drug delivery and their advantages

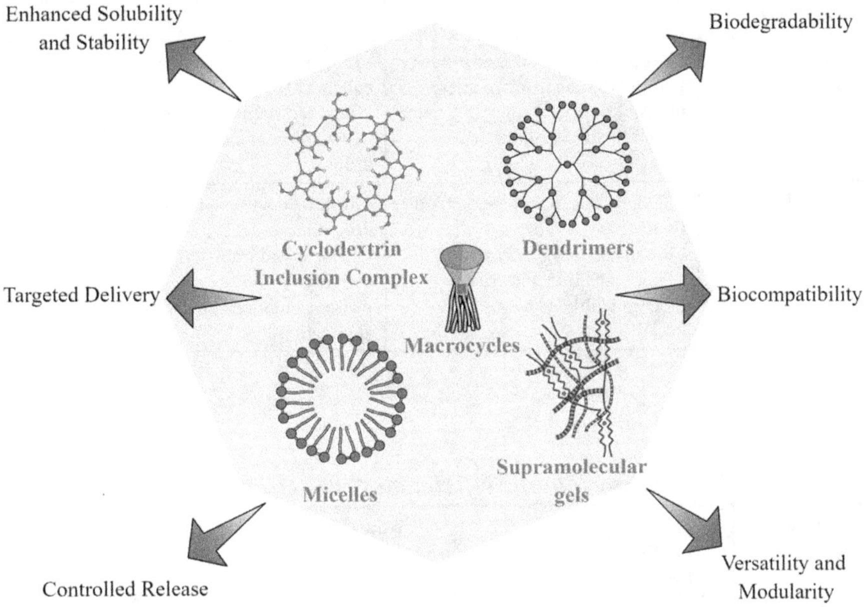

Table 1. Overview of various supramolecular structures used in drug delivery, including their properties and applications

Supramolecular Structures	Properties	Applications	References
Cyclodextrins	- Cyclic oligosaccharides - Hydrophobic cavity and hydrophilic exterior - Ability to form inclusion complexes	- Solubility enhancement - Stability improvement - Taste masking	Kali et al. (2024)
Dendrimers	- Highly branched, tree-like structure - Precisely defined molecular weight - Multiple terminal functional groups	- Targeted drug delivery - Gene therapy - Diagnostic imaging	Khatik et al. (2024)
Micelles	- Amphiphilic (hydrophobic core, hydrophilic shell) - Self-assemble in aqueous solutions - Nanoscale size range	- Solubility enhancement for hydrophobic drugs - Controlled drug release - Reducing systemic toxicity	Xue et al. (2024)

continued on following page

Table 1. Continued

Supramolecular Structures	Properties	Applications	References
Macrocycles	- Large ring-shaped molecules - Ability to encapsulate guest molecules - Chemically versatile	- Drug solubility and stability enhancement - Targeted delivery systems - Protection of sensitive drugs	Lyu et al. (2023)
Supramolecular gels	- 3D network of self-assembled fibers - Responsive to external stimuli - Biocompatible and biodegradable	- Sustained and controlled drug release - Topical and transdermal delivery - Tissue engineering applications	Shang et al. (2022)

Table 2. Comparison of drug loading efficiencies in different supramolecular systems

Supramolecular systems	Typical drug loading efficiency	Notes	References
Cyclodextrins	Moderate to high	Efficiency varies with the size of the cyclodextrin and the nature of the guest molecule. Cyclodextrins are especially effective for small, hydrophobic molecules.	Safari et al. (2023); Sahu et al. (2023)
Dendrimers	High	Dendrimers offer numerous attachment points, allowing for high drug loading. Efficiency depends on dendrimer generation and the type of drug.	Macedo et al. (2023); Pandey et al. (2023)
Micelles	Moderate	Efficient for hydrophobic drugs due to their hydrophobic core. The efficiency is influenced by the core-shell properties of the micelle.	Cai et al. (2022); Shafiq et al. (2023)
Macrocycles	Variable	Depends heavily on the size and chemical compatibility of the macrocycle with the drug. Some macrocycles show selective binding, which can enhance efficiency.	He et al. (2024)
Supramolecular gels	Moderate to high	Efficiency varies based on the network structure of the gel and the physicochemical properties of the drug. Responsive gels can offer controlled release.	Chang Zhang et al. (2024)

Cyclodextrins

Cyclodextrins: An Overview

Cyclodextrins are cyclic oligosaccharides composed of glucose monomers linked by α-1,4 glycosidic bonds. These monomers arrange into a torus-like structure, offering a unique molecular environment characterized by a hydrophobic interior and a hydrophilic exterior (Meenatchi et al., 2024). This distinct configuration allows cyclodextrins to interact effectively with a wide range of drug molecules, making them pivotal tools in drug delivery systems (Zhao et al., 2024). The ability of these compounds to form host-guest complexes is particularly notable, as they enhance the solubility, stability, and bioavailability of poorly water-soluble drugs (Fang et al., 2024). Cyclodextrins are capable of encapsulating various hydrophobic molecules within their central cavity, which is highly valuable in pharmaceutical applications (Kali et al., 2024). The three primary types of cyclodextrins—α, β, and γ—differ in their number of glucose units and the size of their cavities, enabling the encapsulation of diverse hydrophobic drug molecules (Y. Li et al., 2024). In this section, the author presents two key visual elements that are crucial for a comprehensive understanding of cyclodextrins: Figure 2 and Table 3. Figure 2 graphically depicts the molecular structures of α-cyclodextrin (α-CD), β-cyclodextrin (β-CD), γ-cyclodextrin (γ-CD), and hydroxypropyl-β-cyclodextrin (HP-β-CD), providing a visual perspective of their structural differences. Table 3 offers a detailed comparison of the physicochemical properties of these materials, elucidating variations such as the number of glucose units and cavity size and highlighting their impact on their applications and functionalities. Together, these visual aids play an instrumental role in deepening the understanding of the unique properties and roles of different cyclodextrins across various applications.

α–Cyclodextrin. α–CD, a cyclic oligosaccharide composed of six glucopyranose units arranged in a ring, exhibits a range of distinctive chemical properties due to its unique structure. The glycopyrnose units are linked by α-1,4 glycosidic bonds, forming a toroidal shape with a hydrophilic outer surface and a hydrophobic central cavity. This amphiphilic nature allows α–CD to engage in host–guest interactions with various molecules, particularly those that are hydrophobic, leading to the formation of inclusion complexes. The hydrophobic cavity of α–CD, although smaller than that of β–CD or γ–CD, is adept at encapsulating smaller molecules, making it highly effective at enhancing the solubility and stability of hydrophobic drugs. Chemically, α–CD is stable under a wide range of pH levels, which is a critical property for its application in diverse environments, from pharmaceutical formulations to food products. Its solubility in water is moderate and influenced by temperature and the presence of other solutes (Niu et al., 2024). In addition, α–CD can undergo chemical modifications, such as methylation or hydroxypropylation, which can further tailor its solubility, complexation ability, and overall chemical behavior. These modifications expand the range of its applications, allowing for more specific and targeted use in drug delivery, food science, and other fields (Hasanah et al., 2023). The relatively small cavity size of α–CD and its ability to form stable inclusion complexes are key to its role in molecular encapsulation, where it can protect guest molecules from external factors such as oxidation and volatilization, enhance the bioavailability of drugs, or control the release of flavors and fragrances (Yilmaz et al., 2023).

Figure 2. Schematic illustration of the cyclodextrin molecules used in drug delivery systems: (a) α–cyclodextrin, (b) β–cyclodextrin, (c) γ–cyclodextrin, and (d) hydroxypropyl–β–cyclodextrin

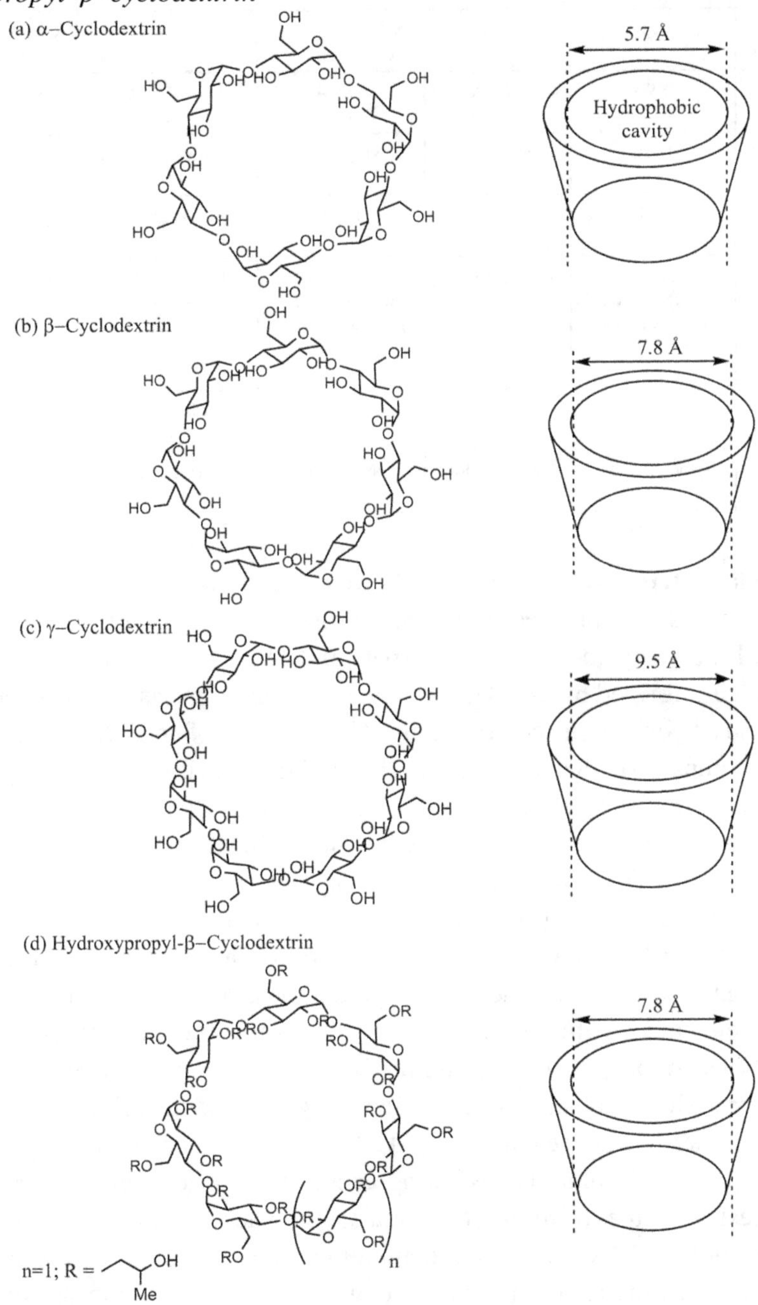

(ChemDraw 20.1.1)

Table 3. Comparative analysis of the physicochemical properties of cyclodextrins

Property	α–Cyclodextrin (α–CD)	β–Cyclodextrin (β–CD)	γ–Cyclodextrin (γ–CD)	Hydroxypropyl–β–cyclodextrin (HP–β–CD)	Reference
Number of glucose units	6	7	8	7 (Modified)	Rubim et al. (2024)
Inner cavity diameter	~5.7 Å	~7.8 Å	~9.5 Å	~7.8 Å (Similar to β–CD)	Y. Li et al. (2024)
Solubility in water	High	Moderate	Very High	Significantly higher than β–CD	Mustafa et al. (2024)
Stability	High	High	High	High	S. Sharma et al. (2023)
Common applications	Flavor encapsulation, pharmaceuticals	Drug delivery, food and fragrance industry	Larger molecule encapsulation, pharmaceuticals	Enhanced drug solubility, reduced nephrotoxicity	Kali et al. (2024)
Binding affinity	Lower than β–CD	Moderate	Higher (for larger molecules)	Similar to β–CD, but with improved properties	Kumar et al. (2024)
Toxicity	Generally, low	Generally, low	Generally, low	Lower than β–CD	Meenatchi et al. (2024)

β–Cyclodextrin. β–CD, composed of seven glucopyranose units linked in a cyclic manner, exhibits a unique set of chemical properties that stem from its molecular structure. Like its cyclodextrin counterpart, β–CD is characterized by its toroidal shape, which is formed by α–1,4 glycosidic bonds and imparts it with a hydrophilic exterior and a hydrophobic central cavity. The presence of an additional glucopyranose unit, compared to that of α–CD, results in a larger cavity size, enabling β–CD to encapsulate a broader range of hydrophobic molecules, a characteristic highly valued in the pharmaceutical and food industries for enhancing the solubility and stability of guest compounds (Tran et al., 2024). Chemically stable across a wide pH range, β–CD maintains its structural integrity, making it suitable for diverse applications, including those in acidic or basic environments. Its solubility in water, although lower than that of α–CD, is adequate for most of its applications, and it can be further modified through chemical substitutions, such as hydroxypropylation or methylation. These modifications not only improve solubility but also reduce the potential for causing irritation or toxicity, enhancing the applicability of these materials, particularly in drug delivery systems. The ability of β–CD to form inclusion complexes is pivotal for its role as a molecular encapsulating agent. It can effectively encapsulate and protect guest molecules from environmental factors such as oxidation, light, and heat, thereby extending the shelf life and maintaining the efficacy of active ingredients (Niu et al., 2024). In the pharmaceutical context, the complexation ability of β–CD significantly contributes to increasing the bioavailability of poorly soluble drugs, thereby improving their therapeutic efficacy (Hasanah et al., 2023). Additionally, its use in food technology for flavor and odor masking and in the cosmetic industry for the stabilization of volatile compounds demonstrates the versatility of its chemical properties. The inherent biocompatibility and low toxicity of β–CD further enhance its desirability for use in human-centric applications. In summary, the chemical properties of β–CD, dictated by its unique structural composition, make it a highly versatile and functional molecule that is integral to various industries because of its ability to modify the physical and chemical characteristics of other compounds through molecular encapsulation (Yilmaz et al., 2023).

γ–Cyclodextrin. γ–CD, composed of eight glucopyranose units, is distinguished within the cyclodextrin family through its specific chemical attributes. The addition of an extra glucopyranose unit compared to that of β–CD results in a notably larger internal cavity, which significantly influences the chemical behavior and application potential of these materials. This enlarged cavity enables γ–CD to form inclusion complexes with larger molecular entities, broadening its utility in encapsulating substances that β–CD and α–CD cannot accommodate. Despite its increased size, γ–CD maintains the characteristic toroidal shape of cyclodextrins, with a hydrophilic outer surface and a hydrophobic core, facilitating the formation of noncovalent interactions with guest molecules. One notable aspect of γ–CD is its superior water solubility relative to both α–CD and β–CD (N. Li et al., 2024). This enhanced solubility is attributable to the increased number of glucopyranose units and the resultant molecular configuration, which allows more efficient interactions with water molecules. The solubility of γ–CD makes it particularly advantageous in aqueous solutions where higher concentrations are needed. In terms of chemical stability, γ–CD exhibits resilience over a broad range of environmental conditions, including varying pH levels and temperatures, a trait that is vital for applications under fluctuating conditions. Its capacity to form stable inclusion complexes with a wide range of substances is pivotal in applications ranging from pharmaceuticals, where it enhances the solubility and stability of drugs, to food and cosmetics, where it is employed for controlled release and protection of flavors, fragrances, and active ingredients (Yilmaz et al., 2023). Additionally, the larger cavity size and inherent chemical properties of γ–CD facilitate its interaction with larger and more structurally complex molecules, offering opportunities for novel applications in industrial and research settings. Its biocompatibility and low toxicity profile, combined with these unique chemical characteristics, underscore the role of γ–CD as an indispensable tool in the development of advanced delivery systems and innovative formulations across multiple domains.

Hydroxypropyl–β–cyclodextrin. HP–β–CD represents a chemically modified variant of β–cyclodextrin, where hydroxypropyl groups are introduced to the glucopyranose units of β–CD. This modification fundamentally alters the chemical profile of the plant, conferring several advantageous properties. One of the most significant changes is the markedly improved water solubility of HP–β–CD compared to that of its unmodified counterpart. This increased solubility is a result of the hydroxypropyl groups enhancing the hydrophilicity of the molecule, making HP–β–CD more amenable to use in aqueous solutions, a crucial factor in pharmaceutical formulations and other applications where high solubility is essential (Jeyadurai et al., 2023). The introduction of hydroxypropyl groups also impacts the complex behavior of HP–β–CD. While it retains the ability to form inclusion complexes with a variety of hydrophobic molecules, akin to native β–CD, the modified structure can potentially alter the affinity and stability of these complexes, expanding its utility in drug delivery and other fields. Moreover, compared with β–CD, HP–β–CD exhibits reduced cytotoxicity and lower hemolytic activity, which broadens its safety profile for use in biological applications (Arruda et al., 2024). This reduced toxicity is particularly important in pharmaceuticals, where HP–β–CD can be used to increase the bioavailability of drugs without eliciting adverse effects associated with the parent cyclodextrin. Additionally, chemical modification imparts molecular flexibility to HP–β–CD, allowing for more tailored and specific complexation with guest molecules, which is crucial in targeted drug delivery systems. Furthermore, HP–β–CD demonstrates improved stability under various environmental conditions, such as at different pH levels and temperatures, making it suitable for diverse applications ranging from pharmaceuticals to food and cosmetics. The unique chemical properties of HP–β–CD, stemming from its hydroxypropyl modification, thus make it a versatile and highly effective derivative of β–CD suitable for advanced applications where the native properties of β–CD are limited (X. Liu et al., 2023).

Enhancing Drug Solubility and Stability

Enhancing drug solubility and stability is a pivotal application of cyclodextrins in pharmaceutical formulations, addressing key challenges in drug delivery. Many APIs exhibit poor water solubility, a significant barrier that often leads to reduced bioavailability and diminished therapeutic efficacy. Several common drugs are known for their poor water solubility, which presents challenges in their formulation and effectiveness. For instance, ibuprofen, a widely used nonsteroidal anti-inflammatory drug, and griseofulvin, an antifungal agent, both suffer from limited solubility in water (Figure 3) (Pinelli et al., 2023). Furosemide, a diuretic prescribed for fluid retention, and licarbazepine, an anticonvulsant for epilepsy, also share this characteristic (Chauhan et al., 2020; Salido-Fortuna et al., 2024). In cancer treatment, paclitaxel, which is used in chemical treatment, has poor water solubility (Pourbadiei et al., 2023). Danazol, a synthetic steroid used for treating endometriosis, and spironolactone, a diuretic for heart failure and liver disease, face similar solubility issues (Fine-Shamir & Dahan; Priya et al., 2023). Diazepam, a benzodiazepine for anxiety and seizures, and simvastatin, used to control high cholesterol, are further examples (Braga et al., 2023; Suárez & Díaz, 2022). Additionally, lipophilic vitamins such as vitamins D, E, and A, which are fat soluble, exhibit limited solubility in water (Burkeev et al., 2021). The poor water solubility of these drugs can significantly impact their bioavailability and effectiveness, posing a considerable challenge in pharmaceutical development. Cyclodextrins, which have unique molecular structures and hydrophobic cavities, provide an innovative solution to this issue. They encapsulate hydrophobic drug molecules within this cavity, effectively increasing their solubility in aqueous environments. This process is driven by the hydrophobic effect, where the nonpolar segments of the drug molecule prefer the environment inside the cyclodextrin cavity, displacing water molecules during the process. This results in the formation of a host-guest complex that enhances the overall solubility of the drug (Ebrahimi et al., 2023).

Moreover, cyclodextrins contribute to the stability of pharmaceutical compounds, safeguarding them against various forms of degradation, such as hydrolysis, oxidation, and photodegradation. This protective action extends the shelf life of drugs and is particularly vital for those susceptible to rapid degradation under standard storage conditions (Su et al., 2024). For example, the use of cyclodextrins in formulations containing drugs such as prostaglandins, which are prone to oxidation and hydrolysis, has been shown to significantly improve their stability (Csige et al., 2022). Similarly, cyclodextrins have been utilized in the stabilization of volatile compounds such as essential oils and flavorings in pharmaceutical and nutraceutical applications (Bai et al., 2022). The ability of cyclodextrins to enhance both solubility and stability not only optimizes drug efficacy but also broadens the range of viable pharmaceutical compounds, enabling the development of new therapies and formulations that were previously limited by solubility and stability issues.

Targeted Drug Delivery and Reduced Side Effects

Cyclodextrins have gained significant attention in the realm of targeted drug delivery due to their unique ability to selectively form complexes with specific drugs. This distinctive property allows them to alter the distribution pattern of these drugs within the human body, consequently optimizing their accumulation at the intended site of action, such as in cancer cells (Tang et al., 2023). This targeted approach is instrumental in diminishing the systemic exposure of the drug, thereby substantially reducing the potential side effects and toxicity that are commonly associated with drug therapies. Particularly noteworthy is the application of cyclodextrins in oncology, where they can be utilized to more precisely target cancerous cells (Akash et al., 2024). This precise targeting significantly mitigates the adverse effects of chemotherapy on healthy cells, enhancing the overall efficacy of cancer treatments. Additionally, the formation of inclusion complexes by cyclodextrins can transform drug release dynamics. This modification is crucial for enabling controlled and sustained release of the medication, a feature that is highly advantageous for maintaining a consistent therapeutic effect over an extended period. Such a sustained release mechanism also plays a pivotal role in enhancing patient adherence to medication regimens by decreasing the frequency of dosages needed, making the treatment process more manageable and less intrusive in patients' daily lives (Kali et al., 2024).

Figure 3. Illustration of selected drugs encapsulated in cyclodextrin inclusion complexes

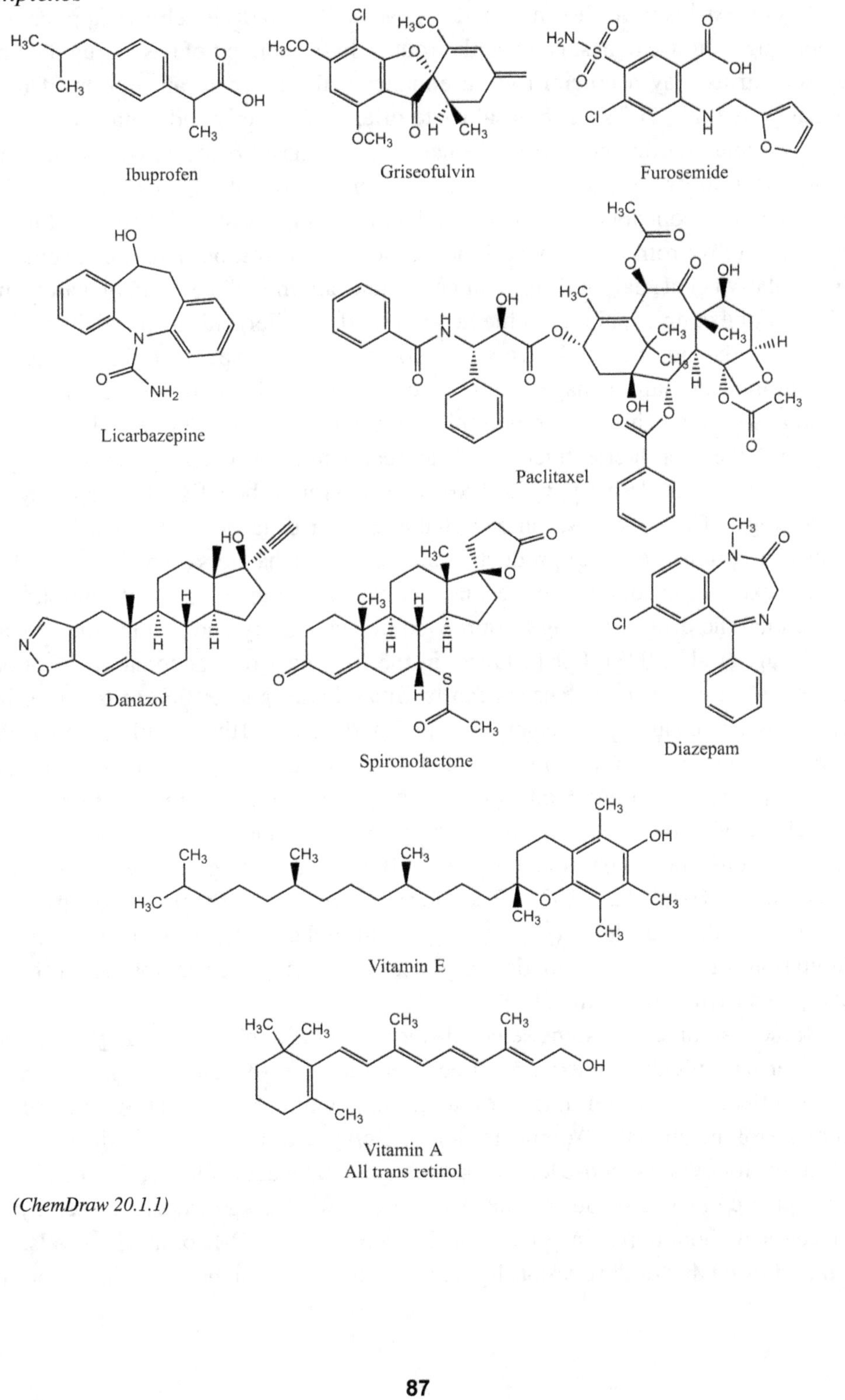

(ChemDraw 20.1.1)

Nasal and Ocular Delivery of Cyclodextrins

Cyclodextrins have shown remarkable versatility in drug delivery applications, extending their use to nasal and ocular routes. In the context of nasal drug delivery, cyclodextrins play a crucial role in enhancing the mucosal absorption of drugs (Changsan et al., 2023). The nasal route offers a distinctive advantage due to the high vascularization and permeable nature of the nasal mucosa, which facilitates rapid drug absorption directly into the systemic circulation (Peng et al., 2024). Cyclodextrins contribute to this process by increasing the solubility and stability of drugs, thereby promoting more efficient mucosal penetration. This enhancement is particularly beneficial for drugs that necessitate an immediate onset of action (Jug, 2020). For example, pain relief medications and antiallergenic agents delivered nasally with the aid of cyclodextrins can provide swift therapeutic effects, which are crucial in acute pain management or in countering sudden allergic reactions. The rapid absorption achieved through this method not only accelerates relief but also improves the overall effectiveness of the treatment (Rassu et al., 2021).

In ocular drug delivery, cyclodextrins offer similar benefits. The anatomy and physiology of the eye pose unique challenges for drug delivery, such as limited drug absorption due to tear production and protective barriers (Dewan et al., 2023). Cyclodextrins can address these challenges by enhancing the solubility and stability of ophthalmic drugs, leading to improved bioavailability and therapeutic efficacy (Q. Wang et al., 2023). For instance, in the treatment of glaucoma or infections, cyclodextrin-enhanced eye drops can facilitate better penetration of the drug into ocular tissues, ensuring greater effectiveness and sustained therapeutic action (Loftsson & Stefánsson, 2023). Moreover, the use of cyclodextrins in ocular formulations can also improve patient comfort by reducing irritation, which is a common issue associated with traditional eye drops. This improvement in patient tolerance can lead to better adherence to treatment regimens, further enhancing the overall success of the therapy (Krstić et al., 2022). Therefore, the incorporation of cyclodextrins into nasal and ocular drug delivery systems exemplifies their potential for overcoming the limitations of traditional drug delivery methods, paving the way for more efficient and patient-friendly treatments.

Table 4 summarizes some selected studies on the applications of cyclodextrins for pharmaceuticals. For example, Lee et al. (2021) investigated the antihyperglycemic effects of α-CD in mice, revealing its potential for regulating glucose and cholesterol metabolism. Administration of 2 g/kg α-CD effectively reduced hyperglycemia following glucose loading by enhancing glucagon-like peptide 1 secretion and hepatic glucose sequestration. A lower dose of 1 g/kg also suppressed hyperglycemia without impacting glucagon-like peptide-1 (GLP-1) or insulin levels. The study showed that α-CD disrupts lipid micelle formation in the gut by incorporating lecithin, a process that, when administered in vitro, negates the antihyperglycemic effect of α-CD in vivo. This nullification was linked to increased intestinal mRNA expression of SREBP2 target genes. These findings suggest that the mode of action of α-CD involves complex interactions in the gut, highlighting its significance in managing hyperglycemia and its dependency on specific gut content interactions.

Mohammad et al. (2024) developed an enhanced oral formulation of lisinopril, a hypertension drug, using β-CD to increase its solubility and bioavailability. The optimized inclusion complex, created through a physical kneading technique and optimized using the Box–Behnken design, showed no adverse interactions and significantly improved drug release and stability. In vivo tests indicated a twofold increase in bioavailability, indicating that this approach is a promising strategy for hypertension treatment.

In a recent study, levo-tetrahydropalmatine (l-THP), which is traditionally used to treat lung injury but is limited by poor solubility and absorption, was encapsulated in γ-CD metal-organic frameworks (CD-MOFs) to create inhalable l-THP@CD-MOF powders. These powders, characterized by methods such as scanning electron microscopy (SEM) and powder X-ray diffractometry (XRD), demonstrated improved solubility and effective pulmonary delivery. Notably, when formulated with lactose, the powders achieved optimal aerosolization with a fine particle fraction of more than 37%. The inhalable form of l-THP was significantly more quickly eliminated in rat liver and lung microsomes than was the orally administered form, and the latter exhibited an 85-fold increase in the area under the curve (AUC) and a 37-fold increase in the maximum concentration (Cmax). Additionally, the inhalation approach enhanced the AUC of active metabolites such as L-isocorypalmine and L-corypalmine. This study indicated that l-THP, when delivered through a dry powder inhaler, has markedly improved bioavailability and promising potential as a treatment for a range of respiratory diseases (Nie et al., 2023).

In addition, C. Zhang et al. (2024) fucoxanthin (Fx) was incorporated into HP-β-CD to form Fx/HP-β-CD nanofibers. This formulation was developed to improve water solubility and bioavailability, overcoming limitations in healthcare applications. These nanofibers, fabricated through electrospinning without additional polymers, exhibited a uniform, beadless morphology with an average diameter of approximately 499 nm. An enhanced thermostability of Fx was achieved postencapsulation. Animal studies revealed significant health benefits, including a 14% reduction in body weight, an 11% decrease in white adipose tissue, a 9% decrease in liver triglyceride levels, and a 23% reduction in total cholesterol in serum, in these mice compared to those in mice treated with Fx alone. Notably, the Fx/HP-β-CD nanofibers also mitigated testicular histopathology in obese mice, demonstrating potential health applications.

Table 4. Summary of recent studies on cyclodextrin-based drug delivery systems

Drug delivered	Cyclodextrin type	Key findings	Remarks	Reference
Lecithin	α-Cyclodextrin (α-CD)	- Studied the anti-hyperglycemic effect of α-CD in mice after glucose and liquid meal loading. - Administration of 2 g/kg α-CD reduced hyperglycemia, increasing glucagon-like peptide 1 secretion and enhancing hepatic glucose sequestration. - 1 g/kg α-CD also suppressed hyperglycemia but without affecting GLP-1 and insulin secretions. - α-CD disrupts lipid micelle formation by including lecithin in the gut luminal fluid. - Prior inclusion of α-CD with lecithin in vitro nullified its anti-hyperglycemic effect in vivo, with increased intestinal mRNA expressions of SREBP2-target genes.	- Demonstrates α-CD's potential in managing hyperglycemia and influencing cholesterol and glucose metabolism through its interaction with dietary components in the gut. - Highlights the importance of dose and interactions with other gut contents in its efficacy.	Lee et al. (2021)
Lisinopril	β-Cyclodextrin	Demonstrated enhanced solubility and stability of the drug. Improved bioavailability observed in in vivo tests.	This study highlights the potential for β-cyclodextrin in improving oral drug delivery.	Mohammad et al. (2024)

continued on following page

Table 4. Continued

Drug delivered	Cyclodextrin type	Key findings	Remarks	Reference
Levo-tetrahydropalmatine (l-THP)	γ-Cyclodextrin metal-organic frameworks (CD-MOFs)	- l-THP encapsulated in γ-CD-MOFs to form inhalable powders, improving solubility and pulmonary delivery. - Achieved optimal aerosolization with a fine particle fraction (FPF) of 37.25 ± 2.80% when formulated with lactose. - l-THP@CD-MOF particles showed faster elimination in rat liver and lung microsomes compared to l-THP alone. - Inhalation increased the AUC_0–24 h and Cmax of l-THP by approximately 85- and 37-folds, respectively, and enhanced the AUC_{0-24h} of metabolites L-ICP and L-CP.	- Demonstrates the potential of l-THP as an effective treatment for respiratory diseases, significantly improving its bioavailability and therapeutic efficacy through inhalation delivery.	Nie et al. (2023)

continued on following page

Table 4. Continued

Drug delivered	Cyclodextrin type	Key findings	Remarks	Reference
Fucoxanthin (Fx)	Hydroxypropyl-β-cyclodextrin (HP-β-CD)	- Fx/HP-β-CD nanofibers fabricated by electrospinning without using polymer. - Molecular docking analysis showed a 1:2 ratio of Fx to HP-β-CD. - Nanofibers are homogeneous, beadless, with ~499 nm diameter. - Improved thermostability of Fx after encapsulation. - In animal studies, Fx/HP-β-CD nanofibers led to body weight, white adipose tissue, liver triglyceride, and total cholesterol reduction in mice. - Attenuated testicular histopathology in obese mice.	- Enhances water solubility and bioavailability of Fx, expanding its potential in food industry applications. - Shows significant health benefits in animal models.	C. Zhang et al. (2024)

Dendrimers

Dendrimers are highly branched, three-dimensional macromolecules with well-defined, symmetrical structures and high degrees of surface functionality (Soltany et al., 2024). Figure 4 illustrates the complex and highly organized structure of a dendrimer, a type of synthetic macromolecule used in advanced drug delivery systems. At the heart of a dendrimer's structure lies a central core, which acts as the foundational element. From this core, a series of branching units radiate outward in a regular, tree-like pattern. These branches are organized in concentric layers, known as generations, each progressively larger than the one before it. As the number of generations increases, the surface area of the dendrimer expands significantly, creating a multitude of terminal groups on the outermost layer. These terminal groups can be chemically modified to serve various functions, such as attaching drug molecules or targeting ligands. The unique architecture of dendrimers, characterized by their branching symmetry and high degree of surface functionality, allows precise control over their size and shape. This structural specificity enables dendrimers to encapsulate or bind drugs efficiently, facilitating targeted delivery and controlled release of therapeutic agents. The ability to tailor the surface properties of dendrimers also makes them ideal candidates for overcoming biological barriers and enhancing the bioavailability of drugs (X. Wang et al., 2023).

Figure 4. Schematic representation of a dendrimer structure - illustrating its highly branched, three-dimensional macromolecule with a distinct central core, multiple branching units, and various terminal functional groups

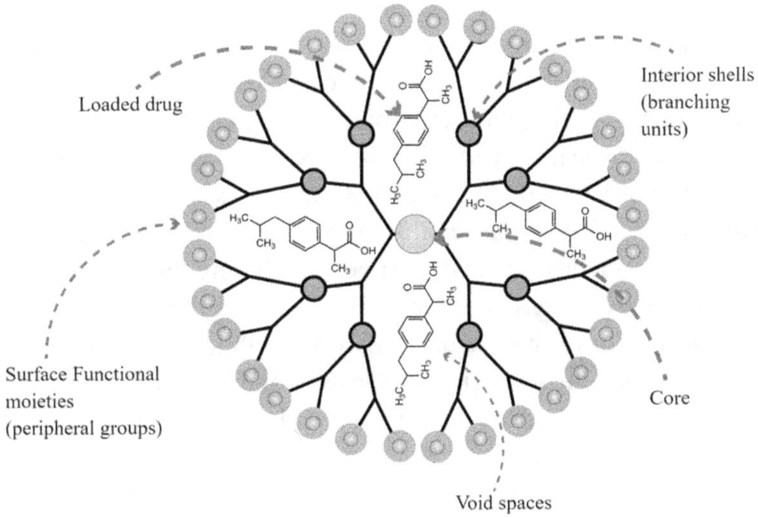

One of the primary advantages of using dendrimers in drug delivery is their ability to encapsulate therapeutic agents within their internal cavities or attach them to their surface functional groups. This encapsulation or conjugation can protect the drug from premature degradation and control its release over time. Dendrimers can be designed to release their payload in response to specific stimuli, such as changes in pH, temperature, or the presence of certain enzymes, which are often characteristic of disease sites. This responsive release ensures that the drug is released in a controlled manner at the target site, maximizing its therapeutic action while minimizing systemic exposure (Khatik et al., 2024). Dendrimers have been extensively explored for targeted delivery in cancer therapy. The surface functional groups of these peptides can be modified with targeting ligands, such as antibodies, peptides, or folic acid, which can recognize and bind to specific receptors overexpressed on cancer cells. This receptor-mediated endocytosis allows dendrimers to deliver their drug payload directly into cancer cells with high precision. Such targeted approaches have shown promising results in reducing the adverse side effects typically associated with chemotherapy, as they limit the exposure of healthy cells to cytotoxic drugs (Michlewska et al., 2023).

In addition to their use in cancer therapy, dendrimers have been applied in the delivery of genetic material, such as DNA and RNA. The cationic nature of certain dendrimers allows them to form complexes with negatively charged nucleic acids, protecting them from degradation and facilitating their entry into cells. This property is particularly valuable in gene therapy and RNA interference (RNAi) approaches, where the delivery of genetic material needs to be highly efficient and targeted. Dendrimers can also be engineered to escape endosomes, ensuring that the genetic material reaches the cytoplasm, where it can exert its therapeutic effect (X.-Y. Liu et al., 2022). Moreover, dendrimers are being investigated for their potential in delivering drugs across challenging biological barriers, such as the blood–brain barrier and the ocular barrier. The ability to cross these barriers opens up new avenues for treating central nervous system disorders and eye diseases (A. Kaur et al., 2023). Scientists are able to improve their ability to permeate biological barriers by adjusting the surface characteristics of dendrimers, including their charge and degree of hydrophobicity. This increase in permeability enables the efficient transport of therapeutic agents to areas that were previously difficult to reach with conventional drug delivery methods. This strategic modification of dendrimer surface properties plays a crucial role in facilitating the targeted and effective delivery of drugs to specific sites within the body.

Table 5 summarizes recent studies on dendrimer-based drug delivery systems and highlights the innovative strides made in this field, revealing a range of drugs and dendrimer types used to improve treatment efficacy and reduce side effects. Mohammad Asghari et al. (2023) developed a unique drug delivery system using nanodiamonds grafted with a thermosensitive polymer and coated with a hyperbranched dendrimer for the controlled release of docetaxel. This system exhibited pH- and temperature-dependent release patterns, with rapid release under near-infrared laser irradiation, making it a promising method for targeted drug delivery. Michlewska et al. (2023) focused on liposomal locked-in dendrimers (LLDs) for doxorubicin delivery and reported that hydrophilic locking was more effective than hydrophobic locking was, as it improved the transfection profile and reduced toxicity. Soltany et al. (2024) introduced a magnetic chitosan nanocomposite (FA–PAMAMG2–MCS) with poly(amidoamine) dendrimer and folic acid for targeted doxorubicin delivery. This system demonstrated remarkable anticancer activity and controlled release in acidic environments. Nori et al. (2023) developed a gold-coated magnetic nanoparticle decorated with a thiol-containing dendrimer for delivering 6-mercaptopurine (6-MP), which was doubled as a hyperthermia treatment and magnetic resonance imaging (MRI) contrast agent, showing significant therapeutic efficacy against breast cancer. Finally, Lee et al. (2022) created hyaluronic acid (HA)-modified dendrimers encapsulating gold nanoparticles (HA-AuDEN-DOX) for targeted doxorubicin delivery to ovarian cancer cells. This system not only provided efficient drug loading and selective delivery but also demonstrated reduced cardiotoxicity and significant tumor growth inhibition in vivo. Overall, these studies demonstrate the diverse potential of dendrimer-based drug delivery systems for enhancing the efficacy and safety of cancer treatments.

Table 5. Summary of recent studies on dendrimer-based drug delivery systems

Drug delivered	Dendrimers	Key findings	Remarks	Reference
Docetaxel	Nanodiamonds with pH/thermo dual responsive polymer and hyperbranched dendrimer	- Developed a drug delivery system using nanodiamonds grafted with thermosensitive polymer and coated with a hyperbranched dendrimer. - The system demonstrated pH and temperature-dependent sustained release of docetaxel (99.78% release at pH 5.6 and 45°C in 6 hours). - Near-infrared laser irradiation accelerated docetaxel release, achieving 90.97% in 20 minutes. - Drug release behavior best modeled by the Higuchi model.	- This study presents an innovative approach for docetaxel delivery, utilizing nanodiamonds with a dual response to pH and temperature, offering a rapid and controllable drug release mechanism.	Mohammad Asghari et al. (2023)
Doxorubicin	Liposomal locked-in dendrimers (LLDs)	- Developed LLDs using two preparation methods, hydrophilic and hydrophobic locking. - The hydrophilic locking method resulted in more efficient transfection profiles and better interaction with erythrocyte membranes. - LLDs showed improved transfection properties and significantly reduced hemotoxicity and cytotoxicity. - Hydrophobic locking was less effective and deemed unsuitable for future use. - Hydrophilic-locked LLDs with doxorubicin demonstrated higher effectiveness against cancer cells compared to normal cells.	- This study highlights the potential of combining dendritic and liposomal carriers for improved drug delivery, with a focus on reducing toxicity and enhancing transfection efficiency.	Michlewska et al. (2023)

continued on following page

Table 5. Continued

Drug delivered	Dendrimers	Key findings	Remarks	Reference
Doxorubicin (DOX)	Folic acid conjugated poly (amidoamine) dendrimer grafted magnetic and pH-responsive nanocarrier (FA–PAMAMG2–MCS)	- Developed a magnetic chitosan nanocomposite (FA–PAMAMG2–MCS) with poly(amidoamine) dendrimer and folic acid for targeted delivery of DOX. - The nanocarrier showed an optimized adsorption capacity for DOX (102.85 mg/gat 298 K). - Demonstrated faster drug release in acidic environments and slower release in neutral environments, following Peppas-Sahlin models. - Exhibited remarkable anticancer activity against the MDA-MB 231 cell line, highlighting its potential for nanomedicine applications.	- This multifunctional system emphasizes the effectiveness of pH-responsive and targeted drug delivery for overcoming limitations of conventional chemotherapy, specifically in cancer treatment.	Soltany et al. (2024)

continued on following page

Table 5. Continued

Drug delivered	Dendrimers	Key findings	Remarks	Reference
6-Mercaptopurine (6-MP)	Gold-coated magnetic nanoparticle decorated with a thiol-containing dendrimer	- Developed a magnetic nanocarrier modified with a thiol-ended dendrimer for delivering 6-MP, effective in hyperthermia and as a magnetic resonance imaging (MRI) contrast agent. - The drug is released in the reductive environment of cancer cells. - Demonstrated cytotoxicity against the MCF-7 breast cancer cell line. - The nanomaterials also served as effective tools for magnetic hyperthermia and contrast enhancement in MRI. - The combined effect of RF hyperthermia and drug release showed significant therapeutic efficiency.	- This study introduces a versatile magneto dendrimer platform that integrates drug delivery with hyperthermia treatment and MRI contrast, offering a multifaceted approach to cancer therapy.	Nori et al. (2023)
Doxorubicin (DOX)	HA-modified dendrimer encapsulating gold nanoparticles (HA-AuDEN-DOX)	- Developed HA-modified dendrimers encapsulating gold nanoparticles for targeted delivery of DOX to ovarian cancer cells. - Achieved efficient DOX loading with high stability. - Showed selective delivery to CD44-overexpressing cancer cells and triggered release in acidic, high GSH tumor microenvironments. - Demonstrated enhanced cytotoxicity against cancer cells, with a superior IC50 value and reduced cardiotoxicity. - In vivo studies revealed significant tumor growth inhibition and reduced cell proliferation in ovarian tumor tissues.	- This study presents HA-AuDEN-DOX as a promising nanocarrier for targeted ovarian cancer treatment, offering enhanced efficacy and reduced side effects compared to free DOX.	Lee et al. (2022)

Micelles

Micelles are a unique form of colloidal particle that play a critical role in various fields, ranging from biology to materials science. Figure 5 depicts the self-assembly of amphiphilic molecules in an aqueous solution, leading to the formation of micelles. These structures are typically formed when amphiphilic molecules, which possess both hydrophobic (water-repelling) and hydrophilic (water-attracting) properties, aggregate in aqueous solutions. The structure of a micelle is characterized by a hydrophobic core and a hydrophilic shell. This core-shell architecture arises because the hydrophobic parts of the molecules avoid water and cluster together, while the hydrophilic parts remain in contact with the surrounding water (Xue et al., 2024). The self-assembly process is driven by the minimization of free energy in the system and is influenced by factors such as the concentration of the amphiphilic molecules, the temperature of the solution, and the nature of the solvent. Micelles can vary in shape, including spherical, rod-like, and lamellar structures, depending on the molecular structure of the amphiphilic molecules and the conditions of the environment (Chen & Liu, 2023).

In practical applications, micelles are highly valued for their ability to encapsulate hydrophobic substances within their cores, making them immensely useful in areas such as drug delivery, detergency medicine, and food science. In drug delivery, micelles can carry poorly soluble drugs, enhancing their solubility and bioavailability. The hydrophilic shell of the micelle can interact with the biological environment, facilitating the transport of the encapsulated drug to the targeted site. This property is particularly important in the development of novel therapeutic agents and in improving the efficacy of existing drugs (Somavarapu et al., 2023). In the field of detergents and cleaning agents, the micellar structure aids in the solubilization of oily and greasy stains, allowing them to be washed away with water (Dhandapani et al., 2022). Furthermore, in food science, micelles are involved in the absorption and transport of lipophilic nutrients and vitamins, underscoring their significance in nutrition (Antuma et al., 2024; Liu et al., 2024). Owing to their unique structural properties, the versatile nature of micelles has led to extensive research and application in various scientific domains.

The design and synthesis of micelles for drug delivery applications have undergone significant advancements. It is possible to tailor the size, shape, and surface characteristics of micelles to optimize drug loading and release profiles by carefully choosing amphiphilic molecules. Block copolymers, consisting of hydrophobic and hydrophilic blocks, are commonly used to form micelles for drug delivery. The versatility of the chemical composition of these copolymers allows for the creation of micelles with the desired properties for specific drugs, enhancing the efficiency of drug delivery systems. Moreover, the stability of these micelles can be controlled, which is crucial for maintaining the integrity of drug-loaded micelles in the physiological environment until they reach the target site (Chaudhuri et al., 2022). Micelles not only are beneficial for increasing drug solubility but also for improving drug biodistribution and reducing side effects. Once administered, micelle-encapsulated drugs can circulate in the bloodstream for extended periods, evading rapid clearance by the reticuloendothelial system. This prolonged circulation time increases the likelihood of the drug reaching the target tissue (M. Zheng et al., 2024). Additionally, the ability of micelles to preferentially accumulate in certain tissues, such as tumor sites, through enhanced permeability and retention further boosts their effectiveness as drug delivery vehicles, particularly in cancer therapy. This targeted accumulation not only enhances therapeutic efficacy but also minimizes the exposure of healthy tissues to the drug, thereby reducing adverse side effects (Lamey et al., 2023).

Figure 5. Illustration of micelle formation and structure

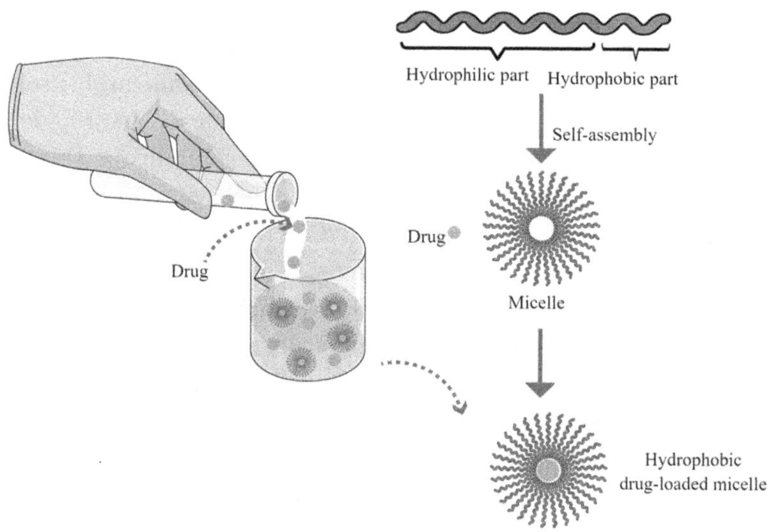

Innovations in micellar drug delivery systems also include the use of stimuli-responsive micelles, which can release their drug payload in response to specific environmental triggers, such as pH changes, temperature fluctuations, or the presence of certain enzymes. For instance, pH-responsive micelles can release drugs in the acidic environment of tumor tissues or inflamed sites, ensuring that the drug is released precisely where needed. Temperature-sensitive micelles can exploit the slightly elevated temperature of inflamed or tumor tissues to release their payload. These stimuli-responsive systems offer a sophisticated approach to controlled drug release, ensuring high drug concentrations at disease sites while minimizing systemic drug levels (Li et al., 2023; Manna et al., 2024). Furthermore, the development of micellar systems for the delivery of a wide range of therapeutic agents, including small molecule drugs, proteins, peptides, and nucleic acids, highlights their versatility. Micelles can protect these biomolecules from enzymatic degradation and immune recognition, enhancing their therapeutic potential. This protective effect is particularly important for the delivery of biologics, which are often susceptible to rapid degradation in the body (Xinjing Wang et al., 2023). The integration of micelles into drug delivery strategies represents a significant step forward in the development of more effective and patient-friendly pharmaceutical formulations. Continued research and development in this field holds the promise of overcoming many of the current limitations in drug solubility and delivery, paving the way for more effective treatments for a variety of diseases.

Recent advancements in micelle-based drug delivery systems, as outlined in Table 6, have led to a range of innovative approaches and significant findings. M. Zheng et al. (2024) explored the effects of polyethylene glycol (PEG) chain lengths in amphiphilic polymer micelles for delivering pyropheophorbide-a (PhA) and revealed that shorter PEG chains led to greater cell internalization, while longer chains resulted in greater tumor accumulation, thus providing insights into optimizing drug delivery to tumors. Li et al. (2023) introduced a novel approach with amphiphilic poly(ethylene glycol)-b-poly(thioketal) (mPEG-b-PTK) copolymer micelles for delivering doxorubicin (DOX), which exhibited reactive oxygen species (ROS) responsiveness and high efficacy in targeted anticancer therapy. Similarly, Qiu et al. (2023) developed hyaluronic acid derivative amphipathic micelles (HATAD-TCSs) that effectively released triclosan (TCS) in response to ROS levels during intracellular bacterial infections, demonstrating significant therapeutic effects in a mouse pneumonia model. Marquet et al. (2024) addressed the challenges of siRNA delivery using linear polyethyleneimine (PEI)-based micellar carriers, which enhance endosomal siRNA release and show potential in treating non-small cell lung cancer. Finally, Aliabadi et al. (2024) synthesized PLA-b-PAPMA diblock copolymers, which formed cationic micelles with high efficacy in nucleic acid condensation and delivery, indicating a promising avenue for therapeutic molecule delivery. Collectively, these studies represent groundbreaking work in the field of drug delivery, highlighting the versatility and potential of micelle-based systems in addressing complex therapeutic challenges.

Table 6. Summary of recent studies on micelle-based drug delivery systems

Drug delivered	Micelle type	Key findings	Remarks	Reference
Pyropheophorbide-a (PhA)	Amphiphilic Polymer with Varying PEG Chain Lengths	- Explored the impact of PEG chain length on cell internalization and tumor targeting in PhA-PEG2000 NPs and PhA-PEG5000 NPs. - PhA-PEG2000 NPs had higher in vitro cell internalization, while PhA-PEG5000 NPs showed greater tumor accumulation in vivo. - Both types exhibited high drug loading and negligible side effects.	- Provides insights into optimizing PEG chain length in drug-PEG conjugate micelles for improved tumor targeting and biodistribution.	M. Zheng et al. (2024)
Doxorubicin (DOX)	Amphiphilic poly(ethylene glycol)-b-poly(thioketal) (mPEG-b-PTK) Copolymer	- Synthesized ROS-responsive, core crosslinked (CCL) micelles via thiol-ene click reaction. - DOX-loaded CCL micelles released nearly 100% of the drug within 32 hours in an ROS environment. - Demonstrated high anticancer activity with low toxicity in HeLa and HUVECs.	- Introduces a novel enzyme-catalyzed polymerization method for synthesizing ROS-responsive dendrimers, with potential for efficient anticancer drug delivery.	Li et al. (2023)
Triclosan (TCS)	Hyaluronic Acid-Derivative Amphipathic Micelle (HATAD-TCS)	- HATAD-TCS micelles incorporated ROS-responsive groups and TCS, releasing drugs effectively in response to intracellular bacterial infections. - Demonstrated significant therapeutic effect in a mouse pneumonia model.	- Highlights the potential of hyaluronic acid-based nanoparticles in treating intracellular bacterial infections with ROS-responsiveness.	Qiu et al. (2023)

continued on following page

Table 6. Continued

Drug delivered	Micelle type	Key findings	Remarks	Reference
siRNA	Linear PEI-based Micellar Carriers	- Shifted from triblock polymeric micelles to linear PEI-based system with orthogonal groups for oligonucleotide delivery. - Effectively enhanced endosomal siRNA release, addressing cytotoxicity and endosomal escape. - Showed promise in targeting STAT3 for NSCLC treatment.	- Underlines the importance of polymer tailoring in overcoming biological barriers for siRNA delivery, with potential in cancer treatment.	Marquet et al. (2024)
Nucleic acids	PLA-b-PAPMA Diblock Copolymers	- Introduced PLA-b-PAPMA diblock copolymers for forming cationic micelles for nucleic acid condensation. - The PLA121-b-PAPMA103 variant showed superior plasmid condensation, stability, and transfection efficiency. - Exhibited spherical morphology and optimal size/polydispersity index for nucleic acid delivery.	- Emphasizes the potential of PLA-b-PAPMA diblock copolymers in nucleic acid delivery, indicating a promising avenue for therapeutic molecule encapsulation and delivery.	Aliabadi et al. (2024)

Macrocycles

Macrocycles are a class of molecules characterized by their large ring structures, typically containing 12 or more atoms. These atoms can be carbon, nitrogen, oxygen, sulfur, or a combination thereof, forming a cyclic backbone. The size of the ring allows for the incorporation of various functional groups and heteroatoms, giving macrocycles a diverse range of chemical properties and potential applications (Garcia Jimenez et al., 2023). The large ring structure of macrocycles often imparts unique physical and chemical properties that are not observed in smaller molecules (Hazarika & Singh, 2023). One of the most fascinating aspects of macrocycles is their ability to interact with other molecules in a highly specific manner. This interaction is often due to the three-dimensional conformation of the macrocycle, which can create a cavity or pocket that is capable of binding other molecules. This feature makes macrocycles extremely useful in supramolecular chemistry, where they are used to create complex structures through noncovalent interactions such as hydrogen bonding, hydrophobic effects, and van der Waals forces (L. Zhang et al., 2023).

In the pharmaceutical industry, macrocycles have gained attention for their potential as therapeutic agents. The large size and structural complexity of miRNAs allow them to bind tightly and specifically to biological targets, such as proteins, making them effective in drug discovery and development. Macrocycles can bridge the gap between small-molecule drugs, which are often limited by their size and the types of targets they can bind to, and larger biomolecules, such as antibodies, which can be highly specific but are more complex and expensive to produce (Yu et al., 2023). This versatility enables macrocycles to encapsulate or bind drugs, thereby influencing their pharmacokinetic properties, including solubility, stability, and bioavailability. Unlike smaller molecules, macrocycles offer a spacious and often flexible framework that can be chemically tailored to suit specific drug delivery needs, allowing for the accommodation of a diverse array of drug molecules.

One of the notable applications of macrocycles in drug delivery is their ability to enhance the solubility of poorly soluble drugs. Solubility issues are a major hurdle in pharmaceutical development, often limiting the clinical efficacy of potent compounds. Macrocycles can form host-guest complexes with such drug molecules, effectively increasing their solubility in aqueous environments. This approach is particularly beneficial for oral drug formulations, where solubility directly impacts the drug's absorption and overall bioavailability. By improving solubility, macrocycles facilitate the efficient absorption of drugs in the gastrointestinal tract, leading to enhanced therapeutic effectiveness (Chowdhury & Goswami, 2023).

In addition to their ability to enhance solubility, the use of macrocycles has also been explored for their potential in targeted drug delivery. The large and often modifiable surface area of these materials allows for the attachment of various targeting ligands or antibodies, enabling macrocycle–drug complexes to specifically home to targeted cells or tissues. This specificity is particularly advantageous in treatments such as cancer therapy, where targeted delivery can minimize the impact on healthy tissues while maximizing the therapeutic effect on tumor cells (Hazarika & Singh, 2023). Such targeted systems can lead to improved treatment outcomes with reduced side effects, a significant advancement over conventional chemotherapy treatments (Pan et al., 2023).

Controlled drug release is another area where macrocycles show promising application potential. It is possible to design drug delivery systems that release their payload over a desired period by manipulating the chemical structure of macrocycles. This controlled release can be achieved through environmental triggers such as pH, temperature, or the presence of specific enzymes, which can induce changes in the macrocycle structure and consequently release the drug. Controlled release is especially important in chronic disease management, where maintaining consistent drug levels in the body can significantly improve treatment efficacy and patient compliance (Attia et al., 2024).

Furthermore, the use of macrocycles in the delivery of biologically active macromolecules such as peptides, proteins, and nucleic acids is being investigated. The inherent stability and large cavity of macrocycles provide an ideal environment for protecting these sensitive molecules from degradation in the biological milieu. This protective capability is crucial for the effective delivery of biologics, which are increasingly important in modern therapeutics (W. Zhang et al., 2023). Additionally, the ability of macrocycles to cross biological barriers, such as the blood–brain barrier, opens new possibilities for treating a range of neurological disorders, offering a pathway to deliver therapeutic agents directly to the brain (Hazarika & Singh, 2023).

Table 7 provides an insightful summary of recent studies of macrocycle-based drug delivery systems, highlighting advancements in the fields of targeted therapy and controlled drug release. Lyu et al. (2023) presented a novel approach using dynamic covalent macrocycles derived from a dithiol monomer for the codelivery of doxorubicin and siRNA, a promising strategy against multiple drug resistance. This study involved formulating nanoparticles and assessing their enhanced anticancer efficacy in vitro, demonstrating the potential of dynamic covalent macrocycles to increase anticancer efficacy. Edilino de Lima et al. (2023) investigated the encapsulation of vitamin E isoforms, α-tocopherol (α-TOC), and α-tocotrienol (α-TCT), into cucurbit[7]uril (CB[7]) using theoretical methods. Their findings revealed the stability of vitamin@CB[7] complexes through extended molecular dynamics simulations and favorable binding energies, suggesting that CB[7] is a promising macrocyclic host for vitamin E, enhancing solubility and potential bioavailability. Z. Liu et al. (2022) explored hyaluronic acid (HA)-based macrocyclic supramolecular assemblies using cyclodextrins and cucurbit[n]urils for diverse applications, including drug delivery and imaging. These HA-based assemblies, capable of responding to various stimuli, highlighted the significant role of macrocyclic host–guest complexes in drug carriers, targeted imaging, and other biomedical applications, highlighting the potential of these assemblies in supramolecular diagnosis and treatment. In addition, Evans et al. (2022) explored the enhancement of antibody–drug conjugates (ADCs) for targeted cancer therapy by incorporating hydrophilic macrocycles, specifically various cyclodextrins and crown ethers, into linker–payload reagent designs. Using bis-sulfone-based disulfide rebridging reagents with Val-Cit-PABC-MMAE linker-payloads, the researchers integrated these macrocycles through a glutamic acid branching point. The resulting ADCs, which included brentuximab as a model antibody, demonstrated comparable in vitro potency to Adcetris® and superior in vivo efficacy, particularly for the 3′-amino-α-cyclodextrin and 1-aza-42-crown-14 macrocycle variants, which outperformed traditional PEG chain architectures. This study underscores the potential of hydrophilic macrocycles for advancing apparent diffusion coefficient (ADC) design, offering enhanced alternatives for targeted cancer treatment.

Thus, macrocycles represent a versatile and powerful tool in the arsenal of drug delivery systems. The unique structural attributes and chemical adaptability of these materials make them suitable for a wide range of applications, from enhancing drug solubility and bioavailability to achieving targeted and controlled drug release. As research in this field continues to evolve, macrocycles are assumed to play a crucial role in the development of more effective, safer, and patient-friendly pharmaceutical therapies.

Table 7. Summary of recent studies on macrocycle-based drug delivery systems

Drug delivered	Macrocycles	Key findings	Remarks	Reference
Doxorubicin and siRNA	Dynamic covalent macrocycles developed from a dithiol monomer	- A protocol for preparing a delivery system using dynamic covalent macrocycles for the codelivery of doxorubicin and siRNA. - The process involves preparing the dithiol monomer and using it to form nanoparticles for codelivery. - Detailed procedures for cell uptake and assessing enhanced anticancer efficacy in vitro.	- Presents a novel approach for combination therapy against multiple drug resistance using codelivery of drug molecules and functional genes. - Demonstrates a potential increase in anticancer efficacy with the use of dynamic covalent macrocycles.	Lyu et al. (2023)
α-Tocopherol (α-TOC) and α-tocotrienol (α-TCT)	Cucurbit[7]uril (CB[7])	- Theoretical methods were used to investigate the encapsulation of α-TOC and α-TCT into CB[7]. - Vitamins@CB[7] complexes were stable over 100 ns molecular dynamics simulations with similar conformations. - Binding free energy is approximately –25 kcal/mol for both complexes, as obtained from the MM/PBSA method. - DFT-D3/B3LYP/6-31G(d,p) calculations show favorable encapsulation with bind energies of –58.25 kcal/mol for α-TOC@CB[7] and –61.08 kcal/mol1 for α-TCT@CB[7]. - Calculated solvation enthalpies indicate enhanced solubility of complexed vitamins compared to their free forms.	- Indicates CB[7] as a promising macrocyclic host for encapsulating vitamin E isoforms, potentially enhancing solubility and stability. - Suggests potential applications in vitamin E delivery and bioavailability enhancement.	Edilino de Lima et al. (2023)

continued on following page

Table 7. Continued

Drug delivered	Macrocycles	Key findings	Remarks	Reference
Various drugs and imaging agents with hyaluronic acid (HA)	Cyclodextrins, cucurbit[n]urils, and HA modified macrocycles	- Explored HA-based macrocyclic supramolecular assemblies for drug delivery and imaging. - HA can assemble with macrocyclic host-guest complexes through hydrogen bonding and electrostatic interactions or can be modified for further assembly. - Macrocycles like cyclodextrins and cucurbit[n]urils encapsulate photosensitizers, drugs, or other functional guest molecules. - The assemblies respond to stimuli like enzymes, light, electricity, and magnetism for controlled delivery, enhancing luminescence and drug loading capacity. - HA-based assemblies improve biocompatibility, reduce drug toxicity, and enhance cell permeability.	- Focuses on the role of macrocyclic host-guest complexes in HA-based assemblies for drug carriers, targeted imaging agents, wound healing, and biomedical soft materials. - Highlights challenges in the field and the potential for supramolecular diagnosis and treatment advancements.	Z. Liu et al. (2022)

continued on following page

Table 7. Continued

Drug delivered	Macrocycles	Key findings	Remarks	Reference
Brentuximab (ADC)	Various cyclodextrins and crown ethers	- Investigated the impact of hydrophilic macrocycles in linker-payload reagent design on ADCs. - Synthesized bis-sulfone based disulfide rebridging reagents with Val-Cit-PABC-MMAE linker-payloads, incorporating cyclodextrins and crown ethers via a glutamic acid branch. - In vitro, ADCs showed similar potency to Adcetris®. - In vivo, cyclodextrin containing ADCs demonstrated greater efficacy than Adcetris®, especially the 3′-amino-α-cyclodextrin variant. - Crown ether containing ADCs also showed enhanced efficacy, with the 1-aza-42-crown-14 macrocycle variant outperforming a 24-unit PEG chain ADC.	- Highlights the potential of hydrophilic macrocycles in enhancing the efficacy of ADCs for targeted cancer therapy. - Offers alternatives to traditional drug-linker architectures in ADC design.	Evans et al. (2022)

Supramolecular Gels

Supramolecular gels are a fascinating class of materials that belong to the broader field of supramolecular chemistry. These gels are formed through the self-assembly of small molecules into a three-dimensional network that immobilizes a liquid, resulting in a gel-like material. Unlike traditional gels, which are often formed by polymerization reactions that create long chains of molecules, supramolecular gels rely on noncovalent interactions between smaller molecules. These interactions can include hydrogen bonding and hydrophobic effects. van der Waals forces, π-π interactions, and metal-ligand coordination (Vandera et al., 2023). The key feature of supramolecular gels is the formation of a network structure through self-assembly. Individual molecules, often referred to as gelators, are designed to spontaneously organize into fibrous or sheet-like structures under certain conditions. When these structures become interconnected, they create a network that can trap and hold solvent molecules, leading to the formation of a gel. This process is reversible and can be triggered by various stimuli, such as changes in temperature, pH, ionic strength, or the presence of specific ions or molecules (Sepulveda et al., 2023). Supramolecular gels, characterized by their network of self-assembled fibrillar structures, have garnered significant attention in the field of drug delivery. These gels are formed through the self-assembly of gelator molecules, which are typically small organic compounds or polymers that can interact through noncovalent interactions such as hydrogen bonding, π-π interactions, and van der Waals forces. The resulting network creates a three-dimensional matrix capable of entrapping drugs, making these gels excellent media for controlled drug release (H. Sharma et al., 2023). The inherent biocompatibility and low toxicity of many supramolecular gelators add to their appeal in pharmaceutical applications, where safety and efficacy are paramount.

Supramolecular gels have several unique properties that make them attractive for various applications. For instance, the mechanical properties of gels can be finely tuned by adjusting the concentration of the gelator or the conditions of gelation. Additionally, since the gel network is formed through reversible interactions, these gels can exhibit self-healing properties. If the gel is disrupted, the network can reassemble, allowing the material to regain its original structure and properties (Sepulveda et al., 2023). One of the primary advantages of using supramolecular gels in drug delivery is their versatility in terms of both composition and functionality. It is possible to tailor the physical properties of a gel, such as stiffness, porosity, and degradability, by altering the chemical structure of the gelator molecules. This flexibility allows for the creation of gels that are specifically suited to the delivery of a wide range of therapeutic agents, from small-molecule drugs to large biomolecules such as proteins and nucleic acids (Ailincai et al., 2023). Additionally, the properties of the gel can be fine-tuned to control the release rate of the encapsulated drug, ensuring sustained release over time, which is particularly beneficial for chronic therapies where consistent drug levels are desired (W. Liu et al., 2023).

Supramolecular gels also exhibit unique responsiveness to external stimuli, such as changes in temperature, pH, or the presence of specific ions or enzymes. This responsiveness can be harnessed to develop smart drug delivery systems that release their payload in response to specific physiological conditions, ensuring that the drug is released at the right place and time. For example, a pH-responsive gel could be designed to release a drug in response to the acidic environment of a tumor or inflamed tissue, providing targeted therapy with minimal side effects (W. Liu et al., 2023). Similarly, temperature-sensitive gels could be used to release drugs in response to the slightly elevated temperatures of infected or inflamed tissues.

Another significant application of supramolecular gels is in topical and transdermal drug delivery. The ability of a gel to form a continuous film on the skin makes it an ideal medium for the delivery of drugs that benefit from localized application, such as anti-inflammatory agents or analgesics. The gel matrix can control drug penetration through the skin, providing sustained release that prolongs the therapeutic effect. Moreover, the ease of application and noninvasive nature of gel-based drug delivery systems enhance patient compliance, particularly in cases where oral administration or injections are not feasible or desirable (Pourbadiei et al., 2023).

In recent years, the potential of supramolecular gels in regenerative medicine and tissue engineering has also been explored. The network structure of these gels can provide a scaffold for cell growth and tissue regeneration, making them suitable for applications such as wound healing and the delivery of growth factors or other bioactive agents. The biocompatibility and biodegradability of many supramolecular gels further enhance their suitability for these applications, as they can degrade and be absorbed by the body over time, reducing the need for surgical removal (Huang et al., 2023). Table 8 summarizes innovative advancements in supramolecular gel-based drug delivery systems, showcasing diverse applications in cancer therapy, wound healing, and antibacterial treatments. C. Zhang et al. (2024) described a rapid-forming polysaccharide injectable hydrogel comprising sulfobutyl ether β-CD (SBE-β-CD), chitosan, and silver ions that effectively encapsulates and retains high concentrations of doxorubicin, showing significant inhibitory effects on bladder tumors. This hydrogel, which regains its shape in less than one hour and maintains body-compatible elasticity, represents a potential in situ treatment for bladder cancer. Monajati et al. (2023) focused on L-asparaginase (ASNase) immobilized in an α-CD dimer (bisCD)-based supramolecular nanogel and revealed its uniform, spherical shape and reduced cytotoxicity. These nanogels significantly increased the circulation half-life of ASNase and reduced its immunogenicity, addressing limitations in cancer chemotherapy. Xu et al. (2023) developed an injectable system for the controlled release of basic fibroblast growth factor (bFGF) and pinocembrin (PNCB), which accelerated wound healing. This system combines anti-inflammatory drugs with growth factors, highlighting a synergistic approach to chronic wound healing. X. Zhang et al. (2023) introduced a guanosine-driven hyaluronic acid-based hydrogel incorporating hemin and glucose oxidase (GOx) for chronic diabetic wound treatment. This hydrogel, which forms rapidly and exhibits peroxidase activity, enhances antibacterial properties and wound healing. Finally, Cheng et al. (2022) described a novel supramolecular hydrogel for bacterial keratitis treatment using guanosine 5′-monophosphate disodium salt and tobramycin. This hydrogel, which forms through guanosine-quartet nanofibers, exhibits thixotropic properties, good biocompatibility, and superior antibacterial activity compared to traditional eye drops, suggesting that it is a new approach for treating bacterial keratitis.

Therefore, supramolecular gels offer a promising and versatile platform for drug delivery because of their ability to encapsulate a wide range of drugs, respond to physiological stimuli, and provide controlled and sustained release. The potential applications of these materials include localized topical treatments, systemic drug delivery, and tissue engineering, highlighting their role as key components in the future of advanced drug delivery systems. As research in this field continues to advance, the development of novel supramolecular gel formulations is expected to open new avenues for effective and patient-friendly therapeutic interventions.

Table 8. Summary of recent studies on supramolecular gel-based drug delivery systems

Drug delivered	Supramolecular gel type	Key findings	Remarks	Reference
Doxorubicin	Sulfobutyl ether β-cyclodextrin (SBE-β-CD)	- Rapid formation of polysaccharide supramolecular injectable hydrogel (CCA hydrogels) by mixing chitosan, SBE-β-CD, and silver ions. - Gel reconstitutes and regains shape in less than 1 hour. - Maintains elasticity suitable for the human body. - High concentration of doxorubicin achieved. - Significant inhibitory effect on bladder tumors in vivo compared to free doxorubicin.	- Injectable hydrogel offers effective retention of chemotherapy drugs in bladder cancer. - Exhibits self-healing properties. - Potential for in situ treatment of bladder cancer.	C. Zhang et al. (2024)
L-asparaginase (ASNase)	α-Cyclodextrin Dimer (bisCD)	- ASNase immobilized in poly hydroxypropyl methacrylamide (P(HPMA))-based supramolecular nanogels formed through host-guest inclusion with P(HPMA-MPEGA) and bisCD, using cystamine as a linker. - Nanogels are discrete, homogeneous, spherical, with an average diameter of 148 ± 41 nm. - Reduced cytotoxicity on HL-60 cell line with IC_{50} of 3 IU/ml. - In vivo studies in rats showed reduced antigenicity and an 8.1-fold increase in circulation half-life compared to native enzyme.	- Addresses limitations of ASNase therapy in cancer chemotherapy, such as poor pharmacokinetics and immunogenicity. - Enhanced therapeutic value through immobilization in supramolecular nanogels.	Monajati et al. (2023)

continued on following page

Table 8. Continued

Drug delivered	Supramolecular gel type	Key findings	Remarks	Reference
Basic fibroblast growth factor (bFGF) and pinocembrin (PNCB)	α-Cyclodextrin (α-CD)	- Injectable host-guest drug delivery system developed for controlled release of hydrophilic bFGF and hydrophobic PNCB. - PNCB loaded into Pluronic F127 micelles (PNCB@F127), then combined with bFGF-mixed α-CD chains to form P/bFGF@F127α-CD hydrogel. - Demonstrated accelerated wound healing in vitro and in vivo, combining therapeutic effects of PNCB and bFGF.	- Addresses the challenge of rapid healing in chronic wounds. - Highlights the synergistic potential of combining anti-inflammatory drugs with growth factors for wound healing.	Xu et al. (2023)

continued on following page

Table 8. Continued

Drug delivered	Supramolecular gel type	Key findings	Remarks	Reference
Hemin and glucose oxidase (GOx)	Phenylboronic acid-functionalized hyaluronic acid (HA-PBA) and guanosine.	- Developed a guanosine-driven hyaluronic acid-based supramolecular hydrogel using phenylboronic acid-functionalized hyaluronic acid (HA-PBA) and guanosine. - Guanosines form G-quartet planes and boronic ester bonds with HA-PBA, inducing rapid hydrogel formation. - Hemin binds to G-quartet via π-π interactions, providing peroxidase activity and bacterial killing by generating hydroxyl radicals with H_2O_2. - Incorporation of GOx into hydrogels enhances antibacterial properties and wound healing in diabetic chronic wounds.	- Novel approach for chronic wound treatment, particularly diabetic wounds. - Guanosine-driven hydrogel formation simplifies gelation and improves biosafety. - G-quartet/hemin complex introduces peroxidase activity for effective wound care.	X. Zhang et al. (2023)

continued on following page

Table 8. Continued

Drug delivered	Supramolecular gel type	Key findings	Remarks	Reference
Tobramycin	Guanosine 5′-monophosphate disodium salt	- Developed a supramolecular hydrogel for bacterial keratitis treatment, assembled from guanosine-5′-monophosphate disodium salt and tobramycin. - The hydrogel forms through guanosine-quartet nanofibers via hydrogen bonding and π-π stacking, with tobramycin crosslinking these nanofibers into gel networks. - Exhibits shear thinning, thixotropic properties, good biocompatibility, and antibacterial activity. - Demonstrated higher therapeutic efficacy in treating *Pseudomonas aeruginosa* induced bacterial keratitis compared to traditional eye drops.	- Offers a novel and efficient antibiotic gel formulation for the clinical treatment of bacterial keratitis, overcoming limitations of low bioavailability and poor efficiency in conventional treatments.	Cheng et al. (2022)

CASE STUDIES AND APPLICATIONS

The environmental impact of drug delivery systems is a significant concern in the pharmaceutical industry, stemming from the extensive use of nonbiodegradable materials, hazardous solvents, and environmentally harmful processes. Traditional drug manufacturing often requires complex chemical synthesis, the generation of substantial waste, and the use of toxic solvents and reagents. A classic example is the production of APIs, which can involve multiple reaction steps, each generating waste and potentially using harmful chemicals. Moreover, conventional drug delivery agents, such as certain pills and injections, frequently utilize synthetic polymers and other nondegradable materials. These materials can accumulate in the environment, posing sustainability issues and potential risks to ecosystems and human health (Idumah et al., 2024; Z. Shi et al., 2024). Another major environmental concern is the accumulation of pharmaceuticals in water bodies. Medications that are excreted by humans or disposed of improperly can persist in aquatic environments. For instance, traces of antibiotics, hormones, and other drugs potentially affecting aquatic life have been detected in rivers and lakes. These drugs can disrupt the hormonal balance of aquatic species, leading to issues such as altered reproductive behaviors and growth patterns. Furthermore, the presence of pharmaceuticals in water bodies raises concerns about their potential re-entry into the human food chain, either through drinking water sources or through the consumption of affected aquatic organisms. This scenario not only highlights the direct environmental impacts of drug delivery systems but also underscores the broader implications for public health and ecological balance (Ali et al., 2024).

In addressing environmental concerns related to drug delivery, supramolecular systems emerge as a promising solution for more sustainable and eco-friendly practices. Supramolecular chemistry, which involves reversible and noncovalent interactions, paves the way for drug delivery systems that are less detrimental to the environment. These systems can be engineered to utilize biodegradable materials, thereby significantly reducing waste and pollution (Idumah et al., 2024). For example, cyclodextrins, which are naturally derived cyclic oligosaccharides, serve as biodegradable alternatives to traditional synthetic drug carriers. These materials breakdown more easily in the environment, mitigating long-term pollution concerns (Zhao et al., 2024). Furthermore, the efficiency of supramolecular systems in drug delivery can minimize the need for harmful solvents and decrease waste production during the manufacturing process. This efficiency is partly due to the ability of these materials to enhance the solubility and bioavailability of drugs. By improving these properties, supramolecular systems can potentially reduce the required dosage of medications, leading to a decrease in the overall quantity of pharmaceuticals released into the environment. This approach is particularly beneficial for reducing the drug load in aquatic ecosystems, where pharmaceutical contamination is a growing issue. Additionally, the precision and targeted delivery capabilities of these systems ensure that the drugs are delivered more effectively to the intended site of action, further reducing the need for higher dosages and minimizing environmental exposure (Zhou et al., 2024). Thus, supramolecular systems offer an innovative and environmentally considerate approach to drug delivery, aligning with the growing need for sustainability in pharmaceutical practices.

Several case studies highlight the successful implementation of green practices in drug delivery through supramolecular systems (Table 9). One notable example is the use of cyclodextrin-based systems for the delivery of anti-inflammatory drugs. These systems not only improve the solubility and efficacy of drugs but also reduce the need for harmful organic solvents in their formulation. Halder et al. (2023) focused on glycyrrhizin combined with HP-β-CD. Despite its therapeutic benefits, glycyrrhizin has limited commercial use due to its poor water solubility. This study addresses this issue by forming a molecular inclusion complex of glycyrrhizin and HP-β-CD, which can be further incorporated into hyaluronic acid nanofibers. This approach significantly enhances glycyrrhizin's solubility and release profile, with a fourfold increase in solubility observed. Compared with those of pure glycyrrhizin, the glycyrrhizin–HP-β-CD nanofibers markedly improved the anti-inflammatory efficacy, as demonstrated by in vitro and in vivo studies, as indicated by a reduction in inflammatory markers and improved inhibition of inflammation. These findings suggest that glycyrrhizin may improve the therapeutic efficacy of this nanofiber-based delivery system.

Another case study involved the use of self-assembling peptide gels for localized drug delivery. This approach eliminates the need for nonbiodegradable polymeric carriers, offering a biocompatible and biodegradable alternative. In their study, Li et al. (2021) developed a supramolecular hydrogel composed of biphenyl-tripeptide for cartilage tissue engineering. This hydrogel, which mimics the nanofibrillar structure of natural collagen fibers, can be manipulated by changing the temperature, ion induction, and pH. It supports cell adhesion and proliferation, particularly by enhancing chondrocyte activity and hyaline cartilage formation, thus showing promise for tissue engineering applications. Additionally, research into dendrimer-based drug delivery systems has demonstrated the potential of these systems for reducing environmental impacts. Dendrimers can be synthesized using greener methodologies, and their precise molecular architecture allows for efficient drug encapsulation, minimizing waste and reducing the overall environmental footprint of drug delivery processes. For example, Zhao et al. (2023) reported that the delivery of isotretinoin occurs via a self-assembled dendrimer-conjugated system, 13cRA-D, which was specifically developed for treating skin diseases such as acne. This novel delivery model enhances the skin permeation and lesion targeting ability of isotretinoin, which is characterized by controlled release, particularly in low-pH environments. The 13cRA-D system exhibited high therapeutic efficacy, better skin penetration than did isotretinoin alone, and minimal skin irritation, suggesting that it is a safe and effective formulation for skin disorders. These case studies exemplify how supramolecular approaches are contributing to the development of greener, more sustainable drug delivery systems, aligning pharmaceutical practices with the principles of environmental stewardship and sustainability.

Table 9. Drugs delivered using supramolecular carriers

Drug	Supramolecular carrier	Indication/area of use	Reference
Glycyrrhizin	Hydroxypropyl β-cyclodextrin incorporated into hyaluronic acid nanofibers	Anti-inflammatory, enhancing solubility and therapeutic potential of glycyrrizin	Halder et al. (2023)
Biphenyl-tripeptide	Supramolecular hydrogel composed of aromatic short peptide gelator	Cartilage tissue engineering, promoting chondrocyte proliferation and extracellular matrix secretion	Li et al. (2021)
Isotretinoin	Self-assembled dendrimer-conjugated system (13cRA-D)	Skin diseases, targeted therapy for acne	Zhao et al. (2023)

CHALLENGES AND FUTURE DIRECTIONS

Current Limitations in Supramolecular Drug Delivery

Scale-Up and Manufacturing Challenges

The scale-up and manufacturing of supramolecular drug delivery systems present a significant and multifaceted challenge, primarily due to the complexity and precision required in their synthesis (Pitingolo & Nastruzzi, 2023). In the field of supramolecular drug delivery, particularly for applications such as targeted cancer therapy, the exactness of the system needed to ensure that it selectively targets cancer cells without impacting healthy cells substantially elevates both the complexity and cost. This precision is not easily translated when moving from a controlled laboratory environment to mass production, where maintaining the same level of accuracy and effectiveness in larger batches becomes a formidable task. The intricate nature of these systems, often involving delicate molecular arrangements and specific environmental responsiveness, adds layers of difficulty to scaling up. This process is further complicated by the need to ensure that these systems remain stable and retain their functional integrity during large-scale manufacturing (Madanayake et al., 2023). Consequently, this aspect of supramolecular drug delivery requires not only substantial financial investment but also considerable innovation in manufacturing processes and technology. Developing methods that can efficiently and reliably scale these complex structures from laboratory-scale prototypes to commercially viable products is crucial for the advancement and practical application of this promising field.

Stability of Supramolecular Assemblies

The stability of supramolecular assemblies in drug delivery is a substantial challenge, as these complex structures often face the risk of disassembly or degradation under physiological conditions. This inherent instability poses a significant hurdle in the effective application of supramolecular drug delivery systems (Arora et al., 2022). The delicacy of these assemblies means that they can be prematurely destabilized by various factors in the human body, such as temperature fluctuations, pH changes, or enzymatic activities, which they are not always designed to withstand (Yan et al., 2023). For instance, a supramolecular system engineered to release a therapeutic agent in response to specific pH changes in the body could be vulnerable to unintended disassembly when exposed to the body's temperature or enzymes, leading to premature drug release. This early release not only diminishes the efficacy of the treatment by preventing the delivery of the drug at the intended site and time but also raises concerns about potential side effects due to the uncontrolled distribution of the drug in the body. Furthermore, the unpredictable nature of these interactions in the dynamic environment of the human body complicates the design and testing of these systems. Addressing the stability of supramolecular assemblies is thus crucial, as meticulous design and extensive testing are required to ensure that these systems can reliably withstand the diverse and often harsh conditions encountered within the body (Ghitman & Voicu, 2023). This aspect underscores the need for advanced materials and design strategies that can bolster the resilience of supramolecular structures, ensuring that they remain intact and functional until they reach their targeted site of action.

Safety and Biocompatibility Concerns

The safety and biocompatibility of supramolecular systems in drug delivery are paramount concerns that demand rigorous scrutiny, as the long-term effects and potential toxicity of these novel materials within the human body remain largely uncharted. Despite being engineered with biocompatibility as a core consideration, the intricate nature of supramolecular systems poses uncertainties regarding their interaction with biological systems over extended periods (Rahman et al., 2023). For example, a newly developed supramolecular material, even if designed to be biodegradable, could exhibit unforeseen long-term effects. These effects might manifest as unexpected interactions with bodily tissues or the immune system, potentially leading to adverse reactions or immune responses. The complexity of these materials, often involving novel molecular structures, raises the possibility of unanticipated biocompatibility issues, as their degradation products or the way they interact with biological molecules and cells are not always predictable.

Therefore, extensive and comprehensive testing is crucial not only for assessing immediate safety concerns but also for understanding long-term implications. This requires a thorough investigation of how these materials behave in the body over time, including their breakdown process, potential accumulation in tissues, and any ensuing physiological responses. The need for such in-depth research underscores the importance of developing robust and reliable testing protocols and models that can effectively simulate the long-term behavior of supramolecular systems in vivo (Mehata & Muthu, 2022). Ensuring the safety and biocompatibility of these systems is a complex yet critical step in their development, necessitating a meticulous and holistic approach to research and testing to guarantee their safe and effective application in medical treatments.

Regulatory Approval Challenges

The path to regulatory approval for innovative supramolecular drug delivery systems is fraught with complexities, primarily because these advanced systems often exhibit properties and pose challenges that are not sufficiently covered by existing regulatory frameworks. Traditional regulatory pathways, as established by bodies such as the Food and Drug Administration, are primarily tailored for conventional drug delivery systems. These pathways may not fully encompass the nuanced and dynamic nature of supramolecular systems (Manjit & Mishra, 2022), which can, for instance, alter their structure in response to specific physiological conditions. This adaptive capability, while therapeutically advantageous, falls outside the scope of current regulatory protocols, which typically focus on more static and predictable drug delivery methods. The dynamic behavior of supramolecular systems, such as their ability to change configuration in response to pH changes, temperature variations, or enzymatic activities within the body, presents a unique set of variables that need thorough evaluation. This situation calls for a significant revision of existing regulatory standards and processes, tailoring them to address the innovative aspects of supramolecular drug delivery. The development of new standards and protocols that can adequately assess the safety, efficacy, and quality of these systems is essential. This includes evaluating the stability, responsiveness, and interaction of these materials with biological systems under various conditions. Moreover, regulatory bodies must also consider the long-term effects and biodegradability of these systems, ensuring that they do not pose unforeseen risks to patients. Such a comprehensive overhaul of regulatory processes is crucial for ensuring that these cutting-edge drug delivery systems can be thoroughly evaluated and safely integrated into medical practice (Musazzi et al., 2023). This is a vital step in unlocking the full potential of supramolecular drug delivery, enabling the transition of these innovative systems from experimental stages to widespread clinical use.

Future Research Trends and Potential Breakthroughs

Future research in supramolecular drug delivery is poised to make significant breakthroughs, addressing current limitations and expanding capabilities in exciting new ways. One of the most promising areas is the development of smart or stimulus-responsive supramolecular systems. These systems are engineered to adapt to changing physiological conditions, enabling them to respond to specific biological triggers, such as pH variations, temperature changes, or enzymatic activities. This adaptability allows more precise and controlled drug release. For example, a supramolecular system could be designed to release a drug only when it reaches a tumor site, as identified by a unique pH, thereby maximizing the efficacy of the drug while minimizing side effects. Such systems could revolutionize treatments for conditions such as cancer, where targeted drug delivery is crucial (R. Kaur et al., 2023).

Another promising trend is the integration of supramolecular systems with nanotechnology, blending the intricate molecular design of supramolecular chemistry with the precision of nanoscale engineering. This combination can lead to more efficient drug targeting and reduced side effects, significantly improving therapeutic outcomes (Ghitman & Voicu, 2023). For instance, nanosized supramolecular carriers could be used to deliver drugs directly to specific cells or tissues, bypassing healthy cells and minimizing collateral damage. This approach could be particularly beneficial in treating diseases such as Alzheimer's disease, where specific brain regions need to be targeted, or in gene therapy, where precise delivery to specific cell types is essential.

Finally, advancements in computational modeling and molecular simulations are expected to play a transformative role in the design and optimization of supramolecular structures. These technologies allow for the virtual testing and refinement of supramolecular designs before they are synthesized, significantly accelerating the development process and reducing costs (Bandara et al., 2023). For example, simulations could predict how a supramolecular system will assemble and behave in the body, allowing researchers to fine-tune the system for optimal performance before it is ever tested in a laboratory. This rapid prototyping and testing could lead to the discovery of new supramolecular structures with unprecedented capabilities, such as systems that can navigate the body to treat hard-to-reach diseases or provide personalized medicine based on an individual's specific physiological makeup. The potential of these technological advancements is immense, indicating a bright future for the field of supramolecular drug delivery, with these trends paving the way for groundbreaking developments that could transform healthcare and medicine.

Overcoming Challenges and Moving Forward

Overcoming the challenges faced by supramolecular drug delivery requires a comprehensive and multifaceted strategy that encompasses continued research and development focused on enhancing the stability, safety, and scalability of these systems (Zakharova et al., 2023). The complexity of supramolecular structures means that improvements in stability could, for example, involve developing new materials that are more resistant to physiological conditions, thereby preventing premature disassembly. Safety enhancements might include extensive biocompatibility testing to ensure that these systems do not elicit adverse immune responses. For scalability, innovative manufacturing techniques that can produce these complex structures on a large scale without compromising their integrity are needed. This task is monumental and necessitates collaboration across various disciplines (Talreja & Tiwari, 2023). Researchers working together can bring diverse perspectives and expertise, leading to innovative solutions that might not emerge within siloed research efforts. For instance, engineers might develop new manufacturing technologies, while biologists and pharmacologists provide insights into physiological interactions and drug efficacy.

Regulatory bodies also play a crucial role in this advancement. The unique characteristics of supramolecular drug delivery systems mean that they may not fit neatly within existing regulatory frameworks. Therefore, regulatory agencies need to adapt and develop new standards and evaluation processes. This adaptation could involve creating guidelines for assessing the stability and biocompatibility of these systems, as well as protocols for clinical testing. A more flexible and dynamic regulatory approach will facilitate the integration of these innovative systems into clinical practice, ensuring that they meet safety and efficacy standards without stifling innovation (Manjit & Mishra, 2022). Furthermore, the advancement of supramolecular drug delivery is heavily dependent on funding. Both public and private investments are essential for supporting the extensive research and development needed. With adequate funding, additional research can be conducted, accelerating the development and refinement of these systems. For instance, funding can support large-scale clinical trials to test the efficacy and safety of these systems, or it can be used to develop new manufacturing processes that make large-scale production feasible.

Overall, despite the considerable challenges it faces, the field of supramolecular drug delivery is poised for significant breakthroughs. These systems have immense potential for use in pharmaceutical development and therapy. With collaborative efforts across disciplines, evolving regulatory frameworks, and sufficient funding, the barriers facing supramolecular drug delivery can be surmounted. This progress holds the promise of significantly impacting healthcare and improving patient outcomes, heralding a new era in medical treatment and drug delivery technologies.

CONCLUSION AND PERSPECTIVES

The exploration of supramolecular approaches for green drug delivery has unveiled a range of innovative strategies that hold great promise for revolutionizing pharmaceutical therapies. One of the key findings of this exploration is the remarkable versatility of supramolecular systems, which can be tailored to address specific challenges in drug delivery, such as solubility enhancement, targeted delivery, and controlled release. The use of environmentally benign materials and processes in these systems underscores the potential of these systems to reduce the ecological footprint of pharmaceutical practices. Advances in cyclodextrins, dendrimers, micelles, macrocycles, and supramolecular gels have demonstrated their efficacy in improving drug solubility, stability, and bioavailability while also offering targeted and controlled drug release. These systems have shown potential in minimizing side effects and enhancing therapeutic outcomes, particularly in the treatment of chronic diseases and cancer.

Looking ahead, the future scope of supramolecular approaches in green drug delivery is vast and full of potential. One area of particular interest is the development of more sophisticated stimuli-responsive systems that can precisely control drug release in response to specific physiological conditions. This precision could lead to highly personalized medicine tailored to the individual patient's needs. Another exciting avenue is the integration of supramolecular systems with emerging technologies such as nanotechnology and biotechnology, which could lead to the development of novel drug delivery platforms with enhanced efficacy and safety. Additionally, there is a growing emphasis on sustainability in pharmaceutical manufacturing, and supramolecular chemistry offers a pathway toward more environmentally friendly production processes. The scalability and commercial viability of these systems are also crucial areas for future development, ensuring that these innovative drug delivery methods are accessible and affordable for widespread clinical use.

As the field moves forward, it will be important to address the current challenges in supramolecular drug delivery, particularly in terms of scalability, stability, and regulatory compliance. Continued interdisciplinary collaboration and innovation are essential to overcome these hurdles. The potential of supramolecular chemistry for creating more effective, safer, and environmentally friendly drug delivery systems is immense, and ongoing research in this area is likely to yield significant benefits for both human health and the environment. The exploration of supramolecular approaches in green drug delivery represents a convergence of scientific inquiry with environmental responsibility. This heralds a new era in pharmaceutical development, one that promises both efficacy and sustainability.

REFERENCES

Ailincai, D., Morariu, S., Rosca, I., Sandu, A. I., & Marin, L. (2023). Drug delivery based on a supramolecular chemistry approach by using chitosan hydrogels. *International Journal of Biological Macromolecules, 248*, 125800. DOI: 10.1016/j.ijbiomac.2023.125800 PMID: 37442500

Akash, B. A., Kanagaraj, S., Sundaravadivelu, S., Varalakshmi, G. S., Manikantan, V., Pillai, A. S., Alexander, A., & Enoch, I. V. M. V. (2024). Cytotoxicity of 5-fluorouracil- loaded, β-cyclodextrin-tethered metallic copper nanoparticles on triple-negative breast cancer cells. *Journal of Molecular Structure, 1295*, 136530. DOI: 10.1016/j.molstruc.2023.136530

Alam, A., Karmakar, R., Rengan, A. K., & Khandelwal, M. (2023). Nanofiber-based systems for stimuli-responsive and dual drug delivery: Present scenario and the way forward. *ACS Biomaterials Science & Engineering, 9*(6), 3160–3184. DOI: 10.1021/acsbiomaterials.3c00363 PMID: 37218599

Ali, I., Jamal, A., Saqib, Z., Ali, M. I., & Masood, A. B. (2024). Biosurfactants: introduction and classification. In Aslam, R., Aslam, J., & Hussain, C. M. (Eds.), *Industrial Applications of Biosurfactants and Microorganisms* (pp. 1–23). Academic Press. DOI: 10.1016/B978-0-443-13288-9.00021-8

Aliabadi, A., Vakili-Azghandi, M., Abnous, K., Taghdisi, S. M., Babaei, M., Ramezani, M., & Alibolandi, M. (2024). Amphiphilic polylactic acid-b-poly(N-(3-aminopropyl) methacrylamide) copolymers: Self-assembly to polymeric micelles for gene delivery. *Journal of Drug Delivery Science and Technology, 91*, 105236. DOI: 10.1016/j.jddst.2023.105236

Antuma, L. J., Braitmaier, S. H., Garamus, V. M., Hinrichs, J., Boom, R. M., & Keppler, J. K. (2024). Engineering artificial casein micelles for future food: Preparation rate and coagulation properties. *Journal of Food Engineering, 366*, 111868. DOI: 10.1016/j.jfoodeng.2023.111868

Arora, K., Singh, A., Javaid, A., & Mudavath, S. L. (2022). Promising Functional Supramolecules in Antiviral Drugs. In Goel, N., & Kumar, N. (Eds.), *Pharmaceutical Applications of Supramolecules* (pp. 135–155). Springer International Publishing. DOI: 10.1007/978-3-031-21900-9_6

Arruda, T. R., Silva, R. R. A., Marques, C. S., Moraes, A. R. F. e., Bernardes, P. C., de Oliveira, T. V., de Oliveira, S. O., Muranyi, P., & Soares, N. F. F. (2024). β-cyclodextrin versus hydroxypropyl-β-cyclodextrin: Is inclusion complexation a suitable alternative to improve the properties of hop-derived β-acids? *Food Hydrocolloids, 149*, 109622. DOI: 10.1016/j.foodhyd.2023.109622

Attia, M. F., Ogunnaike, E. A., Pitz, M., Elbaz, N. M., Panda, D. K., Alexander-Bryant, A., Saha, S., Whitehead, D. C., & Kabanov, A. (2024). Enhancing drug delivery with supramolecular amphiphilic macrocycle nanoparticles: Selective targeting of CDK4/6 inhibitor palbociclib to melanoma. *Biomaterials Science, xx*(3), xx–xx. DOI: 10.1039/D3BM01888A PMID: 38099834

Bai, M.-Y., Zhou, Q., Zhang, J., Li, T., Cheng, J., Liu, Q., Xu, W.-R., & Zhang, Y.-C. (2022). Antioxidant and antibacterial properties of essential oils-loaded β-cyclodextrin-epichlorohydrin oligomer and chitosan composite films. *Colloids and Surfaces. B, Biointerfaces, 215*, 112504. DOI: 10.1016/j.colsurfb.2022.112504 PMID: 35453062

Bandara, C. S., Gamage, N., Gunaratne, J. C., Waruni, K., Nelumdeniya, R. M., & Udayana, R. (2023). Modeling and simulation in drug delivery. In Das, S., Thomas, S., & Das, P. P. (Eds.), *Novel Platforms for Drug Delivery Applications* (pp. 491–536). Woodhead Publishing. DOI: 10.1016/B978-0-323-91376-8.00004-5

Braga, S. S., El-Saleh, F., Oliveira, C., Barbosa, J. S., & Paz, F. A. A. (2023). Co-milling of β-cyclodextrin with simvastatin: Solid-state studies and dissolution profile. *Journal of Drug Delivery Science and Technology, 89*, 105077. DOI: 10.1016/j.jddst.2023.105077

Buddhadev, S. S. C., Garala, K. S. S., Rahamathulla, M., Ahmed, M. M., Farhana, S. A., & Pasha, I. (2024). Quality by design aided self-nano emulsifying drug delivery systems development for the oral delivery of Benidipine: Improvement of biopharmaceutical performance. *Drug Delivery, 31*(1), 2288801. DOI: 10.1080/10717544.2023.2288801 PMID: 38073402

Burkeev, M., Fazylov, S., Bakirova, R., Iskineyeva, A., Sarsenbekova, A., Tazhbaev, E., & Davrenbekov, S. (2021). Thermal decomposition of β-cyclodextrin and its inclusion complex with vitamin E. *Mendeleev Communications, 31*(1), 76–78. DOI: 10.1016/j.mencom.2021.01.023

Cai, Y., Qi, J., Lu, Y., He, H., & Wu, W. (2022). The in vivo fate of polymeric micelles. *Advanced Drug Delivery Reviews, 188*, 114463. DOI: 10.1016/j.addr.2022.114463 PMID: 35905947

Changsan, N., Sawatdee, S., Suedee, R., Chunhachaichana, C., & Srichana, T. (2023). Aqueous cannabidiol β-cyclodextrin complexed polymeric micelle nasal spray to attenuate in vitro and ex vivo SARS-CoV-2-induced cytokine storms. *International Journal of Pharmaceutics*, *640*, 123035. DOI: 10.1016/j.ijpharm.2023.123035 PMID: 37182795

Chaudhuri, A., Ramesh, K., Kumar, D. N., Dehari, D., Singh, S., Kumar, D., & Agrawal, A. K. (2022). Polymeric micelles: A novel drug delivery system for the treatment of breast cancer. *Journal of Drug Delivery Science and Technology*, *77*, 103886. DOI: 10.1016/j.jddst.2022.103886

Chauhan, A., Anton, B., & Singh, M. K. (2020). 3 - Dendrimers for drug solubilization, dissolution and bioavailability. In Chauhan, A., & Kulhari, H. (Eds.), *Pharmaceutical Applications of Dendrimers* (pp. 59–92). Elsevier. DOI: 10.1016/B978-0-12-814527-2.00003-2

Chen, W., & Liu, P. (2023). Design of GSH-triggered disintegrable dendritic polyurethane as bioreducible unimolecular micelles for tumor-specific chemotherapeutic drug delivery. *European Polymer Journal*, *200*, 112521. DOI: 10.1016/j.eurpolymj.2023.112521

Cheng, X., Chen, H., Yang, F., Hong, J., Cheng, Y., & Hu, J. (2022). All-small-molecule supramolecular hydrogels assembled from guanosine 5′-monophosphate disodium salt and tobramycin for the treatment of bacterial keratitis. *Bioactive Materials*, *16*, 293–300. DOI: 10.1016/j.bioactmat.2021.12.024 PMID: 35386321

Chowdhury, A. R., & Goswami, S. (2023). Study of drug delivery using purely organic macrocyclic containers—Cucurbit[7]uril and pillararene. *ACS Omega*, *8*(50), 47340–47366. DOI: 10.1021/acsomega.3c05465 PMID: 38144095

Csige, K., Szabó, J. P., Kálmán-Szabó, I., Dénes, N. S., Szikra, D., Képes, Z., Opposits, G., Méhes, G., Kertész, I., Fenyvesi, F., Trencsényi, G., & Hajdu, I. (2022). In vivo investigation of Gallium-68 and Bismuth-205/206 labeled beta cyclodextrin for targeted alpha therapy of prostaglandin E2 receptor-expressing tumors in mice. *International Journal of Pharmaceutics*, *625*, 122132. DOI: 10.1016/j.ijpharm.2022.122132 PMID: 36028082

Dewan, M., Adhikari, A., Jana, R., & Chattopadhyay, D. (2023). Development, evaluation and recent progress of ocular in situ gelling drug delivery vehicle based on poloxamer 407. *Journal of Drug Delivery Science and Technology*, *88*, 104885. DOI: 10.1016/j.jddst.2023.104885

Dhandapani, G., Wachtel, E., & Patchornik, G. (2022). Detergent micelle conjugates containing amino acid monomers allow purification of human IgG near neutral pH. *Journal of Chromatography. B, Analytical Technologies in the Biomedical and Life Sciences*, *1206*, 123358. DOI: 10.1016/j.jchromb.2022.123358 PMID: 35780745

Ebrahimi, A., Hamishehkar, H., & Amjadi, S. (2023). Development of gelatin-coated nanoliposomes loaded with β-cyclodextrin/vitamin D3 inclusion complex for nutritional therapy. *Food Chemistry*, *424*, 136346. DOI: 10.1016/j.foodchem.2023.136346 PMID: 37201470

Edilino de Lima, F., Giordano Viegas, R., & Vital de Oliveira, O. (2023). In silico study of the encapsulation of α-tocopherol and α-tocotrienol vitamins E into cucurbit[7]uril. *Chemical Physics Letters*, *826*, 140647. DOI: 10.1016/j.cplett.2023.140647

Evans, N., Grygorash, R., Williams, P., Kyle, A., Kantner, T., Pathak, R., Sheng, X. B., Simoes, F., Makwana, H., Resende, R., de Juan, E., Jenkins, A., Morris, D., Michelet, A., Jewitt, F., Rudge, F., Camper, N., Manin, A., McDowell, W., & Bird, M. (2022). Incorporation of hydrophilic macrocycles into drug-linker reagents produces antibody–drug conjugates with enhanced in vivo performance. *Frontiers in Pharmacology*, *13*, 764540. DOI: 10.3389/fphar.2022.764540 PMID: 35784686

Ezike, T. C., Okpala, U. S., Onoja, U. L., Nwike, C. P., Ezeako, E. C., Okpara, O. J., Okoroafor, C. C., Eze, S. C., Kalu, O. L., Odoh, E. C., Nwadike, U. G., Ogbodo, J. O., Umeh, B. U., Ossai, E. C., & Nwanguma, B. C. (2023). Advances in drug delivery systems, challenges and future directions. *Heliyon*, *9*(6), e17488. DOI: 10.1016/j.heliyon.2023.e17488 PMID: 37416680

Fang, F., Zhang, Z., Zhang, P., Zhang, X., Ma, H., & Wei, Y. (2024). Fluorescence detection of amantadine based on competitive β-Cyclodextrin host-guest inclusion process. *Colloids and Surfaces. A, Physicochemical and Engineering Aspects*, *681*, 132801. DOI: 10.1016/j.colsurfa.2023.132801

Fuhrmann, G. (2023). Drug delivery as a sustainable avenue to future therapies. *Journal of Controlled Release*, *354*, 746–754. DOI: 10.1016/j.jconrel.2023.01.045 PMID: 36690037

Gao, J., Karp, J. M., Langer, R., & Joshi, N. (2023). The future of drug delivery. *Chemistry of Materials*, *35*(2), 359–363. DOI: 10.1021/acs.chemmater.2c03003 PMID: 37799624

Garcia Jimenez, D., Poongavanam, V., & Kihlberg, J. (2023). Macrocycles in drug discovery—Learning from the past for the future. *Journal of Medicinal Chemistry*, *66*(8), 5377–5396. DOI: 10.1021/acs.jmedchem.3c00134 PMID: 37017513

Gautam, S., & Kaur, S. (2024). Revolutionizing biomedicine: Metal-organic frameworks combating multidrug resistance. *Materials Letters*, *354*, 135306. DOI: 10.1016/j.matlet.2023.135306

Ghitman, J., & Voicu, S. I. (2023). Controlled drug delivery mediated by cyclodextrin-based supramolecular self-assembled carriers: From design to clinical performances. *Carbohydrate Polymer Technologies and Applications*, *5*, 100266. DOI: 10.1016/j.carpta.2022.100266

Halder, J., Mahanty, R., Rajwar, T. K., Rai, V. K., Kar, B., Ghosh, G., & Rath, G. (2023). Nanofibers of glycyrrhizin/hydroxypropyl-β-cyclodextrin inclusion complex: Enhanced solubility profile and anti-inflammatory effect of glycyrrhizin. *AAPS PharmSciTech*, *24*(7), 196. DOI: 10.1208/s12249-023-02662-0 PMID: 37783948

Hasanah, N., Manurung, R. V., Jenie, S. N. A., Indriyati, , Prastya, M. E., & Andreani, A. S. (2023). The effect of size control of Gold Nanoparticles Stabilized with α-cyclodextrin and β-cyclodextrin and their antibacterial activities. *Materials Chemistry and Physics*, *302*, 127762. DOI: 10.1016/j.matchemphys.2023.127762

Hazarika, B., & Singh, V. P. (2023). Macrocyclic supramolecular biomaterials in anticancer therapeutics. *Chinese Chemical Letters*, *34*(11), 108220. DOI: 10.1016/j.cclet.2023.108220

He, H.-H., Guan, Z.-J., Peng, Y., Liang, Y., Li, J., Zhang, L.-L., & Fang, Y. (2024). Engineering the interactions between metal-organic frameworks and modifying agents: Design, structures, and applications. *Coordination Chemistry Reviews*, *499*, 215515. DOI: 10.1016/j.ccr.2023.215515

Hu, D., Li, Y., Teng, W., Li, F., Li, L., Li, M., Tan, L., & Wu, Z. (2024). pH-responsive mesoporous silica nanorod for high load and oral delivery of insulin. *Journal of Drug Delivery Science and Technology*, *91*, 105256. DOI: 10.1016/j.jddst.2023.105256

Hu, R., Fan, H., Zhou, Y., Tao, K., Tian, Z., & Ma, H. (2024). Electrostatic interactions dominate thermal conductivity and anisotropy in three-dimensional hydrogen-bonded organic frameworks. *International Journal of Heat and Mass Transfer*, *221*, 125071. DOI: 10.1016/j.ijheatmasstransfer.2023.125071

Hu, T., Zhang, Z., & Reches, M. (2024). A self-standing superhydrophobic material formed by the self-assembly of an individual amino acid. *Journal of Colloid and Interface Science*, *655*, 899–908. DOI: 10.1016/j.jcis.2023.11.062 PMID: 37979295

Huang, X., Li, T., Jiang, X., Wang, Z., Wang, M., Wu, X., Li, J., & Shi, J. (2023). Coassembled supramolecular hydrogel of salvianolic acid B and A phosphopeptide for enhanced wound healing. *ACS Applied Materials & Interfaces, 15*(39), 45606–45615. DOI: 10.1021/acsami.3c09219 PMID: 37733024

Idumah, C. I., Nwuzor, I. C., Odera, S. R., Timothy, U. J., Ngenegbo, U., & Tanjung, F. A. (2024). Recent advances in polymeric hydrogel nanoarchitectures for drug delivery applications. *International Journal of Polymeric Materials, 73*(1), 1–32. DOI: 10.1080/00914037.2022.2120875

Jeyadurai, J. A., Muthuselvan, P., Prabhu, A. A. M., Rajendiran, N., & Dasan, A. (2023). Formation of inclusion complexes of 2-hydroxypropyl β-cyclodextrin with 2-hydroxychalcone: Experimental and theoretical study. *Polycyclic Aromatic Compounds, xx*, xx–xx. DOI: 10.1080/10406638.2023.2259565

Jha, A., Kumar, M., & Mishra, B. (2022). Marine Biopolymer-Based Anticancer Drug Delivery Systems. In Jana, S., & Jana, S. (Eds.), *Marine Biomaterials: Drug Delivery and Therapeutic Applications* (pp. 351–401). Springer Nature Singapore. DOI: 10.1007/978-981-16-4787-1_11

Jindal, A. B., Bhide, A. R., Salave, S., Rana, D., & Benival, D. (2023). Long-acting parenteral drug delivery systems for the treatment of chronic diseases. *Advanced Drug Delivery Reviews, 198*, 114862. DOI: 10.1016/j.addr.2023.114862 PMID: 37160247

Joshi, N., Azizi Machekposhti, S., & Narayan, R. J. (2023). Evolution of transdermal drug delivery devices and novel microneedle technologies: A historical perspective and review. *JID Innovations : Skin Science from Molecules to Population Health, 3*(6), 100225. DOI: 10.1016/j.xjidi.2023.100225 PMID: 37744689

Jug, M. (2020). Cyclodextrin-based drug delivery systems. In *Nanomaterials for Clinical Applications* (pp. 29–69). Case Studies in Nanomedicines. DOI: 10.1016/B978-0-12-816705-2.00002-3

Kali, G., Haddadzadegan, S., & Bernkop-Schnürch, A. (2024). Cyclodextrins and derivatives in drug delivery: New developments, relevant clinical trials, and advanced products. *Carbohydrate Polymers, 324*, 121500. DOI: 10.1016/j.carbpol.2023.121500 PMID: 37985088

Kaur, A., Singh, N., Kaur, H., Kakoty, V., Sharma, D. S., Khursheed, R., Babu, M. R., Harish, V., Gupta, G., Gulati, M., Kumar, P., Dureja, H., Alharthi, N. S., Khan, F. R., Rehman, Z., Hakami, M. A., Patel, M., Patel, R., Zandi, M., & Singh, S. K. (2023). Neurodegenerative diseases and brain delivery of therapeutics: Bridging the gap using dendrimers. *Journal of Drug Delivery Science and Technology, 87*, 104868. DOI: 10.1016/j.jddst.2023.104868

Kaur, R., Sharma, S., & Rathore, K. (2023). Stimuli responsive ferrocene-based supramolecular systems: New route for the design of smart devices. *Journal of Organometallic Chemistry*, *997*, 122787. DOI: 10.1016/j.jorganchem.2023.122787

Khatik, A. S., Kurdhane, S., Batheja, S., & Gupta, U. (2024). Dendrimers: promises and challenges in drug delivery. In Gupta, U., & Goyal, A. K. (Eds.), *Molecular Pharmaceutics and Nano Drug Delivery* (pp. 237–267). Academic Press. DOI: 10.1016/B978-0-323-91924-1.00010-1

Kirla, H., Henry, D. J., Jansen, S., Thompson, P. L., & Hamzah, J. (2023). Use of silica nanoparticles for drug delivery in cardiovascular disease. *Clinical Therapeutics*, *45*(11), 1060–1068. DOI: 10.1016/j.clinthera.2023.08.017 PMID: 37783646

Krstić, L., Jarho, P., Ruponen, M., Urtti, A., González-García, M. J., & Diebold, Y. (2022). Improved ocular delivery of quercetin and resveratrol: A comparative study between binary and ternary cyclodextrin complexes. *International Journal of Pharmaceutics*, *624*, 122028. DOI: 10.1016/j.ijpharm.2022.122028 PMID: 35853567

Kumar, P., Bhardwaj, V. K., & Purohit, R. (2024). Highly robust quantum mechanics and umbrella sampling studies on inclusion complexes of curcumin and β-cyclodextrin. *Carbohydrate Polymers*, *323*, 121432. DOI: 10.1016/j.carbpol.2023.121432 PMID: 37940299

Lamey, C. A., Moussa, N., Helmy, M. W., Haroun, M., & Sabra, S. A. (2023). Simultaneous encapsulation of dasatinib and celecoxib into caseinate micelles toward improved in vivo anti-breast cancer efficacy with reduced drug toxicity. *Journal of Drug Delivery Science and Technology*, *87*, 104807. DOI: 10.1016/j.jddst.2023.104807

Lee, C. S., Kim, T. W., Kang, Y., Ju, Y., Ryu, J., Kong, H., Jang, Y.-S., Oh, D. E., Jang, S. J., Cho, H., Jeon, S., Kim, J., & Kim, T. H. (2022). Targeted drug delivery nanocarriers based on hyaluronic acid-decorated dendrimer encapsulating gold nanoparticles for ovarian cancer therapy. *Materials Today. Chemistry*, *26*, 101083. DOI: 10.1016/j.mtchem.2022.101083

Lee, E., Zhang, X., Noda, T., Miyamoto, J., Kimura, I., Tanaka, T., Sakurai, K., Hatano, R., & Miki, T. (2021). Lecithin inclusion by α-cyclodextrin activates SREBP2 signaling in the gut and ameliorates postprandial hyperglycemia. *International Journal of Molecular Sciences*, *22*(19), 10796. DOI: 10.3390/ijms221910796 PMID: 34639136

Li, N., Zhang, T., Xue, W., Zhao, Y., Zhu, B., Pei, X., & Xu, Z. (2024). Tuning composite nanofiltration membranes with γ-cyclodextrin for improved Mg2+/Li+ selectivity. *Separation and Purification Technology*, *330*, 125206. DOI: 10.1016/j.seppur.2023.125206

Li, P., Liu, Q., Xiang, Z., Wang, J., Wu, W.-X., & Yi, W.-J. (2023). ROS-responsive core crosslinked micelles by combination of enzymatic polymerization and thiol-ene click chemistry for anticancer drug delivery. *European Polymer Journal*, *199*, 112473. DOI: 10.1016/j.eurpolymj.2023.112473

Li, X., Bian, S., Zhao, M., Han, X., Liang, J., Wang, K., Jiang, Q., Sun, Y., Fan, Y., & Zhang, X. (2021). Stimuli-responsive biphenyl-tripeptide supramolecular hydrogels as biomimetic extracellular matrix scaffolds for cartilage tissue engineering. *Acta Biomaterialia*, *131*, 128–137. DOI: 10.1016/j.actbio.2021.07.007 PMID: 34245894

Li, Y., Liu, F., Abdiryim, T., & Liu, X. (2024). Cyclodextrin-derived materials: From design to promising applications in water treatment. *Coordination Chemistry Reviews*, *502*, 215613. DOI: 10.1016/j.ccr.2023.215613

Lin, S., Chen, W., Alqahtani, M. S., Elkamchouchi, D. H., Ge, Y., Lu, Y., Zhang, G., & Wang, M. (2024). Exploring the therapeutic potential of layered double hydroxides and transition metal dichalcogenides through the convergence of rheumatology and nanotechnology using generative adversarial network. *Environmental Research*, *241*, 117262. DOI: 10.1016/j.envres.2023.117262 PMID: 37839531

Liu, W., Li, Z., Wang, Z., Huang, Z., Sun, C., Liu, S., Jiang, Y., & Yang, H. (2023). Functional system based on glycyrrhizic acid supramolecular hydrogel: Toward polymorph control, stabilization, and controlled release. *ACS Applied Materials & Interfaces*, *15*(6), 7767–7776. DOI: 10.1021/acsami.2c19903 PMID: 36732699

Liu, X., Niu, W., Liu, J., Cui, Z., Li, J., Zhang, Z., & Ju, J. (2023). Preparation of β-cyclodextrin and hydroxypropyl-β-cyclodextrin inclusion complexes of baicalein and evaluation of their effects on dextran sulfate sodium-induced acute ulcerative colitis in mice. *Journal of Drug Delivery Science and Technology*, *87*, 104714. DOI: 10.1016/j.jddst.2023.104714

Liu, X.-Y., Zhang, X., Yang, J.-B., Wu, C.-Y., Wang, Q., Lu, Z.-L., & Tang, Q. (2022). Multifunctional amphiphilic peptide dendrimer as nonviral gene vectors for effective cancer therapy via combined gene/photodynamic therapies. *Colloids and Surfaces. B, Biointerfaces*, *217*, 112651. DOI: 10.1016/j.colsurfb.2022.112651 PMID: 35759892

Liu, Z., Lin, W., & Liu, Y. (2022). Macrocyclic supramolecular assemblies based on hyaluronic acid and their biological applications. *Accounts of Chemical Research*, *55*(23), 3417–3429. DOI: 10.1021/acs.accounts.2c00462 PMID: 36380600

Liu, Z., Wang, H., Bu, Y., Wu, T., Chen, X., Yan, H., & Lin, Q. (2024). Fabrication of self-assembled micelles based on amphiphilic oxidized sodium alginate grafted oleoamine derivatives via Schiff base reduction amination reaction for delivery of hydrophobic food active ingredients. *International Journal of Biological Macromolecules*, *257*, 128653. DOI: 10.1016/j.ijbiomac.2023.128653 PMID: 38072345

Loftsson, T., & Stefánsson, E. (2023). Topical dexamethasone delivery to the retina: An aqueous cyclodextrin-based microsuspension. *Journal of Drug Delivery Science and Technology*, *81*, 104281. DOI: 10.1016/j.jddst.2023.104281

Lyu, Y., Wu, X., Yang, J., Wang, X., & Li, J. (2023). Protocol for preparing dynamic covalent macrocycles for codelivering genes and drugs to cancer cell lines. *STAR Protocols*, *4*(3), 102350. DOI: 10.1016/j.xpro.2023.102350 PMID: 37314921

Macedo, V. S., Rodrigues, K. G. O., Ciribelli, N. N., Pereira, A. K. S., Cavallini, G. S., & Pereira, D. H. (2023). DFT study of bismuth subsalicylate adsorption on bow-tie dendrimer nanocarrier as drug delivery agents. *Chemical Physics Impact*, *7*, 100381. DOI: 10.1016/j.chphi.2023.100381

Madanayake, S. N., Manipura, A., Thakuria, R., & Adassooriya, N. M. (2023). Opportunities and challenges in mechanochemical cocrystallization toward scaled-up pharmaceutical manufacturing. *Organic Process Research & Development*, *27*(3), 409–422. DOI: 10.1021/acs.oprd.2c00314

Manjit, & Mishra, B. (2022). Research and Development of Supramolecules as Anticancer Drugs. In N. Goel & N. Kumar (Eds.), *Pharmaceutical Applications of Supramolecules* (pp. 55-87). Springer International Publishing. DOI: 10.1007/978-3-031-21900-9_3

Manna, K., Dey, S., Phukan, A., Dey, S. K., & Pal, S. (2024). Alginic acid-based pH and thermo responsive reversible switched polymeric micelle via RAFT polymerization. *Polymer*, *291*, 126581. DOI: 10.1016/j.polymer.2023.126581

Marquet, F., Hagen, H., Stanchieri, M., Beinier, V. S., Grasso, G., Danani, A., Patrulea, V., & Borchard, G. (2024). Clickable polyethyleneimine incorporated into triblock copolymeric micelles as an efficient platform in the delivery of siRNA to NSCLC cells. *International Journal of Pharmaceutics*, *649*, 123632. DOI: 10.1016/j.ijpharm.2023.123632 PMID: 38000648

Meenatchi, V., Sood, A., Bhaskar, R., & Soo Han, S. (2024). Electrospun hydroxypropyl-β-cyclodextrin nanofibers loaded with anti-inflammatory drugs for potential therapeutic wound dressings. *Journal of Molecular Liquids, 393*, 123558. DOI: 10.1016/j.molliq.2023.123558

Mehata, A. K., & Muthu, M. S. (2022). Development of Supramolecules in the Field of Nanomedicines. In Goel, N., & Kumar, N. (Eds.), *Pharmaceutical Applications of Supramolecules* (pp. 211–239). Springer International Publishing. DOI: 10.1007/978-3-031-21900-9_9

Michlewska, S., Garaiova, Z., Šubjakova, V., Hołota, M., Kubczak, M., Grodzicka, M., Okła, E., Naziris, N., Balcerzak, Ł., Ortega, P., de la Mata, F. J., Hianik, T., Waczulikova, I., Bryszewska, M., & Ionov, M. (2023). Lipid-coated ruthenium dendrimer conjugated with doxorubicin in anticancer drug delivery: Introducing protocols. *Colloids and Surfaces. B, Biointerfaces, 227*, 113371. DOI: 10.1016/j.colsurfb.2023.113371 PMID: 37244201

Mohammad, A., Singh, S., Swain, S., & Ghose, D. (2024). Quality by design enabled β-cyclodextrin complexes of lisinopril by kneading method: Improved solubility and bioavailability [Article]. *Current Materials Science, 17*(2), 135–147. DOI: 10.2174/2666145416666230407100318

Mohammad Asghari, B., Samadi Zadeh, M., Ahmad Panahi, H., Hesami Tackallou, S., & Safaeijavan, R. (2023). Surface modification of nanodiamond with pH/thermo dual responsive polymer and hyperbranched dendrimer as a near-infrared photothermal-triggered drug delivery for cancer therapy. *Journal of Molecular Liquids, 390*, 123155. DOI: 10.1016/j.molliq.2023.123155

Monajati, M., Tamaddon, A. M., Abolmaali, S. S., Yousefi, G., Javanmardi, S., Borandeh, S., Heidari, R., Azarpira, N., & Dinarvand, R. (2023). L-asparaginase immobilization in supramolecular nanogels of PEG-grafted poly HPMA and bis(α-cyclodextrin) to enhance pharmacokinetics and lower enzyme antigenicity. *Colloids and Surfaces. B, Biointerfaces, 225*, 113234. DOI: 10.1016/j.colsurfb.2023.113234 PMID: 36934612

Musazzi, U. M., Franzè, S., Condorelli, F., Minghetti, P., & Caliceti, P. (2023). Feeding next-generation nanomedicines to Europe: Regulatory and quality challenges. *Advanced Healthcare Materials, xx*(30), 2301956. DOI: 10.1002/adhm.202301956 PMID: 37718353

Mustafa, W. W., Khoder, M., Abdelkader, H., Singer, R., & Alany, R. G. (2024). Interactions of cyclodextrins and their hydroxyl derivatives with etodolac: Solubility and dissolution enhancement. *Current Drug Delivery*, *21*(1), 126–139. DOI: 10.2174/1567201820666230320164210 PMID: 36959150

Nie, Q., Wang, C., Xu, H., Mittal, P., Naeem, A., Zhou, P., Li, H., Zhang, Y., Guo, T., Sun, L., & Zhang, J. (2023). Highly efficient pulmonary delivery of levo-tetrahydropalmatine using γ-cyclodextrin metal-organic framework as a drug delivery platform. *Journal of Drug Delivery Science and Technology*, *82*, 104296. DOI: 10.1016/j.jddst.2023.104296

Niu, H., Wang, J., Jia, Q., Wang, Q., Zhao, J., & Yan, F. (2024). QSPR models for complexation performance of α-cyclodextrin and β-cyclodextrin complexes by norm indices. *Chemical Engineering Science*, *284*, 119484. DOI: 10.1016/j.ces.2023.119484

Nori, Z. Z., Bahadori, M., Moghadam, M., Tangestaninejad, S., Mirkhani, V., Mohammadpoor-Baltork, I., Jafari, S. S., Emamzadeh, R., & Alem, H. (2023). Synthesis and characterization of a new gold-coated magnetic nanoparticle decorated with a thiol-containing dendrimer for targeted drug delivery, hyperthermia treatment and enhancement of MRI contrast agent. *Journal of Drug Delivery Science and Technology*, *81*, 104216. DOI: 10.1016/j.jddst.2023.104216

Pan, Z., Zhao, X., Li, Q., Zhang, Z., & Liu, Y. (2023). Recent advances in supramolecular-macrocycle-based nanomaterials in cancer treatment. *Molecules (Basel, Switzerland)*, *28*(3), 1241. DOI: 10.3390/molecules28031241 PMID: 36770907

Pandey, G., Phatale, V., Khairnar, P., Kolipaka, T., Shah, S., Famta, P., Jain, N., Srinivasarao, D. A., Rajinikanth, P. S., Raghuvanshi, R. S., & Srivastava, S. (2024). Supramolecular self-assembled peptide-engineered nanofibers: A propitious proposition for cancer therapy. *International Journal of Biological Macromolecules*, *256*, 128452. DOI: 10.1016/j.ijbiomac.2023.128452 PMID: 38042321

Pandey, M., Lang, H. X., Loh, J. S., Chai, Y. L., Tee, H. L., Mayuren, J., Candasamy, M., Gorain, B., Jain, N., Gupta, G., & Dua, K. (2023). Dendrimer platform against prostate cancer: Recent update on new horizon of treatment. *Journal of Drug Delivery Science and Technology*, *85*, 104589. DOI: 10.1016/j.jddst.2023.104589

Parveen, N., Sheikh, A., Abourehab, M. A. S., Karwasra, R., Singh, S., & Kesharwani, P. (2023). Self-nanoemulsifying drug delivery system for pancreatic cancer. *European Polymer Journal*, *190*, 111993. DOI: 10.1016/j.eurpolymj.2023.111993

Peng, G., Cai, J., Wang, Z., Zhang, W., Xu, J., Zhang, D., & Gong, D. (2024). Facile fabrication of diatomite biosilica-based nasal drug delivery vehicle for enhanced treatment of allergic rhinitis. *Colloids and Surfaces. B, Biointerfaces, 234*, 113715. DOI: 10.1016/j.colsurfb.2023.113715 PMID: 38134821

Pinelli, F., Ponti, M., Delleani, S., Pizzetti, F., Vanoli, V., Vangosa, F. B., Castiglione, F., Haugen, H., Nogueira, L. P., Rossetti, A., Rossi, F., & Sacchetti, A. (2023). β-Cyclodextrin functionalized agarose-based hydrogels for multiple controlled drug delivery of ibuprofen. *International Journal of Biological Macromolecules, 252*, 126284. DOI: 10.1016/j.ijbiomac.2023.126284 PMID: 37572821

Pitingolo, G., & Nastruzzi, C. (2023). Production of supramolecular aggregates by microfluidic platforms: application to food industries. In Anandharamakrishnan, C., & Dutta, S. (Eds.), *Liposomal Encapsulation in Food Science and Technology* (pp. 169–187). Academic Press. DOI: 10.1016/B978-0-12-823935-3.00009-6

Pourbadiei, B., Adlsadabad, S. Y., Rahbariasr, N., & Pourjavadi, A. (2023). Synthesis and characterization of dual light/temperature-responsive supramolecular injectable hydrogel based on host-guest interaction between azobenzene and starch-grafted β-cyclodextrin: Melanoma therapy with paclitaxel. *Carbohydrate Polymers, 313*, 120667. DOI: 10.1016/j.carbpol.2023.120667 PMID: 37182982

Priya, A. S., Sivakamavalli, J., Vaseeharan, B., Rajamohan, R., Lee, Y. R., & Thambusamy, S. (2023). Interaction of torsemide with native cyclodextrin through inclusion complexation: In vitro drug release, antibacterial and antibiofilm activities. *Journal of Molecular Structure, 1286*, 135624. DOI: 10.1016/j.molstruc.2023.135624

Puhlmann, N., Vidaurre, R., & Kümmerer, K. (2024). Designing greener active pharmaceutical ingredients: Insights from pharmaceutical industry into drug discovery and development. *European Journal of Pharmaceutical Sciences, 192*, 106614. DOI: 10.1016/j.ejps.2023.106614 PMID: 37858896

Qiu, Y., Shang, K., Xu, N., Chen, P., Gao, H., Mu, H., Feng, W., & Duan, J. (2023). Clearance of intracellular bacterial infections by hyaluronic acid-based ROS responsive drug delivery micelles. *International Journal of Biological Macromolecules, 245*, 125506. DOI: 10.1016/j.ijbiomac.2023.125506 PMID: 37356691

Rahman, M., Almalki, W. H., Afzal, O., Alfawaz Altamimi, A. S., Najib Ullah, S. N. M., Abul Barkat, M., & Beg, S. (2023). Chiral-engineered supraparticles: Emerging tools for drug delivery. *Drug Discovery Today, 28*(1), 103420. DOI: 10.1016/j.drudis.2022.103420 PMID: 36309193

Rajaram, R., Angaiah, S., & Lee, Y. R. (2023). Polymer supported electrospun nanofibers with supramolecular materials for biological applications – a review. *International Journal of Polymeric Materials*, *72*(13), 1042–1058. DOI: 10.1080/00914037.2022.2075871

Rassu, G., Sorrenti, M., Catenacci, L., Pavan, B., Ferraro, L., Gavini, E., Bonferoni, M. C., Giunchedi, P., & Dalpiaz, A. (2021). Versatile nasal application of cyclodextrins: Excipients and/or actives? *Pharmaceutics*, *13*(8), 1180. DOI: 10.3390/pharmaceutics13081180 PMID: 34452141

Rubim, A. M., Rubenick, J. B., Vendrame, L. O., Zanella, I., Rolim, C. M. B., & Rhoden, C. R. B. (2024). Formulation and characterization of amiodarone-methyl-beta-cyclodextrin inclusion complexes: A molecular modeling perspective. *Journal of Molecular Graphics & Modelling*, *126*, 108639. DOI: 10.1016/j.jmgm.2023.108639 PMID: 37774479

Safari, J. B., Mona, L. B., Sekaleli, B. T., Avudi, B. K.-N., Isamura, B. K., Mukubwa, G. K., Salami, S. A., Mbinze, J. K., Lobb, K. A., Krause, R. W. M., & Nkanga, C. I. (2023). Inclusion complexation and liposomal encapsulation of an isoniazid hydrazone derivative in cyclodextrin for pH-dependent controlled release. *Journal of Drug Delivery Science and Technology*, *81*, 104302. DOI: 10.1016/j.jddst.2023.104302

Saha, S., Ali, M. R., Khaleque, M. A., Bacchu, M. S., Aly Saad Aly, M., & Khan, M. Z. H. (2023). Metal oxide nanocarrier for targeted drug delivery toward the treatment of global infectious diseases: A review. *Journal of Drug Delivery Science and Technology*, *86*, 104728. DOI: 10.1016/j.jddst.2023.104728

Sahu, K. M., Patra, S., & Swain, S. K. (2023). Host-guest drug delivery by β-cyclodextrin assisted polysaccharide vehicles: A review. *International Journal of Biological Macromolecules*, *240*, 124338. DOI: 10.1016/j.ijbiomac.2023.124338 PMID: 37030461

Salido-Fortuna, S., Salgado, A., Castro-Puyana, M., & Marina, M. L. (2024). Study on the interactions between RS-licarbazepine and carboxyethylated cyclodextrins by electrokinetic chromatography and nuclear magnetic resonance. *Microchemical Journal*, *196*, 109621. DOI: 10.1016/j.microc.2023.109621

Sanphui, P., Fayaz, T. K. S., Chanduluru, H. K., Lal, P., Ghosh, A., & Chernyshev, V. V. (2024). Structural analysis of anti-retroviral drug raltegravir and its potential impurity C: Investigation of solubility and stability. *CrystEngComm*, *xx*(4), xx–xx. DOI: 10.1039/D3CE01071F

Sepulveda, A. F., Kumpgdee-Vollrath, M., Franco, M. K. K. D., Yokaichiya, F., & de Araujo, D. R. (2023). Supramolecular structure organization and rheological properties modulate the performance of hyaluronic acid-loaded thermosensitive hydrogels as drug-delivery systems. *Journal of Colloid and Interface Science, 630*, 328–340. DOI: 10.1016/j.jcis.2022.10.064 PMID: 36327735

Shafiq, M., Rafique, M., Cui, Y., Pan, L., Do, C.-W., & Ho, E. A. (2023). An insight on ophthalmic drug delivery systems: Focus on polymeric biomaterials-based carriers. *Journal of Controlled Release, 362*, 446–467. DOI: 10.1016/j.jconrel.2023.08.041 PMID: 37640109

Shang, Q., Su, Y., Leslie, F., Sun, M., & Wang, F. (2022). Advances in peptide-drug conjugate-based supramolecular hydrogel systems for local drug delivery. *Medicine in Drug Discovery, 14*, 100125. DOI: 10.1016/j.medidd.2022.100125

Sharma, H., Kalita, B. K., Pathak, D., & Sarma, B. (2023). Low molecular weight supramolecular gels as a crystallization matrix. *Crystal Growth & Design, xx*, xx–xx. DOI: 10.1021/acs.cgd.3c01211

Sharma, S., Naman, S., & Baldi, A. (2023). Cyclodextrin-Based Molecular Inclusion by Grinding: Quality by Design in Green Chemistry. *Sustainable Approaches in Pharmaceutical Sciences*, 217-236. DOI: 10.1002/9781119889878.ch9

Shi, W., Zhang, D., Han, L., Shao, W., Liu, Q., Song, B., Yan, G., Tang, R., & Yang, X. (2024). Supramolecular chitin-based hydrogels with self-adapting and fast-degradation properties for enhancing wound healing. *Carbohydrate Polymers, 323*, 121374. DOI: 10.1016/j.carbpol.2023.121374 PMID: 37940271

Shi, Z., Hu, Y., & Li, X. (2024). Polymer mechanochemistry in drug delivery: From controlled release to precise activation. *Journal of Controlled Release, 365*, 259–273. DOI: 10.1016/j.jconrel.2023.10.042 PMID: 39491171

Soltany, P., Miralinaghi, M., & Pajoum Shariati, F. (2024). Folic acid conjugated poly (amidoamine) dendrimer grafted magnetic chitosan as a smart drug delivery platform for doxorubicin: In vitro drug release and cytotoxicity studies. *International Journal of Biological Macromolecules, 257*(127564), 127564. Advance online publication. DOI: 10.1016/j.ijbiomac.2023.127564 PMID: 37865361

Somavarapu, S., Sornsute, A., Trill, J., Abdelhakim, H. E., Goonatilaka, D., Pannala, A. S., & Taylor, K. M. G. (2023). Engineering artesunate-loaded micelles using spray drying for pulmonary drug delivery. *Journal of Drug Delivery Science and Technology, 86*, 104641. DOI: 10.1016/j.jddst.2023.104641

Su, Q., Su, W., Xing, S., & Tan, M. (2024). Enhanced stability of anthocyanins by cyclodextrin–metal organic frameworks: Encapsulation mechanism and application as protecting agent for grape preservation. *Carbohydrate Polymers*, *326*, 121645. DOI: 10.1016/j.carbpol.2023.121645 PMID: 38142106

Suárez, D., & Díaz, N. (2022). Amphiphilic cyclodextrins: Dimerization and diazepam binding explored by molecular dynamics simulations. *Journal of Molecular Liquids*, *349*, 118457. DOI: 10.1016/j.molliq.2021.118457

Sun, S., Wang, Y. H., Gao, X., Wang, H. Y., Zhang, L., Wang, N., Li, C., & Xiong, S. Q. (2023). Current perspectives and trends in nanoparticle drug delivery systems in breast cancer: Bibliometric analysis and review. *Frontiers in Bioengineering and Biotechnology*, *11*, 1253048. DOI: 10.3389/fbioe.2023.1253048 PMID: 37771575

Talreja, S., & Tiwari, S. (2023). Supramolecular chemistry: Unveiling the fascinating world of noncovalent interactions and complex assemblies. *Journal of Pharmacy and Pharmacology Research*, *7*(3), 133–139. DOI: 10.26502/fjppr.075

Tang, X., Wen, Y., Zhang, Z., Zhu, J., Song, X., & Li, J. (2023). Rationally designed multifunctional nanoparticles as GSH-responsive anticancer drug delivery systems based on host-guest polymers derived from dextran and β-cyclodextrin. *Carbohydrate Polymers*, *320*, 121207. DOI: 10.1016/j.carbpol.2023.121207 PMID: 37659810

Tran, C. T. H., Wang, H.-M. D., Anh, L. T. H., Lin, C., Huang, C.-Y., & Kuo, C.-H. (2024). Evaluate the effect of β-cyclodextrin on the sensory and physicochemical properties of bitter gourd extract during thermal processing. *Food Chemistry*, *433*, 137394. DOI: 10.1016/j.foodchem.2023.137394 PMID: 37690136

Vandera, K. K. A., Pague, C., Omar, J., González-Gaitano, G., Ways, T. M. M., Khutoryanskiy, V. V., & Dreiss, C. A. (2023). Formation of supramolecular gels from host–guest interactions between PEGylated chitosan and α-cyclodextrin. *Macromolecular Materials and Engineering*, *308*(6), 2200646. DOI: 10.1002/mame.202200646

Wang, Q., Yan, X.-T., Fan, J.-W., Xu, S.-H., Yao, H., & Yan, C.-G. (2024). Design of a serum albumin sensitive probe for cell imaging and drug delivery by modifying a fluorescent agent. *Journal of Molecular Structure*, *1297*, 136973. DOI: 10.1016/j.molstruc.2023.136973

Wang, Q., Zhang, A., Zhu, L., Yang, X., Fang, G., & Tang, B. (2023). Cyclodextrin-based ocular drug delivery systems: A comprehensive review. *Coordination Chemistry Reviews*, *476*, 214919. DOI: 10.1016/j.ccr.2022.214919

Wang, X., Wu, Y., Shang, H., Sun, X., An, K., Zhang, Q., & Qiao, N. (2023). Preparation of pH/Light dual-responsive biocompatible polymer micelles: Application to curcumin delivery. *Journal of Drug Delivery Science and Technology*, *86*, 104652. DOI: 10.1016/j.jddst.2023.104652

Wang, X., Zhang, M., Li, Y., Cong, H., Yu, B., & Shen, Y. (2023). Research status of dendrimer micelles in tumor therapy for drug delivery. *Small*, *19*(50), 2304006. DOI: 10.1002/smll.202304006 PMID: 37635114

Wang, Z., Chen, J., Gao, R., Jiang, L., Zhang, G., Zhao, Y., Miao, Y.-B., & Shi, Y. (2024). Spatiotemporal manipulation metal–organic frameworks as oral drug delivery systems for precision medicine. *Coordination Chemistry Reviews*, *502*, 215615. DOI: 10.1016/j.ccr.2023.215615

Wu, X., Chen, X., Wang, X., He, H., Chen, J., & Wu, W. (2024). Paclitaxel-lipid prodrug liposomes for improved drug delivery and breast carcinoma therapy. *Chinese Chemical Letters*, *35*(2), 108756. DOI: 10.1016/j.cclet.2023.108756

Xiang, S., Guilbaud-Chéreau, C., Hoschtettler, P., Stefan, L., Bianco, A., & Ménard-Moyon, C. (2024). Preparation and optimization of agarose or polyacrylamide/amino acid-based double network hydrogels for photocontrolled drug release. *International Journal of Biological Macromolecules*, *255*, 127919. DOI: 10.1016/j.ijbiomac.2023.127919 PMID: 37944737

Xu, J., Wang, K., Li, Y., Li, Y., Li, B., Luo, H., Shi, H., Guan, X., Zhang, T., Sun, Y., Chen, F., He, H., Zhang, J., Cai, L., Song, W., Wu, J., & Li, X. (2023). Injectable host-guest supramolecular hydrogel codelivers hydrophobic and hydrophilic agents for enhanced wound healing. *Chemical Engineering Journal*, *454*, 140027. DOI: 10.1016/j.cej.2022.140027

Xue, H., Ju, Y., Ye, X., Dai, M., Tang, C., & Liu, L. (2024). Construction of intelligent drug delivery system based on polysaccharide-derived polymer micelles: A review. *International Journal of Biological Macromolecules*, *254*, 128048. DOI: 10.1016/j.ijbiomac.2023.128048 PMID: 37967605

Yan, M., Wu, S., Wang, Y., Liang, M., Wang, M., Hu, W., Yu, G., Mao, Z., Huang, F., & Zhou, J. (2023). Recent progress of supramolecular chemotherapy based on host–guest interactions. *Advanced Materials*, *xx*(21), 2304249. DOI: 10.1002/adma.202304249 PMID: 37478832

Yilmaz, A. S., Ozturk, S., Salih, B., Ayyala, R. S., & Sahiner, N. (2023). ESI-IM-MS characterization of cyclodextrin complexes and their chemically cross-linked alpha (α-), beta (β-) and gamma (γ-) cyclodextrin particles as promising drug delivery materials with improved bioavailability. *Colloids and Surfaces. B, Biointerfaces*, *230*, 113522. DOI: 10.1016/j.colsurfb.2023.113522 PMID: 37657404

Yong Joon Kim, J., Sang, Z., Xiang, Y., Shen, Z., & Shi, Y. (2023). Nanobodies: Robust miniprotein binders in biomedicine. *Advanced Drug Delivery Reviews*, *195*, 114726. DOI: 10.1016/j.addr.2023.114726 PMID: 36754285

Yu, H., Song, Z., Yu, J., Ren, B., Dong, Y., You, Y., Zhang, Z., Jia, C., Zhao, Y., Zhou, X., Sun, H., & Zhang, X. (2024). Supramolecular self-assembly of EGCG-selenomethionine nanodrug for treating osteoarthritis. *Bioactive Materials*, *32*, 164–176. DOI: 10.1016/j.bioactmat.2023.09.020 PMID: 37822916

Yu, L., Liu, S., Jia, S., & Xu, F. (2023). Emerging frontiers in drug delivery with special focus on novel techniques for targeted therapies. *Biomedicine and Pharmacotherapy*, *165*, 115049. DOI: 10.1016/j.biopha.2023.115049 PMID: 37364480

Zakharova, L. Y., Maganova, F. I., Sinyashin, K. O., Gaynanova, G. A., Mirgorodskaya, A. B., Vasilieva, E. A., & Sinyashin, O. G. (2023). Supramolecular strategy for the design of nanocarriers for drugs and natural bioactives: Current state of the art (a review). *Russian Journal of General Chemistry*, *93*(7), 1867–1899. DOI: 10.1134/S1070363223070253

Zhang, C., Niu, J., Li, J., Zhang, H., Yu, Q., Chen, Y., & Liu, Y. (2024). Polysaccharide based supramolecular injectable hydrogels for in situ treatment of bladder cancer. *Chinese Chemical Letters*, *35*(1), 108556. DOI: 10.1016/j.cclet.2023.108556

Zhang, C., Niu, J., Li, J., Zhang, H., Yu, Q., Chen, Y., & Liu, Y. (2024). Polysaccharide based supramolecular injectable hydrogels for in situ treatment of bladder cancer. *Chinese Chemical Letters*, *35*(1), 108556. DOI: 10.1016/j.cclet.2023.108556

Zhang, L., Xu, Y., & Wei, W. (2023). Water-soluble organic macrocycles based on dye chromophores and their applications. *Chemical Communications*, *59*(91), 13562–13570. DOI: 10.1039/D3CC04159J PMID: 37901908

Zhang, W., Yang, W., & Zhou, J. (2023). Biphenarenes, versatile synthetic macrocycles for supramolecular chemistry. *Molecules (Basel, Switzerland)*, *28*(11), 4422. DOI: 10.3390/molecules28114422 PMID: 37298898

Zhang, X., Ren, K., Xiao, C., & Chen, X. (2023). Guanosine-driven hyaluronic acid-based supramolecular hydrogels with peroxidase-like activity for chronic diabetic wound treatment. *Acta Biomaterialia, 172*, 206–217. DOI: 10.1016/j.actbio.2023.10.014 PMID: 37839631

Zhang, Y.-J., Liang, J.-X., Xu, Y.-S., Liu, Y.-X., Cui, Y., Qiao, Z.-Y., & Wang, H. (2024). Covalent drugs based on small molecules and peptides for disease theranostics. *Biomaterials Science*, *xx*(3), xx–xx. DOI: 10.1039/D3BM01138K PMID: 37975197

Zhang, Z., Xie, B., Lu, X., Xiong, L., Li, X., Zhang, Y., Li, C., & Wang, C. (2024). Intracellular self-aggregation of biomimetic Fe_3O_4 nanoparticles for enhanced ferroptosis-inducing therapy of breast cancer. *Nanoscale*, *xx*(2), xx–xx. DOI: 10.1039/D3NR04305C PMID: 38108145

Zhao, T., Zhou, M., Wu, R., Wang, H., Zouboulis, C. C., Zhu, M., & Lee, M. (2023). Dendrimer-conjugated isotretinoin for controlled transdermal drug delivery. *Journal of Nanobiotechnology*, *21*(1), 285. DOI: 10.1186/s12951-023-02052-5 PMID: 37605256

Zhao, Y., Zheng, Z., Yu, C., & Wei, H. (2024). Engineered cyclodextrin-based supramolecular hydrogels for biomedical applications. *Journal of Materials Chemistry. B*, *12*(1), 39–63. DOI: 10.1039/D3TB02101G PMID: 38078497

Zheng, H.-y., Liu, Y., Zhang, Y., Shi, Q., Hou, X., & An, L. (2024). Camptothecin-loaded supramolecular nanodelivery system based on amphiphilic calix[4]arene for targeted tumor therapy. *New Journal of Chemistry*, *xx*(3), xx. DOI: 10.1039/D3NJ03289B

Zheng, M., Mei, T., Deng, C., Xin, J., Aras, O., An, F., & Chen, N. (2024). The impact of PEG chain length on the in vivo tumor accumulation of self-assembled amphiphilic drug-PEG polymeric micelles. *Colloids and Surfaces. A, Physicochemical and Engineering Aspects*, *682*, 132823. DOI: 10.1016/j.colsurfa.2023.132823

Zhou, J., Li, Y., Wang, L., Lv, P., Chen, M., Xiao, F., Si, T., Tao, J., & Yang, B. (2024). Bifunctional drug delivery system with carbonic anhydrase IX targeting and glutathione-responsivity driven by host-guest amphiphiles for effective tumor therapy. *Carbohydrate Polymers*, *326*, 121577. DOI: 10.1016/j.carbpol.2023.121577 PMID: 38142063

Zhuo, S., Liang, Y., Wu, Z., Zhao, X., Han, Y., & Guo, B. (2024). Supramolecular hydrogels for wound repair and hemostasis. *Materials Horizons*, *xx*(1), xx–xx. DOI: 10.1039/D3MH01403G PMID: 38018225

ABBREVIATIONS

API: Active pharmaceutical ingredient
AUC: Area under the curve
CDs: Cyclodextrins
Cmax: Maximum concentration
DDS: Drug delivery system
ROS: Reactive oxygen species

Chapter 5
Exploring Thiazoline, Thiophane, and Thiazole Advances in Sulphur-Based Heterocyclic Compounds

Nandan Chamanlal Pomal
https://orcid.org/0000-0001-5890-3163
Sigma University, Vadodara, India

Keyur D. Bhatt
https://orcid.org/0000-0001-9301-576X
Mehsana Urban Institute of Science, India

Ageetha Vanaamudan
Sigma University, India

ABSTRACT

Sulphur-containing heterocycles, including thiazoline, thiophane, and thiazole, have garnered significant attention in medicinal chemistry due to their diverse biological activities and crucial roles in the biosynthesis of essential compounds. This review highlights the pivotal roles of these sulfur-containing heterocycles' structural diversity and biological significance. The synthesis section delves into the methods and mechanisms behind the formation of these heterocycles. Thiazoline plays essential role in the biosynthesis of thiamine (vitamin B1) and is synthesized in living organisms. Thiophane is a key structural motif in various biologically active natural products, while thiazole is widely prevalent in antimicrobial peptides and

DOI: 10.4018/979-8-3693-2774-6.ch005

antibiotics. The biological evaluation aspect explores the diverse pharmacological activities of these compounds. Thus, the chapter features insights into synthesis and biological activities of these moieties, provides a valuable resource for researchers in medicinal chemistry and chemical biology.

INTRODUCTION

In organic chemistry, compounds which incorporate minimum one hetero-atom in the ring contains carbon atoms are said to be heterocyclic. Among the number of hetero-atoms nitrogen, oxygen, and sulphur are fairly common hetero atoms, a report indicating the complex formation by interaction between H-atom of DNA and these N-,O- and S-heterocycles (Laxmikeshav, Kumari, & Shankaraiah, 2022) pointing the eminence of them. Additionally, these rings occasionally also have phosphorus, selenium, magnesium, iron, etc. (Kabir & Uzzaman, 2022). Heterocyclic compounds have been earned a special status in the organic chemistry due to their medicinal and biological significance (Dua, Shrivastava, Sonwane, & Srivastava, 2011). Pharmaceutical sector investing their large funds and times on R & D of these moieties due to their structure, properties and applications apt to therapeutic means.

Precisely speaking that, because of their bioactive properties, these compounds are acclaimed for their roles in a variety of sectors, notably agrochemicals, pharmaceuticals, biology, veterinary medicines, etc. (Lamberth & Dinges, 2012). They have anticancer (Gomha, Riyadh, Huwaimel, Zayed, & Abdellattif, 2022), antiviral (De, Sarkar, & Majee, 2021), antibacterial (Borah et al., 2023), antioxidant (El-Helw, Sallam, & Elgubbi, 2019), anti-HIV (Kurapati, Atluri, Samikkannu, Garcia, & Nair, 2016), analgesic (Ahirwar et al., 2018), antihistamine (Rosa Rodes Solanes, 2016), and many more bioactive libraries of chemicals accessible in the literature. In a reservoir of heterocycles, the sulfur-containing heterocycles obtained a unique place due to their pharmacology actions. Recent report on US FDA approved drugs indicated the share of sulphur-containing drugs from 2015 to 2020 was 39 in numbers, which exhibited the sulfur-containing molecules are significant in medicinal terms. In this context, in the large pool of the heterocyclic molecules, sulphur-containing heterocyclic chemicals (S-heterocycles) have an exceptional position bearing in the mind that, they have extreme potential to fight against the diseases. Widely used S-heterocycles possess three to seven membered ring known for their therapeutic performance. Generally, thiirane, thietane, thiazole, thiophene, thiophane, thiopyran, thiepine, and other than that thiazide, thiazepine, thiazolidine, β-lactams etc. are used in broad manner by means of natural or synthetic products (Feng, Tang, Liang, & Jiang, 2015). Some of the US FDA approved drugs containing S-heterocycles are Thiabendazole is familiar name for the treatment of strongyloids and various larva migrans used as antihelmintic drug (Vardanyan & Hruby, 2006), Bacitracin is utilized as topical antimicrobial comprising of thiazoline and peptide side chain (O'Donnell, Gelone, & Safdar, 2015), Biotin (Vitamin B7) containing fused rings of thiophane and imidizalone employed in distinct cellular processes such as Kreb cycle, formation of fatty acids and metabolism of fats etc. (Ledesma-Amaro, Santos, Jiménez, & Revuelta, 2013), Articaine is the most popular dental anesthetic encompassing thiophene derivative which aid to prevent the ion channel to enter (W. Luo et al., 2022). Till date more than 360 drugs are approved by FDA that contains the sulfur atom (Feng et al., 2015).

In this review, the narrow line of sulphur containing heterocyclic compounds in particular thiazoline, thiophane and thiazole with updates are discussed.

DISCUSSION

Thiazoline

Thiazoline is a class of five-membered heterocycles featuring three isomers, incorporating 2-thiazoline, 3-thiazoline, and 4-thiazoline, which have a sulfur and a nitrogen atom in an opposite position to the carbon atom and one double bond, respectively. Sulfur and nitrogen are next to each other in the fourth isomer.

Since its first ever synthesis by Richard Willstatter in 1909, from dialkylation and thioamides, it is the ever growing and unstoppable molecule (Willstätter & Wirth, 1909).

Recently, a method was developed to design the thiazoline derivatives from aldehydes or ketones by reacting with N^4-substituted thiosemicarbazide in MeOH, in glacial acetic acid catalyzed conditions, which gave an intermediate thiosemicarbazone derivatives. These thiosemicarbazones were reacted in absolute EtOH, with 4-substituted phenacyl bromide and anhydrous CH_3COONa which gave different 36 thiazoline derivatives (Figure 1a) out of which, 8 moieties were found to be active aldehyde reductase (ALR1) and aldol reductase (ALR2) (anti-diabatic) inhibitor enzymes, moreover 7 compounds were selective ALR2 inhibitors. Additionally, in the present study, *In-silico* assay was carried out and found in good agreement with experimental outcomes (Shehzad et al., 2021).

El-Helw *et al.* have synthesized a thiazoline derivative bearing -Ph, $-NH_2$ and -CONH-, functionalities (Figure 1b) which exhibited the cytotoxicity against human tumors HepG2 and MCF-7 cell lines. A compound was formed by Et_3N catalyzed condition, reacting hydrazide-hydrazone derivative with sulfur and phenyl isothiocyanate in dioxane at 60 °C, that formed the thiazoline derivative in mixture with another compound. In the antiproliferative study thiazoline compound was found to be active agent with IC_{50} values of 21.65 ± 2.9 μM for HepG2, and 23.53 ± 3.1 μM for MCF-7 cell lines. The anticancer potency was also assayed by virtue of structure activity relationship (SAR) (El-Helw & El-Badawy, 2020).

In another foundation work of creating thiazoline derivative by utilizing cascade protocol, in which domino reaction was approached through N-substituted thiourea, thiobenzamide and thiosemicarbazones with methyl-4-bromocrotonate as electrophile and other reagent was sodium acetate in hexfluoroisopropanol (HFIP) as a promotor and reaction medium in reflux for 8 hours. A library of thirty nine new thiazoline derivatives (Figure 1c) were prepared with aim of applications in drug discovery and biological study. The key of success of the current research was good yield due to HFIP, a strong H-bond donor and polar solvent, metal free and mild conditions, no work-up and no column (Alsharif & Alam, 2017).

Eleven new thiazoline-tetralin combination of derivatives (Figure 1d) were prepared through a three steps synthesis, for that three intermediates were equipped i.e., tetrahydronapthalen-2-yloxy acetate, acetohydrazide of it and third was hydrazine carbothioamide, which further converted to 11 compounds by reacting with phenacyl bromide derivatives. All compounds were assessed against cancer cell lines MCF-7, A549, and NIH/3T3 in accordance with MTT assay, along with that DNA synthesis inhibition was also undertook for the cell lines. Moreover, acetylcholinesterase (AChE) and butyrylcholinesterase (BuChE) enzyme inhibition was conducted. As supported by the outcomes, these compounds were manifested greater results for A549 lines, even some of them showed apoptosis results higher then the cisplatin which was used as a standard drug (Turan-Zitouni et al., 2018).

A latest research was reported for the hybrid of thiazoline-coumarin scaffolds (Figure 1e). Preparation of final product was transmitted from various strides; i.e., started from salicylaldehyde and phenol under catalytic milieu to coumarins, which then coupled with thiosemicarbazide of D-glucose and D-galactose in MeOH in acid catalysis in microwave synthesis to give thiosemicarbazones, in advance to that the thiosemicarbazones turned into commensurate combo of thiazoline-coumarin derivatives in anhydrous sodium acetate with ω-bromoacetophenone. A plausible mechanism of formation of combo was well explained in the report. Different human carcinoma cell lines namely, HepG2, MCF-7, HeLa, SK-Mel-2 and LU-1 were screened for all the derivatives and displayed very promising cytotoxicity in terms of IC_{50} values. Above that, Epidermal growth factor receptors (EGFR) and HER2 tyrosine kinase were examined and found good results as well as molecular modeling study was performed that endorsed the experimental data (Ngoc Toan & Dinh Thanh, 2021).

Ibrayev and co-workers have synthesized eight 1,3-thiazoline derivatives. Three of them were prepared using cysteine, salsoline and anabasine alkaloids in conc. HCl with N-allylisothiocyanate which experienced intramolecular cyclization after heating in ampoule for 5 to 6 hours, other two moieties were formed as a result of pyrazole derivative reacted with dibromoisothiocyanate that provide an intermediate and that underwent cyclization by addition of excess of base, three other compounds were prepared from cysteine, salsoline and pyrazole derivative and dibromoisothiocyanate under basic conditions in good yields (Figure 1f). Among them last five molecules were selected for molecular docking with Penicillin binding proteins 4 (PBP4) *E.Coli* and *S.Aureus* “Homo sapiens”. It was concluded that thiazoline bearing natural alkaloids have enhanced the affinity for PBP4 (Ibrayev et al., 2022).

Figure 1. Different thiazoline molecules prepared for their biological significance

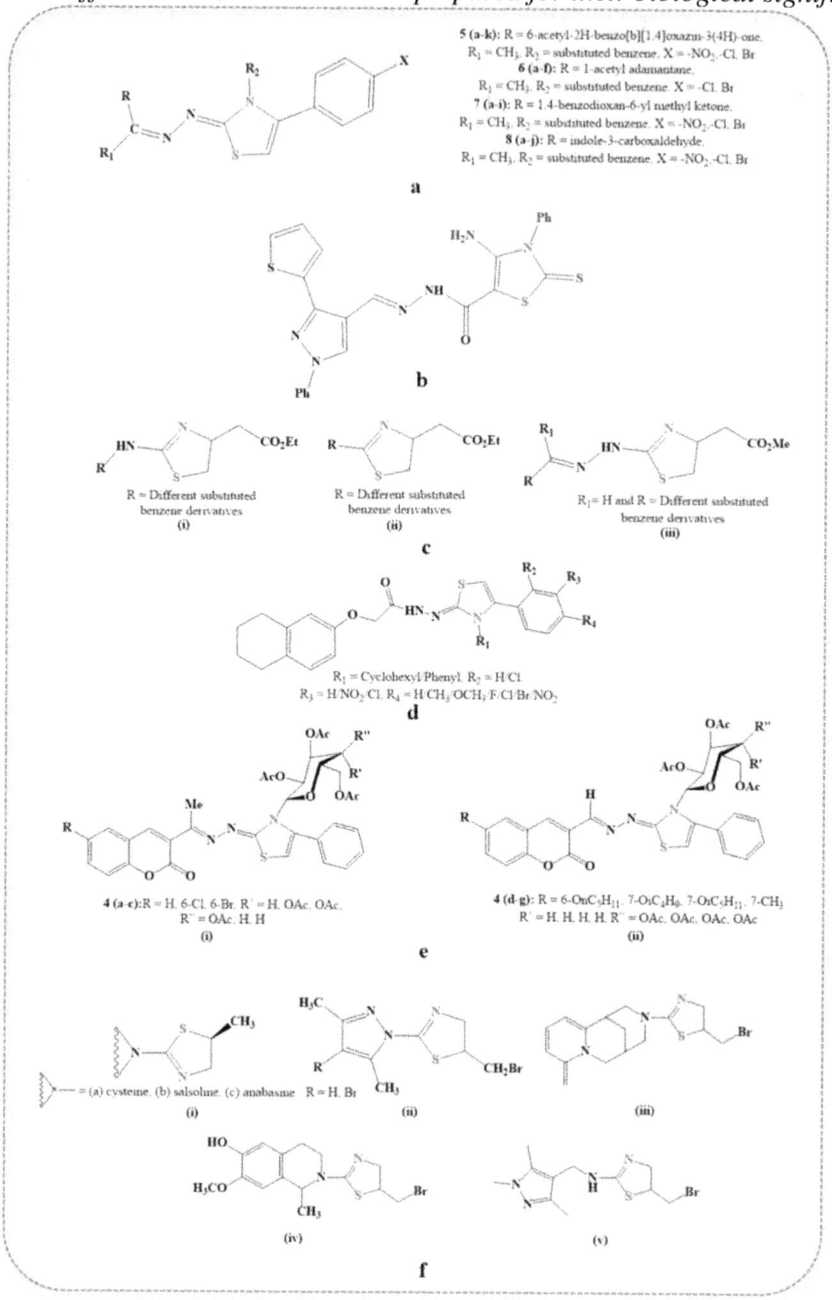

Alom and group has developed the thiazoline series exploiting various alkenes and thioamides. Here in a synthesis, a styrene experienced halogenation and turned into 1,2-dihaloalkane in ACN at ambient temperature which then underwent nucleophilic addition by thioamides to yield substituted thiazoline compounds in same solvent at 80° C, in pot procedure, further to improve the % of yield various acids and inorganic bases were utilized. Product yield was discovered to increase on addition of trifluoroacetic acid of 2 equivalent and 3 equivalents of $NaHCO_3$ as well as 3 equivalents of halide source. In progressive work, variety of alkenes substrates were employed with thioamides, in which investigation established that substitution at para position of benzene by electron donating and neutral species gave high yield then that of withdrawing species. On the other side, thioamides substrates also evaluated and found higher yields at ortho or para for methyl substitution and efficiency increased for halogen at ortho, meta and para substitution (Figure 2). After the substantial work of different substrate scope, a gram scale synthesis of thiazoline derivative was achieved in 61% yield, which on further hydrolysis with 5M HCl and refluxing gave β-aminothiol (92%) and oxidation of thiazoline with DDQ in DCM at 50° C formed 95% of thiazole derivative (Alom, Wu, & Li, 2017).

Figure 2. Synthesis of thiazoline derivatives (a) from alkene and thioamide, (b) styrene and thioamide

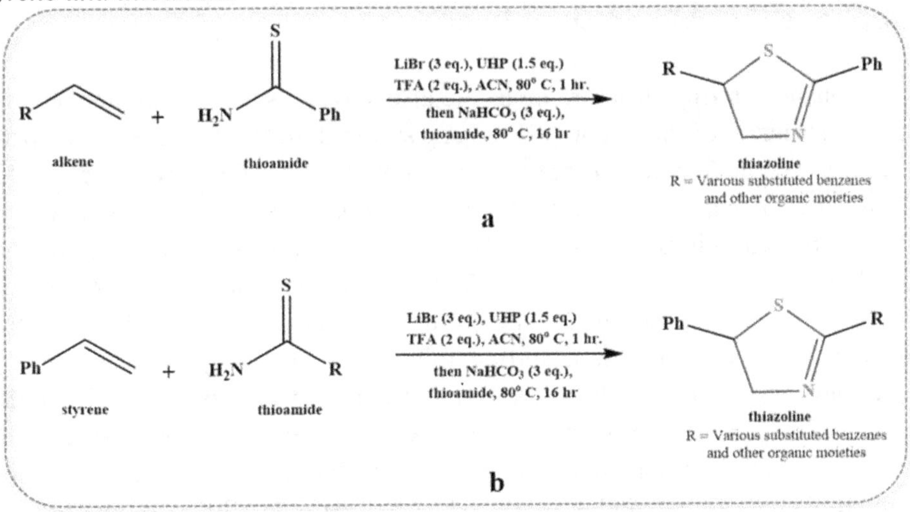

Figure 3. Thiazoline derivatives containing various groups

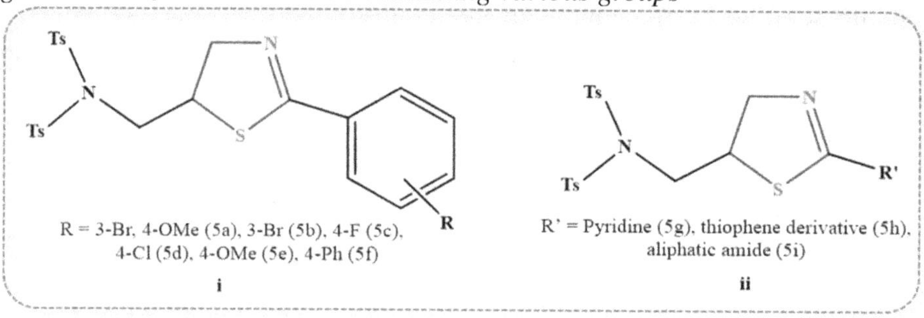

On other report Jeon *et al.* have prepared derivatives of thiazoline (Figure 3) along with oxazolines. Thiazoline furnished in better yield by reaction of benzothioamide with various electron-donating and -withdrawing groups. In a typical process 2 equivalents of benzothioamide mixed with 1 equivalent of Ts_2NH and 2 equivalent of $PhI(OAc)_2$ in 2.5 mL dichloroethane at 50° C, for 4 hours, resulting species were purified by flash chromatography (Jeon et al., 2018).

Thiophane (Tetrahydrothiophene)

Thiophane is a ring structure molecule also known as tetrahydrothiophene is a saturated structure of thiophene possess one sulfur atom and four methylene parts. It is available in nature in *Eruca sativa* (a family of mustard) (Aissani et al., 2015), *Allium sativum* (garlic) (Block et al., 2018), *Allium cepa* (onion) (Aoyagi et al., 2011), biotin (vitamin B7 or vitamin H) (Bitsch, 2003), natural antibiotic Albomycins (Wang, Zhang, Lv, Kong, & Niu, 2022) and other alkaloids. It is colourless, volatile substance in liquid form with strong foul smell and insoluble in water. It is also known as tetramethylene sulfide, thiolane, THT, thiacyclopentane, etc. Its major applications in fuel odorant, solvent, intermediate compounds of biological importance, in synthesis of some natural products, insecticides and repellent (Services, 1999). THT demonstrate the better biological actions such as anti-cancer, antimicrobial, antifungal and anti-HIV, etc. (El-Aasr et al., 2011; Winter, Rupcic, Stadler, & Trauner, 2019). Having more light on this moiety can be fruitful for the future aspects and for that reason some of the latest synthesis are overviewed here.

Chen *et al.* have reported the successful domino reactions to synthesize the novel thiophane derivatives (Figure 4) using 3.0 eq. sulfur powder (S_8), 1.0 eq. 2-acetylfuran, 1.1 eq. benzaldehyde in 3.0 eq. sodium hydroxide and solvent used 5.0 mL methanol at ambient temperature furnished medicinally useful thiophane derivative in good yield in 48 hours. In the further optimization, 2-acetylfuran, 2-acetylthiophene were used with substituted and unsubstituted benzaldehyde as well as alkyl ketones and formaldehyde in same reagents discussed above which produced decreased to higher % yields thiophanes. A gram scale reaction was also carried out and yielded in more 70% product with best plausible mechanism explained (Chen et al., 2022).

Figure 4. Structure of thiophane synthesized

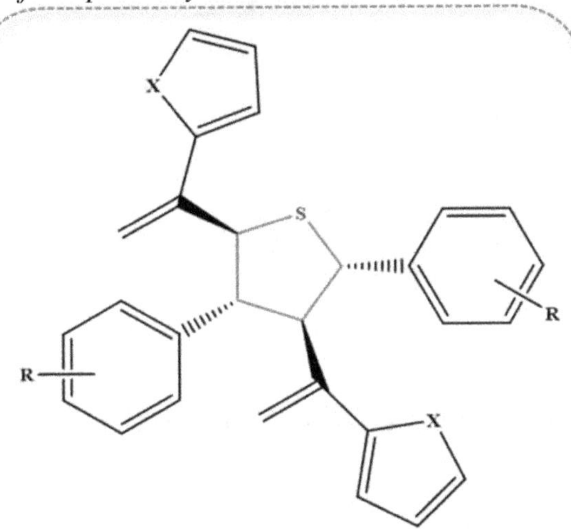

R = Substituted and unsubstituted benzaldehyde, X = O, S

(Chen et al., 2022)

An attempt of bacterial resistant for quaternary ammonium compounds (QASc) was addressed by providing alternative compounds along with amphiphilic profiles i.e., trivalent sulfonium compounds (TSCs) using THT as starting materials made by Feliciano and group. They synthesized ammonium based N-methyl pyrrolidines (MPs) and alkylated tetrahydrothiophenes (THTs) (Figure 5). In a preparation of 8 mono THTs, 0.1 eq. dodecyl iodide was mixed to 10-fold extra THT in MeCN, in reflux at 80° C for 24 hours and then bolaTSCs (THT-n-THTs) were formed by heating α,ω-diiodoalkanes with 5-fold oversupply of THT in same conditions gave yield up to 87%. THT-n yield (73% for monoTHTs and bolaTHTs) was enhanced by employing trifluoroacetic acid. From the synthesized pool of the MPs and THTs, only THT-18 having 18 carbons offered minimum MIC magnitudes for Gram-positive species, this therapeutic numbers of THT-18 were equivalent to commercially available products which opened the doors for future hopes for the THT-18 (Feliciano et al., 2021).

Figure 5. Synthesis of mono and bola tetrahydrothiophenes (THTs)

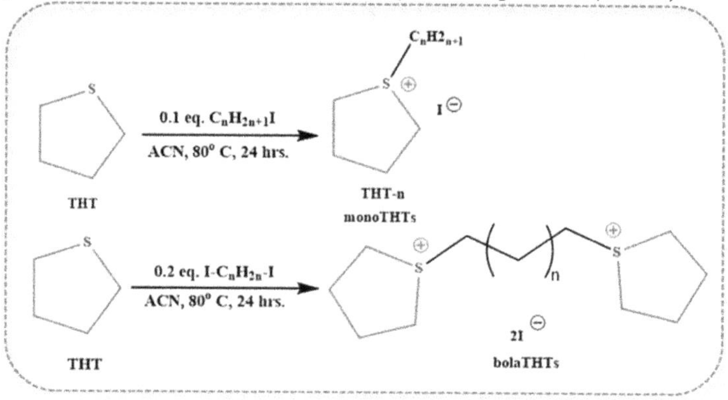

Figure 6. Preparations of isoxazole-THT and spiro oxindole-isoxazole-THT derivatives

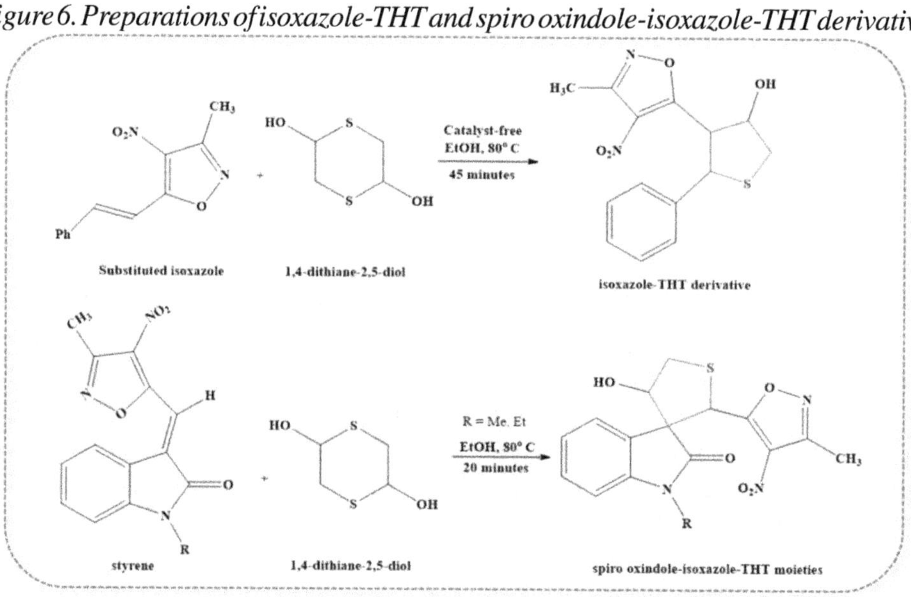

Figure 7. Highly functionalized thiophane diastereomers

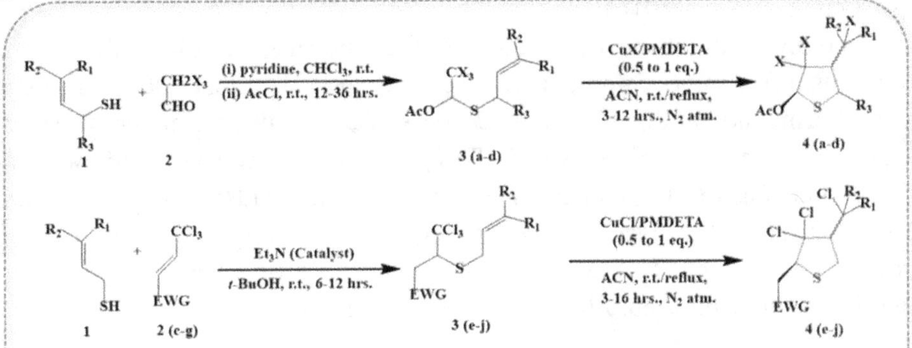

1a: R_1=R_2=Me, R_3=H, **1b:** R_2 = Ph, R_1=R_3=H, **1c:** R_2-R_3 = -$(CH_2)_3$-, R_1=H, **1d:** R_1=R_2=R_3=H

2a: X=Cl, 2b: X=Br, **2c-g:** EWG: (c) COPh, (d) COMe, (e) CO_2Me, (f) NO_2, (g) CN

In another cascade reaction under thia-Michael and intramolecular aldol reaction, Nagaraju *et al.* have prepared a biologically essential hybrid of isoxazole-spirooxindole-THT in catalyst-extricate environment in regioselective manner (Figure 6). Initially isoxazole-THT hybrid was formed in ethanol at 80° C, in 45 minutes by reacting isoxazole derivative with 1,4-dithiane-2,5-diol without catalyst, on success of that, thereafter to synthesize the desired hybrid, styrenes were designed from substituted isoxazole and oxindole derivative using vinylogous backed with mesylation and dehydration stages. These styrene were then infused with 1,4-dithiane-2,5-diol in ethanol at 80° C which gave corresponding hybrid in 20 minutes which were 1,4-thia-Michael addition products which was generated possibly due to lack of electron releasing ability of isoxazoles. In addition to this, complex moieties containing butadiene isoxazole and few familiar compounds were treated with 1,4-dithiane,2,5-diol to yield such hybrids (Nagaraju, Sathish, Satyanarayana, Paplal, & Kashinath, 2020).

Exceedingly functionalized diastereomer THTs were furnished by Cu^{I}/PMDETA-fostered HATRC (halogen atom transferred radical cyclization) of β-haloalkyl sulfides by Ram *et al.* (Figure 7) In the first step, in $CHCl_3$, at 10° C, prenyl mercaptan and anhy. chloral with pyridine yielded thioether derivative. After that, the reaction conditions were optimized and 81% THT derivative was prepared from substituted thioether using 100 mol% of CuCl in Acetonitrile with CuCl to PMDETA mol ratio of 1:1 at room temperature in 12 hours in Nitrogen atmosphere. Optimized temperature conditions were used to determine the diastereomers obtained, i.e., *cis* and *trans* of functionalized THTs. High ratio of diastereomers offer good isolation of *cis* and *trans* products. Once, the reaction conditions were set, the precursor were altered and distinct biologically useful highly functionalized THTs were obtained (Ram, Gupta, & Soni, 2016).

(Z)-tetrahydrothiophene was produced by transition metal free radical cyclization of sulfur using 0.5 mmol chloropentynyl benzene and 1.0 mmol $EtOCS_2K$ in 2.0 mL dimethylformamide at 80° C within 12 hours. Various alkynyl halide of benzene compounds were used to achieve (Z)-THT, possessing electron-withdrawing and electron-donating species, had given good to excellent yields (Figure 8). Synthesis work was extended to pyridine, pyrazole, quinoline, thiophene substrates which produced good yields (X. Luo, Li, Yibiao, Chen, Xiuwen, Song, Zhiyan, Liang, Junyi, Liao, Chunshu, Zhu, Zhongzhi, Chen, Lu, 2019).

Figure 8. Synthesis of Z/E-thiophane derivatives using alkynyl halide and $EtOCS_2K$

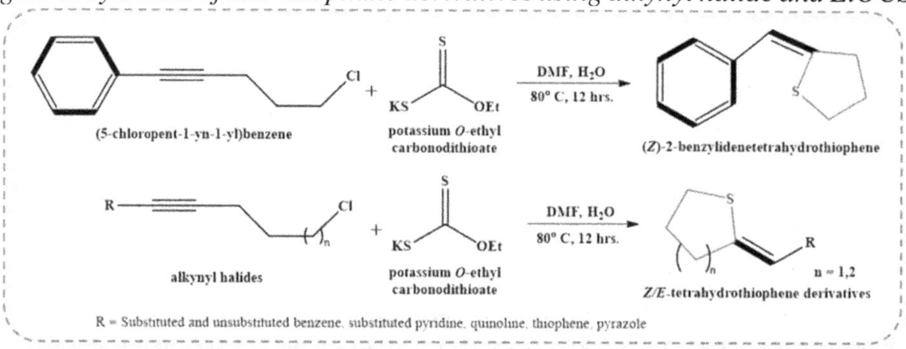

Jain *et al.* have portrayed the stereoselective preparation of tri-substituted THTs, in which second position of THT was occupied by ester functional group that was yet another thia-Michael-Aldol cascade reaction (Figure 9). In attainment of the final product the care was also taken for diastereoselective ratio (dr) of the product along with the yield. Different base, solvent and substrate scopes were inspected in controlled conditions. A reaction was proceeded with a phenylbutenoate and 1,4-dithiane-2,5-diol in toluene confined in 1,4-diazabicyclo[2.2.2.]octane (DABCO) in 5 mol% as a base that provided 94% tri-substituted THT in 2 hours with dr >20:1, and in solvent scope assessment, an ethyl acetate was confirmed as the appropriate solvent. Reaction was further optimized with several aryl rings, wherein methoxy substituted aryl moieties have taken considerably prolonged duration then nitro species. Moreover, di-halo aryl rings, naphthyls, heterocyclic and alkyl groups were evaluated with 1,4-dithiane-2,5-diol offered the notable outcomes. Enantioselective adaption was also checked in chiral squaramide catalyzed reaction with anhydrous magnesium sulphate given moderate to good yields of tri-substituted THTs. *In silico*

bioactivity towards Mtb enzyme MptpB inhibitor was inspected that advocated this THT as future anti-mycobacterium molecule (Jain et al., 2022).

Figure 9. Stereoselective preparation of tri-substituted THT

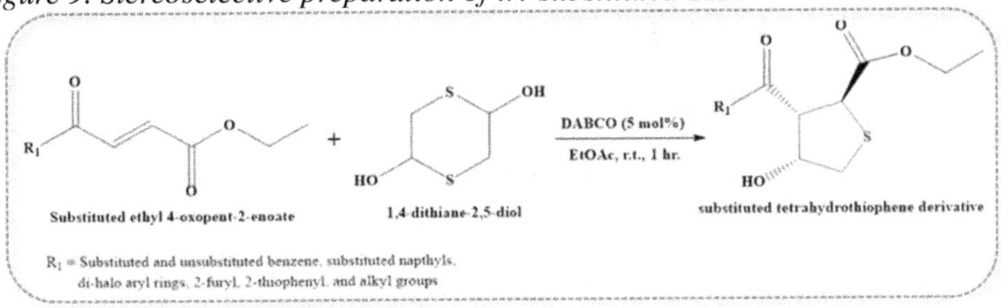

A [3+2]-(1,3-dipolar) cycloaddition reaction was demonstrated between 1.0 eq. *trans*-alkenes and *in situ* produced thiocarbonyl species (from 1.5 eq. chloromethyl trimethylsilylmethyl sulfide in 2.0 eq. CsF) in acetonitrile under reflux in 5 hours, that synthesized 3,4-disubstituted THT (Figure 10). Among the prepared compounds 4d compound was proven by X-ray diffraction. 4 a, b, d, and e structures received the endowment of NMR spectroscopy, and 4c was possessed two diastereomers of cycloadduct. From the above thiolanes, further S-oxide, S,S-dioxide, sulfoximines, carboxylic acid derivative prepared. On the positive side, the researchers have characterized all the compounds using distinct spectroscopy techniques (Markitanov, Timoshenko, Rudenko, Rusanov, & Shermolovich, 2019).

Figure 10. Synthesis of disubstituted thiophane by [3+2] cycloaddition reaction

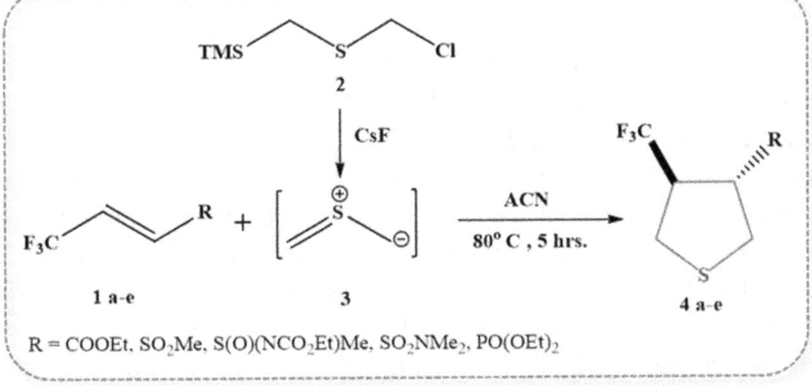

Figure 11. Formation of chiral 3-amine thiophane based on asymmetric sulfa-Michael/Aldol cascade reaction

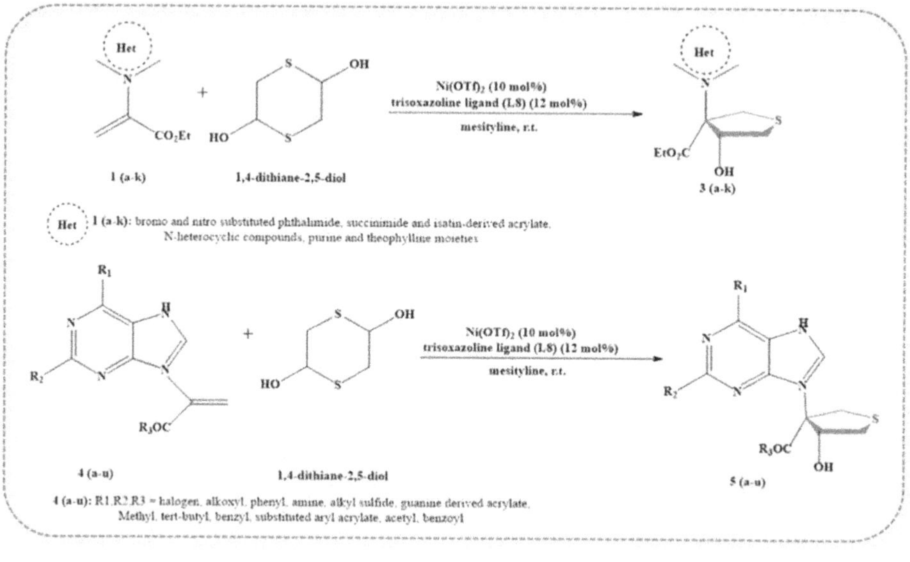

Huan *et al.* has constructed a new chiral 3-amine THT using 10 mol% $Ni(OTf)_2$ catalyst in asymmetric sulfa-Michael/Aldol cascade reaction (Figure 11). In the model reaction 0.05 mmol phthalimide substituted acrylate was made to react with 0.03 mmol 1,4-dithiane-2,5-diol in the presence of 12 mol% ligand trisoxazoline gave 91% yield in mesitylene with dr >20:1 and 90% enantiomeric excess (ee) at room temperature under nitrogen atmosphere. Subsequently, the substrate scope was assessed using bromo and nitro substituted phthalimide, succinimide and isatin-derived acrylate plus N-heterocyclic compounds, purine and theophylline moieties (1 a-k) that produced THT products (3 a-k). Various purine based derivatives and acrylate substrates (4 a-u) were checked which produced good to excellent yields (5 a-u), dr and ee factors in the other products (Huang, Xie, Sang, Qu, & Guo, 2020).

Borthakur *et al.* produced THT in Aldol-Michael kind of reaction using 30 mol% of $Bi(OTf)_3$ catalyst with β,γ-unsaturated ketone and 1,4-dithiane-2,5-diol in dichloromethane in 60% yield in 60 hours at room temperature of diastereoselective substituted THT derivatives (Figure 12). Above reaction was optimized in distinct ways, including altered the mol% of $Bi(OTf)_3$, modify the catalysts, temperatures, time of reaction and solvents but eventually successful in the way described above. Reaction proceeded to hydroxy ketone in first step in aldol way, which converted to THT in Michael fashion in the next step. It was stated in mechanism that high diastereoselectivity occurred in the product due to aldehyde group in the reaction (Borthakur & Saikia, 2019).

Figure 12. Preparation of diastereoselective substituted thiophane derivatives

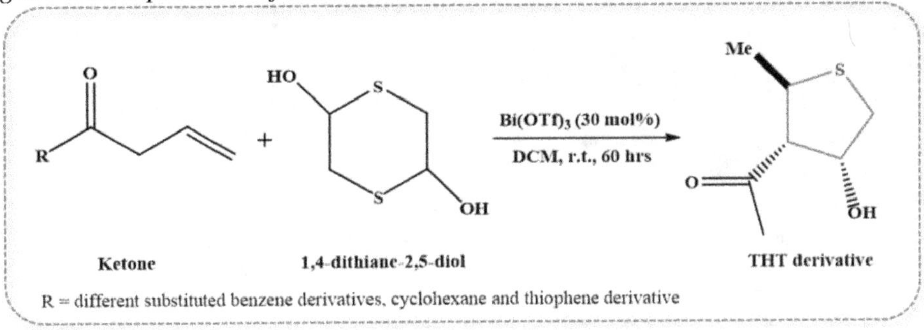

Thiazole

Thiazole, also known as 1,3-thiazole belongs to azole family, its structure is planar, and holding higher aromaticity compared to oxazole. It possess sulfur and nitrogen atoms in five membered ring of heterocyclic compounds, its general chemical form is C_3H_3NS. It is colourless or pale yellow liquid, having foul smell or pyridine like odor. Main source of thiazole are marine microbiology, plants, some foods including peanuts, coffee, potato chips, special products, vitamin B_1, etc. (Dahiya et al., 2020; Eskin, Ho, & Shahidi, 2013). The major applications of these scaffolds are in agricultural pesticides (Mak, Xu, & Fairlie, 2017), vulcanization process, fungicides, therapeutic agents (Chhabria, Patel, Modi, & Brahmkshatriya, 2016), and many more, and as a consequence these motifs are very significantly influence the field of research and development. To embedded the thiazole moieties some of the recent synthetic methods were outlined here. Such a rich thiazole derivatives were reportedly occurred utilizing various methods incorporating, Hantzsch, Robinson-Gabriel process, Cook-Heilbron, Herz reaction, etc. comprising different organic and inorganic reagents in various conditions (Panda & Hansen, 2023).

Figure 13. Formation of thiazole derivative through [3+1+1] cyclization between

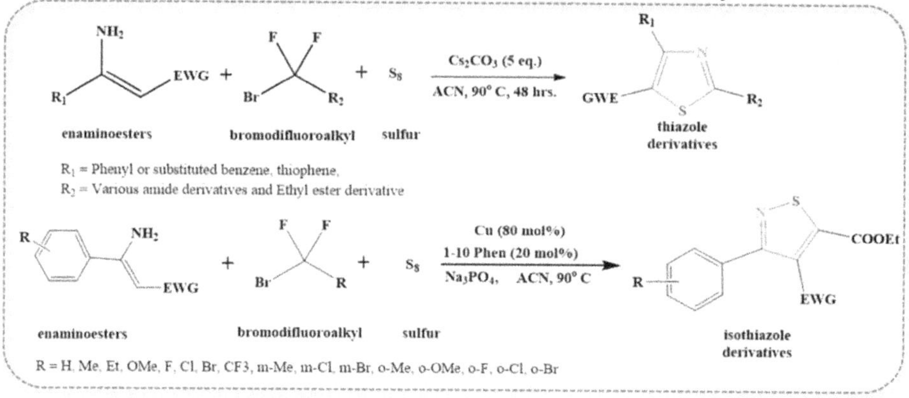

In order to create the different thiazoles and isothiazoles, Ma and his colleagues used three ingredients as building blocks; enaminoester, halodifluoroalkyl (bromodifluoroamide), and sulfur (S_8) in [3+1+1] cyclization decorum (Figure 13). In this context, in ACN at 90° C, in 48 hours along with 5 eq. of cesium carbonate as a base, diverse thiazole derivatives were prepared by acquiring multifarious substrate range of both enaminoesters and bromodifluoroalkyl with S_8. Bromodifluoroacetate was considered for the reaction but reluctantly produced a low yield; this can be attributed since the acetate peer is considerably more unstable than the amide equivalent. Besides that, diethyl (bromdifluoromethyl) phosphonate did not construct the thiazole moiety. Harmoniously, in examining the steric hindrance on the aryl ring, as well as distinctly substituted aryl ring offered high yield of thiazoles. X-ray crystallographic proved thiazole structure explicitly. Furthermore, the work was extended to originate the isothiazoles which encompass the same substrates by regulating the reaction conditions in modest to good yield of isothiazoles. The authors' delicacy was determining control experiments in combination with plausible mechanisms (Ma, Yu, Huang, Zhou, & Song, 2020).

Figure 14. Different thiazole derivatives synthesis from 2-aminothiazole and 2-chloro-N-(thiazol-2-yl)acetamide

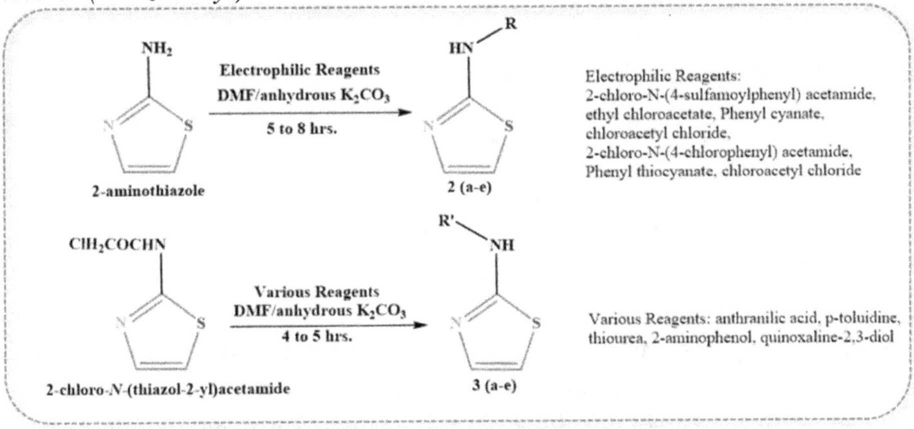

Abdel-Sattar *et al.* (2017) have provided the synthesis of six thiazole derivatives (Figure 14) in first phase of work employing 10 mmol of 2-aminothiazole with 10 mmol of various electrophiles in anhydrous potassium carbonate under reflux for five to eight hours that brought forth the divergent substituted thiazole compounds on quenching with water and recrystallization. Amongst the six compounds generated in first half of work, 2e was the significant one owing to its further use as chemical precursor with anthranilic acid, p-toluidine, thiourea, 2-aminophenol, quinoxaline-2,3-diol in anhydrous K_2CO_3 and DMF for 4-5 hours that offered other six thiazole moieties; viz. oxazipin-one, amide derivative, thiadiazolamine, thiazine, and ether derivative respectively. All the molecules were confirmed by IR, NMR and X-Ray crystallography spectroscopies. Density Functional Theory (DFT) was deployed for structure confirmation by means of theoretical aspects. Additionally, molecular reactivity parameters such as chemical potential, hardness, electrophilicity, nucleophilicity were calculated for the compounds (Abdel-Sattar, El-Naggar, & Abdel-Mottaleb, 2017).

Figure 15. Synthesis of hydrazinyl thiazole employing phenacyl bromide, thiosemicarbazide, and carbonyl compounds

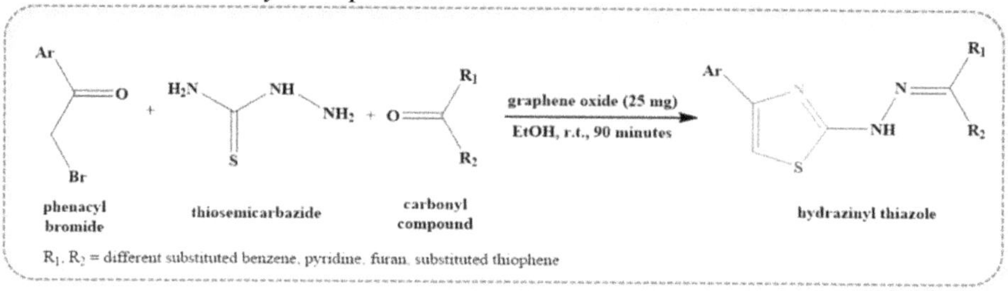

Das and group has prepared the hydrazinyl thiazole scaffolds that possess the inhibitory influences on human pancreatic cells α-amylase and glucosamine 6-phosphate synthase. Three constituent parts; i.e. 1 mmol carbonyl compound, 1 mmol thiosemicarbazide, and 1 mmol phenacyl bromide were mixed together in graphene oxide (synthesized) catalyzed reaction in a one-pot procedure in the ethanol, that imparted various hydrazinyl thiazoles (Figure 15), which were confirmed by ^{1}H and ^{13}CNMR spectroscopy. In addition to that, molecular docking assessment embarked for the further evaluation of the products (Das et al., 2021).

Khidre and Radini have materialized the thiazole derivatives (Figure 16) embracing the pyridine that exemplified the antibacterial and antifungal activities as well as DNA gyrase inhibitory action was assessed using *in silico* molecular docking. Primarily N-aminopyridone was prepared adopting acetylacetone, cyanoacetohydrazide in ethanol at reflux which was further treated with conc.HCl and ammonium isothyocyanate resulted in thiourea derivative. This thiourea moiety was then made to react with different reagents including hydrazonyl chlorides, chloroacetone, equimolar quantities of bromoketones, ethyl chloroacetate and 2,3-dichloroquinoxaline in EtOH at reflux and few drops of Et_3N base that entailed thiazole derivative holding pyridine species. All these derivatives were docked *in silico* and found hydrophobic interactions and H-bonding with some binding sites. Alongside this, these compounds were manifested the promising antibacterial results for both Gram-positive and Gram-negative bacteria and antifungal outcomes (Khidre & Radini, 2021).

Figure 16. Preparation of thiazole embracing pyridine holding antimicrobial and antifungal activities

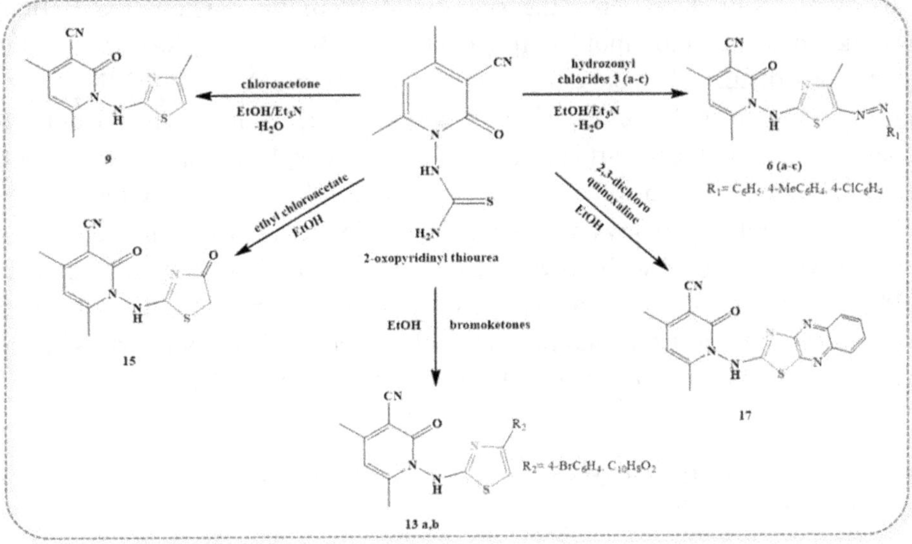

Figure 17. Synthesis of thiazole compounds A,B, and C utilized as coating on poly-urethane as antimicrobial and antifungal activities

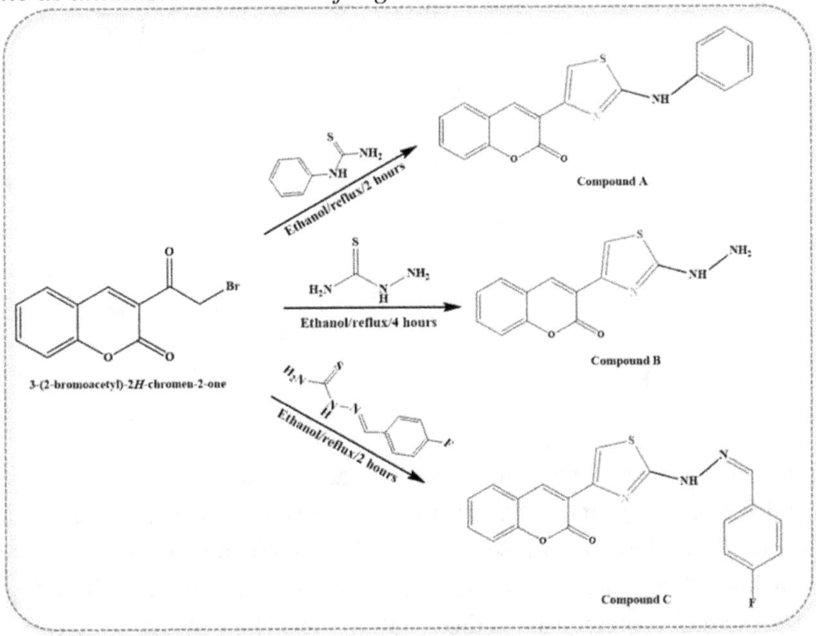

Bashandy *et al.* have reported the synthesis if thiazoles in which, in a reflux conditions, 0.01 mol of bromoacetyl substituted 2-chromanone was reacted in absolute ethanol with different reagents; incorporating 0.01 mol phenylthiourea, 0.01 mol carbothioamide, and 0.01 mol fluorobenzylidene substituted hydrazine carbothioamide to get distinct thiazole derivatives i.e. compound A, B and C (Figure 17). These compounds were then assessed for antimicrobial action in mixed and blank modes with polyurethane varnish. In three compounds, C has proven to be highly antimicrobial for Gram-positive species, moderately active against Gram-negative species and mildly active towards the fungi. Besides that, C was appeared as flame retardant when blended with polyurethane, having Limiting Oxygen Index (LOI) of 48 which proved to be less appealing towards the burning. Additionally, all the compounds were physically and mechanically tested particularly for gloss, scratch hardness, cross-hatch adhesion and flexibility (M. S. Bashandy, 2016).

Figure 18. Synthesis of indenyl-thiazole combinations assessed against cancer cell lines SNU-16 and Colo205

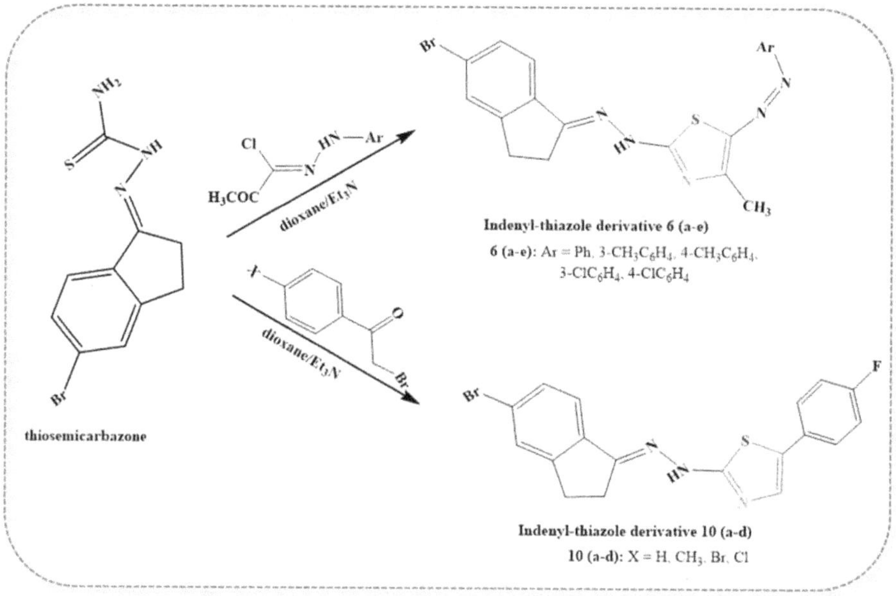

Alfaifi and associates have epitomized the indenyl-thiazole (Figure 18) along with the indenyl-formazan compounds. In primary phase they have synthesized thiosemicarbazone utilizing 0.01 mol thiosemicarbazide and 0.01 mol 5-bromo-indan-1-one in reflux using EtOH. Forenamed thiosemicarbazone was enabled to react with five various hydrazonoyl chloride using dioxane and triethylamine as base in reflux state, lead to indenyl-thiazole combinations. Ahead of this, four phenacyl bromide moieties were treated with same thiosemicarbazone, which fabricated four indenyl-thiazole derivatives. These moieties were then adopted to examine the anticancer activity against the gastric cancer cell line SNU-16 and colon cancer cell lines Colo205, with that the standard Cisplatin and Sunitinib were used as reference drugs. IC_{50} values of some of the indenyl-thiazole derivatives were found to have beneficial effect over standard references. Plus, to that *in silico* docking was performed for the said combination that binds with 2 tested proteins of cancers. Drug-likeness character for some inhibitors were also investigated and given promising results. Apart from these, researchers have derived indenyl-formazan compounds and examined all aforementioned parameters for the same (Alfaifi, Farghaly, & Magda, 2023).

Oliveira *et al.* has portrayed 22 derivatives based on thiazole (Figure 19). To synthesize these compounds a starting material phthalimido-thiosemicarbazide (1a and 1b) were obtained in beginning, following that a series of phthalimido-1,3-thiazole derivatives 2 (a-i), 3 (a, b), 4 (a-d), 5 (a-g) was prepared using distinct synthetic protocols. In synthesizing 2 (a-i), 2.26 mmol α-haloketones and 25 mL isopropanol were sonicated which blended to sodium acetate and substituted thiourea and underwent ultrasonic for 2 hours yielded the desired derivatives. For 3 (a-b), 2.26 mmol phthalimido-thiosemicarbazide intermediates (1a/1b) were made soluble in propanone, then 1.2:1 ratio of CH_3COONa:chlorine substituted pentanedione dissolved that produced preferred products in 90 minutes. To generated 4 (a-d), compound 3a or 3b were solubilized in ethanol and five drops of H_2SO_4 was added and sonicated for half an hour, after that1.5 mmol of thiosemicarbazide was added to that mixture and further leave for sonication for 3 hours, which on work-up given essential outcomes 4 (a-d). At the last synthetic steps, compounds 5 (a-g) were developed by addition of 4a (or 4b for 5g) with α-bromoacetophenone, sodium acetate and isopropanol which sonicated for 4 hours at room temperature, provided the products. Including the synthesis, an anticancer activeness of all the compounds were tested, choosing MOLT-4, TOLEDO, HL-60-MX1, K562, T47-D, SK-MEL-28, HL60, and MCF-7 cell lines. From the group of moieties, 5b and 5e exhibited cytotoxicity for all cell lines, on the other hand 5d displayed for 6, 5c has 5, 5f and 4c demonstrated against 4 and 5a indicated for 3 cell lines (Oliveira et al., 2021).

Figure 19. Phthalimido-1,3-thiazole derivatives investigated for various cancer cell lines

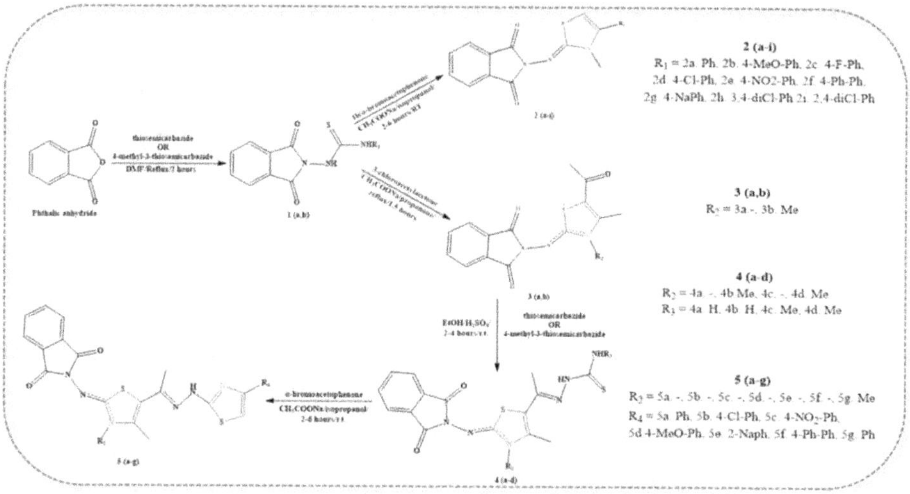

Figure 20. Biologically active trisubstituted thiazole structures examined against Mycobacterium Tuberculosis

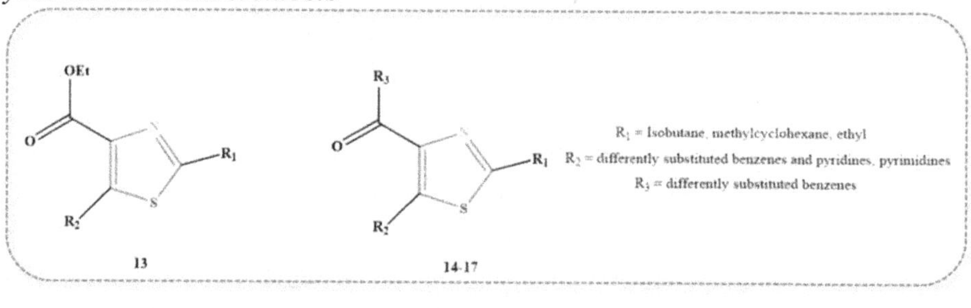

Karale and group have synthesized the number of trisubstituted thiazole derivatives, some of which were active against *Mycobacterium Tuberculosis*. Various thiazole scaffolds 12 (a-c), 13 (a-r) and 14a-17f were synthesized (Figure 20). The process of preparing the product started with crude thioamide, which was worked-up and given 12 (a-c) motifs in 30 minutes. This reaction was completed with 11.42 mmol α-bromopyruvate in ethanol. 0.5 mmol of 12a was reportedly combined with 0.45 mmol of 4-bromo-1-2-(methylenedioxy) benzene in 5 mol% dichlorobis(triphenyl-phosphine)palladium (II), and potassium acetate in dry dimethylacetamide at 150° C, following final work-up and purification, this mixture produced 13a, and in a similar manner, corresponding 13 (b-r) were produced. Then, using tetrahydrofuran and methanol in a 3:1 ratio at 0° C, 13a was utilized to produce 15a. After adding two eq. of NaOH to the mixture gradually over the course of three hours at room temperature, the mixture was dried under vacuum, and the solid acidic product that resulted from the summation of 4M hydrochloric acid was solvated in DMF. Next, 1 mmol of EDC hydrochloride and 1 mmol HOBt were added as reagents, and after half an hour, 0.8 milligrams of (S)-1-phenylethan1-amine and 1.33 mmol triethyl-amine were added. The mixture was then allowed to whip at room temperature for twelve hours, at which point the reaction was completed and the product worked up to yield crude 15a and, in accordance, products 14 to 17 were also got prepared. Amongst the large series of trisubstituted thiazole derivatives, 13h and 13p were found to resist dormant *TB* strains H37Ra and H37Rv moreover they were benign to Chinese hamster ovary cells (CHO-K1). It was concluded in structural activity relationship (SAR)study that, at 2nd carbon hydrophobic, ester functional group at 4th carbon and H-bond acceptor part at carbon 5th give desired outcomes (Karale et al., 2019).

Figure 21. thiazole scaffolds have an inhibitory influence against human Pin1 (Peptidyl prolyl cis-trans isomerase)

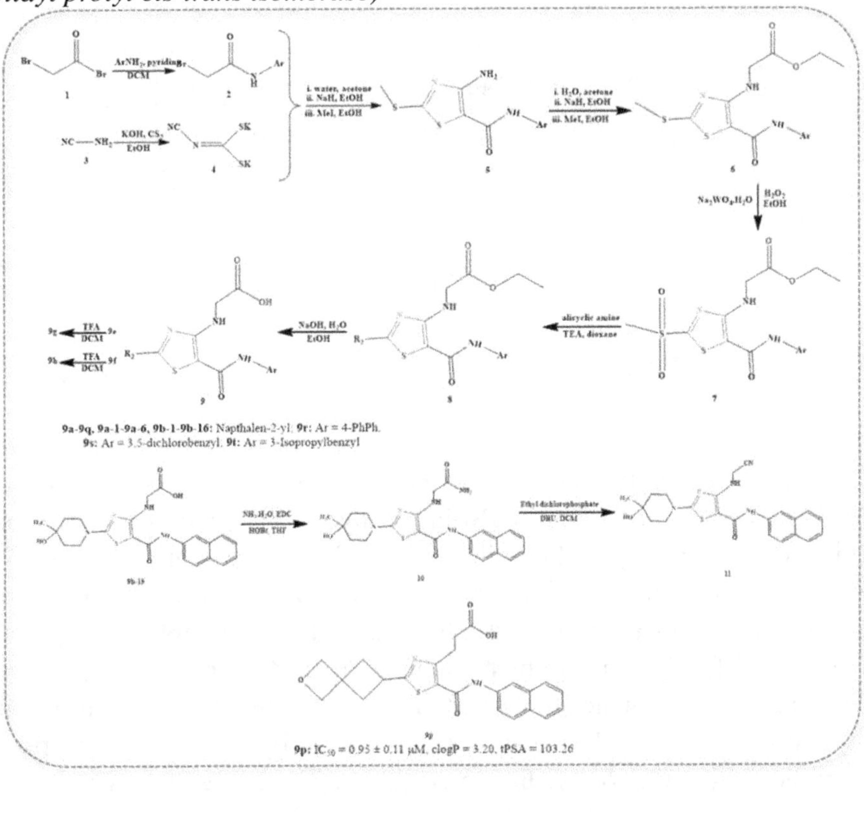

According to a biological evaluation by Du *et al.*, new thiazole scaffolds have an inhibitory influence against human Pin1 (Peptidyl prolyl cis-trans isomerase). A number of chemical reactions were performed for the intended use (Figure 21). Using the acylation of 2-napthylamine and bromoacetyl bromide, 2 was created. This reaction offered 5, a thiazole intermediate, when it reacted with potassium cyanocarbonimidodithiate. After subjecting 5 to ethyl glyoxylate, 6 was generated. 6 experienced treatment with hydrogen peroxide and sodium tungstate dihydrate to yield another thiazole moiety i.e. 7. Following a reaction between this thiazole motif 7 and alicyclic amines 8, which produced 9e and 9f upon hydrolysis, 9g and 9h in trifluoroacetic acid (TFA) in dichloromethane (DCM) were the final products of these reactions.

Figure 22. Synthesis of dihydropyrazole-thiazole derivatives studied for their anti-inflammatory activity

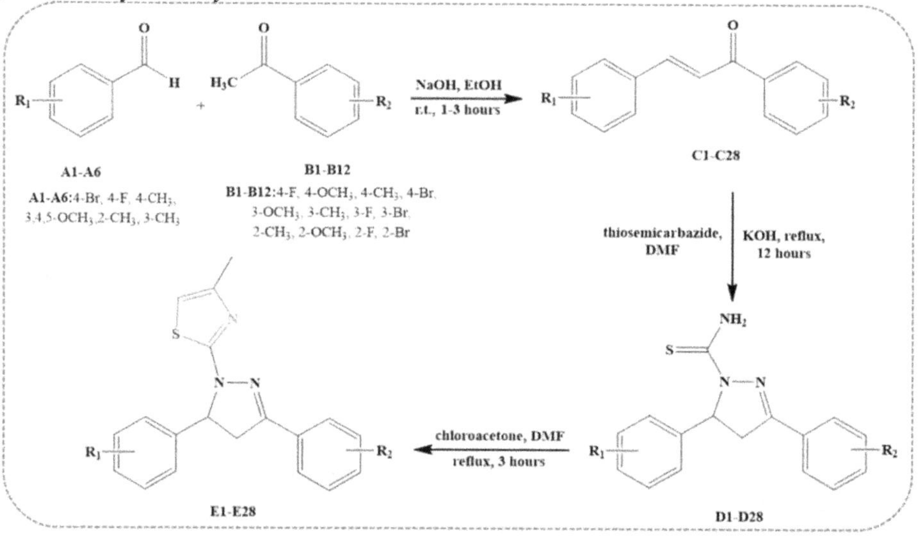

This work extended further to create the 10 by exposing the compounds 9b-15 towards Ammonia-water, EDC, HOBt in tetrahydrofuran, a 10 was embraced for the formation of 11, to ethyl dichlorophosphate, 1,8-diazabicyclo[5.4.0]undac-7-ene in DCM. Next to it, on occlusion of ring 4 equipped interim derivative 12, which on Sandmeyer reaction modified to 13. In virtue of hydrolysis followed by condensation provided 15. Reacting 15 by pursuing Suzuki-Miyaura coupling shaped the principal derivative 16, which translated to 17 using reduction by means of sodium borohydride and nickel chloride. 19 was originated by underwent oxidation and nucleophilic substitution of 17. In due course they succeeded to construct the 20 on hydrolysis by sodium hydroxide in ethanol and THF in 67% amount. SAR suggested the 2nd position of thiazole was suitable place to combat against Pin1, and in the huge task of biological evaluation, 9p was as drug-like specie with 0.95 μM IC_{50}, 3.20 calculated logP and 103.26 of topological polar surface area (tPSA) (Du, Wang, Cui, & Xu, 2021).

The dihydropyrazole-thiazole compounds that were engineered by Zhang and team were examined for their anti-inflammatory properties (Figure 22). Aldol condensation of benzaldehydes (A1-A6) and acetophenone (B1-B12) was performed in the first step of the synthesis, A1-A6 were made soluble in pure EtOH, and 10% sodium hydroxide (5 mL) was included gradually to it, following that B1-B12 summed up in the above mixture in 1 eq., and then mixture was whisked at ambient temperature for approximately one to three hours, that offered chalcone compounds C1-C28. In second step of ring closure reaction of thiosemicarbazide, one equivalent of chalcones were dissolved in DMF with 1.5 equivalents of thiosemicarbazide and potassium hydroxide, that refluxed for twelve hours at 100° C to yield the carbothioamide D1-D28 species. In the last step, 2 mmol of D1-D28 was dissolved in DMF subsequently, 10 mmol chloroacetone was blended and next to that whole mixture swirled at 100° C for 180 minutes produced compounds E1-E28. Thereupon, survival of LPS processed RAW264.7 macrophages was assessed using E1-E28, among them, E3, E6, E9, E11 demonstrated valuable inherent results on 1μM amount of these substances. Moreover, in primary SAR study of inhibition of all E moieties on emergence of nitric oxide (NO) caused by LPS. It was found that 1μM concentrations, $-CH_3$ at R_1 and $-OCH_3$ at R_2 given far better inhibition against NO production, and at 1 μM amount E26 has provided the finest effect. At 10 μM, E26 has imparted the activity more than two times which is superior then standard positive control, and hence in advance SAR mechanism, E26 turned out to be reduce the impacts of cyclooxygenase-2 (COX-2), tumor necrosis factor (TNF-α), interleukin-1β (IL-1β), inducible nitric oxide synthase (iNOS), by obstructing MAPKs signaling routes. E26 also exhibited *In vivo* study in C57BL-6J mice, enhanced results (Zhang et al., 2021).

CONCLUSION

Thiazoline, thiophane, and thiazole are highly important scaffolds with respect to their vital biological significance. Owing to resistance against antibiotics by microorganisms it is the primary challenge to scientific community to design such motifs that not only eradicate the microorganisms but at the same time, other diseases can be prevented. Among the diverse heterocycles the sulfur containing species have substantial roles as anti-microbial, anticancer, antihypertension, antiviral, anti-inflammatory, etc. The need of S-containing heterocycles are due to their medicinal character, and curtailed toxicity draft, bioavailability, metabolic stability and binding affinity over others. Synthesis of these moieties using various routes and utilizing numerous chemical functional groups under environmentally benign conditions made them accessible in large scale, another aspect observed was the preparation of such compounds through cascade reactions that are very crucial mode to synthesize them with good to excellent yield which offers large atom economy, stereo- and regioselectivity, acquiring intermediates without isolation, provide diverse chemical structure and complex framework, and tenders the complex molecular architectures in biosynthetic pathways as well as those are cost effective and having functional group tolerance ability. Though classical synthesis approaches also quotes large scale production. Thus, it is concluded that synthesizing these compounds in excellent yields require advanced environment friendly routes that fosters high efficiency in reduced energy and time.

REFERENCES

Abdel-Sattar, N. E. A., El-Naggar, A. M., & Abdel-Mottaleb, M. S. A. (2017). Novel Thiazole Derivatives of Medicinal Potential: Synthesis and Modeling. *Journal of Chemistry, 4102796*, 1–11. Advance online publication. DOI: 10.1155/2017/4102796

Ahirwar, J., Ahirwar, D., Lanjhiyana, S., Jha, A. K., Dewangan, D., & Badwaik, H. (2018). Analgesic and Anti-inflammatory Potential of Merged Pharmacophore Containing 1,2,4-triazoles and Substituted Benzyl Groups via Thio Linkage. *Journal of Heterocyclic Chemistry, 55*(9), 2130–2141. DOI: 10.1002/jhet.3258

Aissani, N., Urgeghe, P. P., Oplos, C., Saba, M., Tocco, G., Petretto, G. L., Eloh, K., Menkissoglu-Spiroudi, U., Ntalli, N., & Caboni, P. (2015). Nematicidal Activity of the Volatilome of Eruca sativa on Meloidogyne incognita. *Journal of Agricultural and Food Chemistry, 63*(27), 6120–6125. DOI: 10.1021/acs.jafc.5b02425 PMID: 26082278

Alfaifi, G. H., Farghaly, T. A., & Magda, H. A. (2023). Indenyl-thiazole and indenyl-formazan derivatives: Synthesis, anticancer screening studies, molecular-docking, and pharmacokinetic/ molin-spiration properties. *PLoS One, 18*(3), e0274459. DOI: 10.1371/journal.pone.0274459 PMID: 36857383

Alom, N.-E., Wu, F., & Li, W. (2017). One-Pot Strategy for Thiazoline Synthesis from Alkenes and Thioamides. *Organic Letters, 19*(4), 930–933. DOI: 10.1021/acs.orglett.7b00079 PMID: 28177643

Alsharif, Z. A., & Alam, M. A. (2017). Modular synthesis of thiazoline and thiazole derivatives by using a cascade protocol. *RSC Advances, 7*(52), 32647–32651. DOI: 10.1039/C7RA05993K PMID: 29170713

Aoyagi, M., Kamoi, T., Kato, M., Sasako, H., Tsuge, N., & Imai, S. (2011). Structure and Bioactivity of Thiosulfinates Resulting from Suppression of Lachrymatory Factor Synthase in Onion. *Journal of Agricultural and Food Chemistry, 59*(20), 10893–10900. DOI: 10.1021/jf202446q PMID: 21905712

Bashandy, M. S. (2016). Synthesis and Characterization of Thiazole Derivatives and Evaluating their Performance as Additives in Polyurethane Coating. *Egyptian Journal of Chemistry, 59*(1), 115–129. DOI: 10.21608/ejchem.2016.1375

Bitsch, R. (2003). BIOTIN | Properties and Determination. In Caballero, B. (Ed.), *Encyclopedia of Food Sciences and Nutrition* (2nd ed., pp. 506–516). Academic Press. DOI: 10.1016/B0-12-227055-X/00101-2

Block, E., Dethier, B., Bechand, B., Cotelesage, J. J. H., George, G. N., Goto, K., Pickering, I. J., Mendoza Rengifo, E., Sheridan, R., Sneeden, E. Y., & Vogt, L. (2018). Ajothiolanes: 3,4-Dimethylthiolane Natural Products from Garlic (Allium sativum). *Journal of Agricultural and Food Chemistry*, *66*(39), 10193–10204. DOI: 10.1021/acs.jafc.8b03638 PMID: 30196701

Borah, P., Hazarika, S., Chettri, A., Sharma, D., Deka, S., Venugopala, K. N., & Deb, P. K. (2023). Heterocyclic compounds as antimicrobial agents. In Bagchi, D., Das, A., & Downs, B. W. (Eds.), *Viral, Parasitic, Bacterial, and Fungal Infections* (pp. 781–804). Academic Press. DOI: 10.1016/B978-0-323-85730-7.00068-0

Borthakur, U., & Saikia, A. K. (2019). Bismuth(III)-Triflate-Catalyzed Highly Diastereoselective Synthesis of Substituted Tetrahydrothiophene via Tandem Isomerization, Michael and Aldol Reactions. *ChemistrySelect*, *4*(37), 11136–11139. DOI: 10.1002/slct.201902871

Chen, D., Bai, Y., Cheng, Q., Li, J., Tong, Z., Hou, J., Liu, T., Guo, Y., Tang, X., Yang, X., & Yang, X. (2022). Domino synthetic strategy for tetrahydrothiophene derivatives from 2-acetylfuran/2-acetylthiophene, benzaldehydes, and sulfur powder. *Arabian Journal of Chemistry*, *15*(9), 104097. DOI: 10.1016/j.arabjc.2022.104097

Chhabria, T. M., Patel, S., Modi, P., & Brahmkshatriya, S. P. (2016). Thiazole: A Review on Chemistry, Synthesis and Therapeutic Importance of its Derivatives. *Current Topics in Medicinal Chemistry*, *16*(26), 2841–2862. DOI: 10.2174/15680 26616666160506130731 PMID: 27150376

Dahiya, R., Dahiya, S., Fuloria, N. K., Kumar, S., Mourya, R., Chennupati, S. V., Jankie, S., Gautam, H., Singh, S., Karan, S. K., Maharaj, S., Fuloria, S., Shrivastava, J., Agarwal, A., Singh, S., Kishor, A., Jadon, G., & Sharma, A. (2020). Natural Bioactive Thiazole-Based Peptides from Marine Resources: Structural and Pharmacological Aspects. *Marine Drugs*, *18*(6), 329. Advance online publication. DOI: 10.3390/md18060329 PMID: 32599909

Das, A., Dey, S., Chakraborty, S., Barman, A., Naresh Yadav, R., Gazi, R., Jana, M., & Firoj Hossain, M. (2021). Metal-Free One-Pot Synthesis of 2-(2-Hydrazinyl) Thiazole Derivatives Using Graphene Oxide in a Green Solvent and Molecular Docking Studies. *ChemistrySelect*, *6*(36), 9552–9558. DOI: 10.1002/slct.202102642

De, A., Sarkar, S., & Majee, A. (2021). Recent advances on heterocyclic compounds with antiviral properties. *Chemistry of Heterocyclic Compounds*, *57*(4), 410–416. DOI: 10.1007/s10593-021-02917-3 PMID: 33994556

Du, L., Wang, X., Cui, G., & Xu, B. (2021). Design, synthesis and biological evaluation of novel thiazole-based derivatives as human Pin1 inhibitors. *Bioorganic & Medicinal Chemistry*, *29*, 115878. DOI: 10.1016/j.bmc.2020.115878 PMID: 33246256

Dua, R., Shrivastava, S., Sonwane, S., & Srivastava, S. (2011). Pharmacological significance of synthetic heterocycles scaffold: a review. Academic Press.

El-Aasr, M., Fujiwara, Y., Takeya, M., Ono, M., Nakano, D., Okawa, M., Kinjo, J., Ikeda, T., Miyashita, H., Yoshimitsu, H., & Nohara, T. (2011). Garlicnin A from the Fraction Regulating Macrophage Activation of Allium sativum. *Chemical & Pharmaceutical Bulletin*, *59*(11), 1340–1343. DOI: 10.1248/cpb.59.1340 PMID: 22041068

El-Helw, E. A. E., & El-Badawy, A. A. (2020). Synthesis of chromenone, pyrimidinone, thiazoline, and quinolone derivatives as prospective antitumor agents. *Journal of Heterocyclic Chemistry*, *57*(6), 2354–2364. DOI: 10.1002/jhet.3948

El-Helw, E. A. E., Sallam, H. A., & Elgubbi, A. S. (2019). Antioxidant activity of some N-heterocycles derived from 2-(1-(2-oxo-2H-chromen-3-yl)ethylidene) hydrazinecarbothioamide. *Synthetic Communications*, *49*(20), 2651–2661. DOI: 10.1080/00397911.2019.1638938

Eskin, N. A. M., Ho, C.-T., & Shahidi, F. (2013). Browning Reactions in Foods. In Eskin, N. A. M., & Shahidi, F. (Eds.), *Biochemistry of Foods* (3rd ed., pp. 245–289). Academic Press. DOI: 10.1016/B978-0-08-091809-9.00006-6

Feliciano, J. A., Leitgeb, A. J., Schrank, C. L., Allen, R. A., Minbiole, K. P. C., Wuest, W. M., & Carden, R. G. (2021). Trivalent sulfonium compounds (TSCs): Tetrahydrothiophene-based amphiphiles exhibit similar antimicrobial activity to analogous ammonium-based amphiphiles. *Bioorganic & Medicinal Chemistry Letters*, *37*, 127809. DOI: 10.1016/j.bmcl.2021.127809 PMID: 33516911

Feng, M., Tang, B., Liang, S., & Jiang, X. (2015). Sulfur Containing Scaffolds in Drugs: Synthesis and Application in Medicinal Chemistry. *Current Topics in Medicinal Chemistry*, *16*. Advance online publication. DOI: 10.2174/1568026615666150915111741 PMID: 26369815

Gomha, S. M., Riyadh, S. M., Huwaimel, B., Zayed, M. E. M., & Abdellattif, M. H. (2022). Synthesis, Molecular Docking Study, and Cytotoxic Activity against MCF Cells of New Thiazole–Thiophene Scaffolds. *Molecules (Basel, Switzerland)*, *27*(14), 4639. Advance online publication. DOI: 10.3390/molecules27144639 PMID: 35889511

Huang, K.-X., Xie, M.-S., Sang, J.-W., Qu, G.-R., & Guo, H.-M. (2020). Asymmetric Synthesis of 3-Amine-tetrahydrothiophenes with a Quaternary Stereocenter via Nickel(II)/Trisoxazoline-Catalyzed Sulfa-Michael/Aldol Cascade Reaction: Divergent Access to Chiral Thionucleosides. *Organic Letters*, *23*(1), 81–86. DOI: 10.1021/acs.orglett.0c03747 PMID: 33332122

Ibrayev, M. K., Nurkenov, O. A., Rakhimberlinova, Z. B., Takibayeva, A. T., Palamarchuk, I. V., Turdybekov, D. M., Kelmyalene, A. A., & Kulakov, I. V. (2022). Synthesis, Structure and Molecular Docking of New 4,5-Dihydrothiazole Derivatives Based on 3,5-Dimethylpyrazole and Cytisine and Salsoline Alkaloids. *Molecules (Basel, Switzerland)*, *27*(21), 7598. Advance online publication. DOI: 10.3390/molecules27217598 PMID: 36364423

Jain, A., Maji, S., Shukla, K., Kumari, A., Garg, S., Metre, R. K., Bhattacharyya, S., & Rana, N. K. (2022). Stereoselective synthesis of tri-substituted tetrahydrothiophenes and their in silico binding against mycobacterial protein tyrosine phosphatase B. *Organic & Biomolecular Chemistry*, *20*(15), 3124–3135. DOI: 10.1039/D2OB00052K PMID: 35343552

Jeon, H., Kim, D., Lee, J. H., Song, J., Lee, W. S., Kang, D. W., Kang, S., Lee, S. B., Choi, S., & Hong, K. B. (2018). Hypervalent Iodine-Mediated Alkene Functionalization: Oxazoline and Thiazoline Synthesis via Inter-/Intramolecular Aminohydroxylation and Thioamination. *Advanced Synthesis & Catalysis*, *360*(4), 779–783. DOI: 10.1002/adsc.201701087

Kabir, E., & Uzzaman, M. (2022). A review on biological and medicinal impact of heterocyclic compounds. *Results in Chemistry*, *4*, 100606. DOI: 10.1016/j.rechem.2022.100606

Karale, U. B., Krishna, V. S., Krishna, E. V., Choudhari, A. S., Shukla, M., Gaikwad, V. R., Mahizhaveni, B., Chopra, S., Misra, S., Sarkar, D., Sriram, D., Dusthackeer, V. N. A., & Rode, H. B. (2019). Synthesis and biological evaluation of 2,4,5-trisubstituted thiazoles as antituberculosis agents effective against drug-resistant tuberculosis. *European Journal of Medicinal Chemistry*, *178*, 315–328. DOI: 10.1016/j.ejmech.2019.05.082 PMID: 31195172

Khidre, R. E., & Radini, I. A. M. (2021). Design, synthesis and docking studies of novel thiazole derivatives incorporating pyridine moiety and assessment as antimicrobial agents. *Scientific Reports*, *11*(1), 7846. DOI: 10.1038/s41598-021-86424-7 PMID: 33846389

Kurapati, K. R. V., Atluri, V. S., Samikkannu, T., Garcia, G., & Nair, M. P. N. (2016). Natural Products as Anti-HIV Agents and Role in HIV-Associated Neurocognitive Disorders (HAND): A Brief Overview. *Frontiers in Microbiology*, *6*, 6. DOI: 10.3389/fmicb.2015.01444 PMID: 26793166

Lamberth, C., & Dinges, J. (2012). The Significance of Heterocycles for Pharmaceuticals and Agrochemicals. In *Bioactive Heterocyclic Compound Classes* (pp. 1-20). Academic Press.

Laxmikeshav, K., Kumari, P., & Shankaraiah, N. (2022). Expedition of sulfur-containing heterocyclic derivatives as cytotoxic agents in medicinal chemistry: A decade update. *Medicinal Research Reviews*, *42*(1), 513–575. DOI: 10.1002/med.21852 PMID: 34453452

Ledesma-Amaro, R., Santos, M. A., Jiménez, A., & Revuelta, J. L. (2013). 21 - Microbial production of vitamins. In McNeil, B., Archer, D., Giavasis, I., & Harvey, L. (Eds.), *Microbial Production of Food Ingredients, Enzymes and Nutraceuticals* (pp. 571–594). Woodhead Publishing. DOI: 10.1533/9780857093547.2.571

Luo, W., Zheng, K., Kuang, H., Li, Z., Wang, J., & Mei, J. (2022). The potential of articaine as new generation of local anesthesia in dental clinics. *Revista de Medicina (São Paulo)*, *101*(48). PMID: 36482533

Luo, X., Li, Y., Chen, X., Song, Z., Liang, J., Liao, C., Zhu, Z., & Chen, L. (2019). (Z)-Tetrahydrothiophene and (Z)-tetrahydrothiopyran synthesis through nucleophilic substitution and intramolecular cycloaddition of alkynyl halides and EtOCS2K. *Organic & Biomolecular Chemistry*, *17*(31), 7315–7319. DOI: 10.1039/C9OB01370A PMID: 31342046

Ma, X., Yu, X., Huang, H., Zhou, Y., & Song, Q. (2020). Synthesis of Thiazoles and Isothiazoles via Three-Component Reaction of Enaminoesters, Sulfur, and Bromodifluoroacetamides/Esters. *Organic Letters*, *22*(14), 5284–5288. DOI: 10.1021/acs.orglett.0c01275 PMID: 32603128

Mak, J. Y. W., Xu, W., & Fairlie, D. P. (2017). Thiazoles in Peptides and Peptidomimetics. In Lubell, W. D. (Ed.), *Peptidomimetics I* (pp. 235–266). Springer International Publishing.

Markitanov, Y. M., Timoshenko, V. M., Rudenko, T. V., Rusanov, E. B., & Shermolovich, Y. G. (2019). 3-Functional substituted 4-trifluoromethyl tetrahydrothiophenes via [3+2]-cycloaddition reactions. *Journal of Sulfur Chemistry*, *40*(6), 629–640. DOI: 10.1080/17415993.2019.1633326

Nagaraju, S., Sathish, K., Satyanarayana, N., Paplal, B., & Kashinath, D. (2020). Regioselective synthesis of spiro isoxazole-oxindole-tetrahydrothiophene hybrids via cascade reactions under catalyst-free conditions. *Journal of Heterocyclic Chemistry, 57*(1), 469–476. DOI: 10.1002/jhet.3803

Ngoc Toan, V., & Dinh Thanh, N. (2021). Novel thiazoline–coumarin hybrid compounds containing sugar moieties: Synthesis, biological evaluation and molecular docking study as antiproliferative agents. *New Journal of Chemistry, 45*(24), 10636–10653. DOI: 10.1039/D1NJ00680K

O'Donnell, J. A., Gelone, S. P., & Safdar, A. (2015). 37 - Topical Antibacterials. In Bennett, J. E., Dolin, R., & Blaser, M. J. (Eds.), *Mandell, Douglas, and Bennett's Principles and Practice of Infectious Diseases* (8th ed., pp. 452–462.e452). W.B. Saunders. DOI: 10.1016/B978-1-4557-4801-3.00037-0

Oliveira, A. R., dos Santos, F. A., Ferreira, L. P. L., Pitta, M. G. R., Silva, M. V. O., Cardoso, M. V. O., Pinto, A. F., Marchand, P., de Melo Rêgo, M. J. B., & Leite, A. C. L. (2021). Synthesis, anticancer activity and mechanism of action of new phthalimido-1,3-thiazole derivatives. *Chemico-Biological Interactions, 347*, 109597. DOI: 10.1016/j.cbi.2021.109597 PMID: 34303695

Panda, S. S., & Hansen, D. W. (2023). Microwave-assisted synthesis of five-membered azaheterocyclic systems using nanocatalysts. In Bhunia, S., Kumar, B., Singh, P., Oraon, R., & Kim, K.-H. (Eds.), *Nanoparticles in Green Organic Synthesis* (pp. 255–278). Elsevier. DOI: 10.1016/B978-0-323-95921-6.00007-X

Ram, R. N., Gupta, D. K., & Soni, V. K. (2016). Copper(I)-Promoted Synthesis of Highly Substituted and Functionalized Tetrahydrothiophenes. *European Journal of Organic Chemistry, 2016*(20), 3434–3440. DOI: 10.1002/ejoc.201600497

Rosa Rodes Solanes, Herrero, Royo, & Gòmez. (2016). Spain Patent No. EP 3 040 334 A1. European Patent Office: E. P. Application.

Services, N. J. D. o. H. a. S. (1999). Tetrahydrothiophene. *Hazardous Substance Fact Sheet*, 1-6.

Shehzad, M. T., Imran, A., Hameed, A., Rashida, M., Bibi, M., Uroos, M., Asari, A., Iftikhar, S., Mohamad, H., Tahir, M. N., Shafiq, Z., & Iqbal, J. (2021). Exploring synthetic and therapeutic prospects of new thiazoline derivatives as aldose reductase (ALR2) inhibitors. *RSC Advances, 11*(28), 17259–17282. DOI: 10.1039/D1RA01716K PMID: 35479726

Turan-Zitouni, G., Yurttaş, L., Tabbi, A., Akalın Çiftçi, G., Temel, H. E., & Kaplancıklı, Z. A. (2018). New Thiazoline-Tetralin Derivatives and Biological Activity Evaluation. *Molecules (Basel, Switzerland), 23*(1), 135. Advance online publication. DOI: 10.3390/molecules23010135 PMID: 29320423

Vardanyan, R. S., & Hruby, V. J. (2006). 38 - Antihelmintic Drugs. In Vardanyan, R. S., & Hruby, V. J. (Eds.), *Synthesis of Essential Drugs* (pp. 583–593). Elsevier. DOI: 10.1016/B978-044452166-8/50038-8

Wang, M., Zhang, Y., Lv, L., Kong, D., & Niu, G. (2022). Biosynthesis and Chemical Synthesis of Albomycin Nucleoside Antibiotics. *Antibiotics (Basel, Switzerland), 11*(4), 438. Advance online publication. DOI: 10.3390/antibiotics11040438 PMID: 35453190

Willstätter, R., & Wirth, T. (1909). Über Thioformamid. *Berichte der Deutschen Chemischen Gesellschaft, 42*(2), 1908–1922. DOI: 10.1002/cber.19090420267

Winter, N., Rupcic, Z., Stadler, M., & Trauner, D. (2019). Synthesis and biological evaluation of (±)-hippolachnin and analogs. *The Journal of Antibiotics, 72*(6), 375–383. DOI: 10.1038/s41429-019-0176-x PMID: 30979942

Zhang, Z., Cao, P., Fang, M., Zou, T., Han, J., Duan, Y., Xu, H., Yang, X., & Li, Q.-S. (2021). Design, synthesis, and SAR study of novel 4,5-dihydropyrazole-Thiazole derivatives with anti-inflammatory activities for the treatment of sepsis. *European Journal of Medicinal Chemistry, 225*, 113743. DOI: 10.1016/j.ejmech.2021.113743 PMID: 34403978

Chapter 6
Functionalized Gold Nanoparticles Enable Colorimetric Biosensing of Pesticides for Advanced Applications:
Detection of Pesticides Using Functionalized Gold Nanoparticles

Keyur Bhatt
https://orcid.org/0000-0001-9301-576X
Ganpat University, India

Nandan C. Pomal
https://orcid.org/0000-0001-5890-3163
Sigma University, India

Dhruvansh Patni
https://orcid.org/0009-0007-7321-5693
Sigma University, India

Nihal Patel
Ganpat University, India

ABSTRACT

The growing use of pesticides in agriculture poses significant risks to food safety and environmental health. Traditional detection methods are accurate but require

DOI: 10.4018/979-8-3693-2774-6.ch006

complex instruments and skilled personnel. Biosensing approaches using functionalized gold nanoparticles (AuNPs) have emerged as rapid, cost-effective, and portable alternatives. This mini-review explores the synthesis, surface modification, and characterization of AuNPs, emphasizing their surface plasmon resonance (SPR) properties for colorimetric biosensing. Pesticide exposure induces aggregation of ligand-functionalized AuNPs, leading to visible color changes detectable without advanced equipment. Recent studies demonstrate the successful detection of pesticides such as paraquat, glyphosate, and deltamethrin across diverse matrices. These biosensors offer advantages like visual output, simplicity, and fast response. Despite some limitations, functionalized AuNPs show strong potential for developing efficient biosensing systems, supporting safer agriculture and environmental monitoring.

INTRODUCTION

Growing population on the earth today is on its peak and world has witnessed the serious issues in last few years like COVID-19, climate crisis, economic troubles, and war etc., which influence the world as a whole. Among the various kinds of challenges, environmental crisis severely impacted the mankind. The greatest concerns for the future are the contaminated and unhygienic air, water, and food. About 50% of the world population depends on different agriculture products and above all the existing population growth trend on globe will increased by 10 billion by 2050 that rises the risk of food security, (Alavi & Rai, 2019). Population demands large amount of foodstuff, to encounter the worldwide demand of food the effective and sustainable strategies has to be implemented for high agricultural yield, (Amaraneni, 2018). To gain more annual yield of agricultural commodities, the pest and insect can be controlled efficiently by use of pesticides because global food supply affected by pests and insects, (Arshad, Hassan, & Naikoo, 2023). Pesticide-free fruits, vegetables, and crops are either a thing of a bygone era that mankind values greatly or are simply illusions of the present. These farming chemicals are adsorbed and absorbed by the surfaces of fruits, vegetables, leaves, and other parts of crops that causes them to disseminate across the ecosystem by entering through bio-chains. The majority of such insecticides and pesticides displays a strong link with human-health and the environment; some are even prohibited because of their potentially hazardous nature, (Bala, Sharma, & Wangoo, 2018; Beyene et al., 2017). Consuming these substances in whatsoever proportion via any of the route causes one's health at risk for conditions including gastrointestinal, dermatological, lung problems, cancer, neurological, and endocrine disruptions, among others and high exposure can even be fatal, (Bhagyaraj, Oluwfemi, & Kalarikkal, 2018; Bhatt, Vyas, & Modi, 2014). In order to mitigate this possible threat, the agro-chemicals incorporating pesticides, insecticides, herbicides, and fungicides monitoring is indispensable, (Chadha, Sharma, & Maiti, 2022), therefore the scientific community has developed the different kinds of techniques and devices which are simple, quick, reliable, requires less sample that are able to give on-site results, (Chandrakala, Aruna, & Angajala, 2022).

Figure 1. Graphical abstract

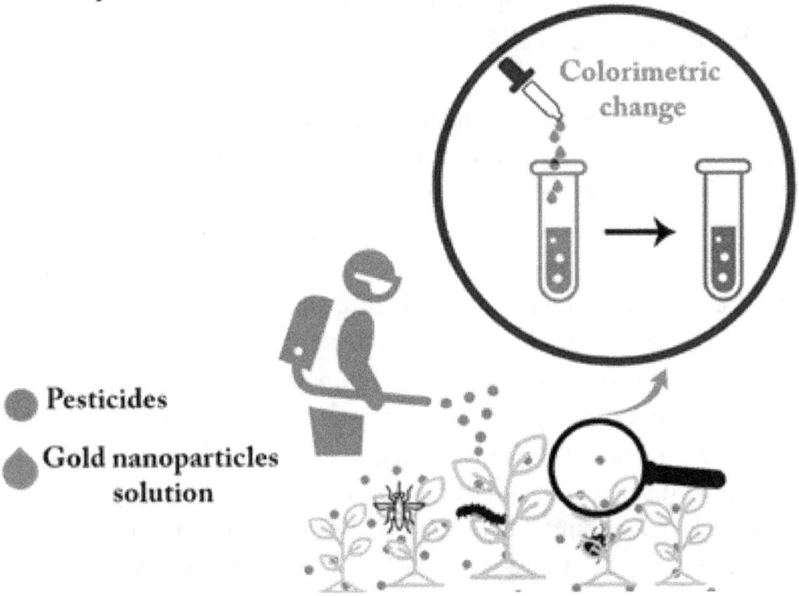

Metal-based nanomaterials have emerged as one of the most prominent and widely studied classes of nanomaterials due to their unique physicochemical properties, tunable functionalities, and broad application spectrum. These materials, which include metal nanoparticles, metal oxides, and metal-organic frameworks, exhibit enhanced catalytic, electronic, magnetic, and optical characteristics at the nanoscale. This chapter provides a comprehensive overview of metal-based nanomaterials, focusing on their synthesis methods, structural features, characterization techniques, and current and emerging applications in diverse sectors such as medicine, energy, environmental remediation, and electronics. It concludes with a discussion of challenges and future directions in this dynamic research field. Nanotechnology has revolutionized materials science by enabling the manipulation of materials at the atomic and molecular scales. Among the various classes of nanomaterials, metal-based nanomaterials (MBNMs) hold a central place due to their exceptional surface-area-to-volume ratio, quantum size effects, and superior reactivity. These materials are composed of metals or metal compounds and are engineered at dimensions typically below 100 nm. As particle size decreases, the properties of metals significantly change, giving rise to new phenomena that are not observed in their bulk counterparts.

CARBON-BASED NANOMATERIALS FOR PESTICIDE DETECTION: STRUCTURES, MECHANISMS, AND SENSING STRATEGIES

The increasing use of pesticides in agriculture to improve crop yield has raised significant concerns regarding their residual presence in food, soil, and water. Rapid, sensitive, and field-deployable detection methods are crucial for ensuring food safety and environmental sustainability. In this context, carbon-based nanomaterials (CBNs) have emerged as highly effective platforms for pesticide biosensing. Their outstanding electrical conductivity, large surface area, chemical tunability, and biocompatibility make them ideal candidates for developing advanced sensors capable of detecting trace levels of harmful pesticide compounds.

CBNs include a diverse range of nanostructures such as graphene and graphene oxide (2D), carbon nanotubes (1D), carbon quantum dots and fullerenes (0D), and 3D carbon architectures like mesoporous carbon and carbon aerogels. These materials enable multiple sensing mechanisms—electrochemical, optical, field-effect transistor (FET)-based, and more—tailored for detecting various classes of pesticides, including organophosphates, carbamates, pyrethroids, and neonicotinoids.

Roles of Carbon Nanomaterials in Pesticide Biosensing

Graphene and Graphene Oxide (GO)

Graphene's high surface area and excellent electrical conductivity make it a preferred material for electrochemical pesticide biosensors. Its surface can be functionalized with enzymes such as acetylcholinesterase (AChE), which is commonly used to detect organophosphate and carbamate pesticides by measuring their enzyme inhibition effects. Graphene oxide, rich in oxygenated groups, provides enhanced dispersibility in aqueous solutions and easy anchoring sites for antibodies, DNA probes, or aptamers. GO is also widely used in fluorescence-based pesticide sensors due to its quenching capabilities via Förster Resonance Energy Transfer (FRET).

Carbon Nanotubes (CNTs)

CNTs offer excellent mechanical strength, electrical conductivity, and surface area, making them effective in fabricating sensitive pesticide sensors. Functionalized CNTs can immobilize enzymes, aptamers, or molecularly imprinted polymers (MIPs), enabling specific recognition of pesticide molecules. Their 1D structure supports rapid electron transport, improving signal transduction in electrochemical sensing. CNT-based sensors have demonstrated detection of pesticides like malathion, chlorpyrifos, and paraquat at nanomolar or even lower concentrations.

Carbon Quantum Dots (CQDs)

CQDs exhibit size-tunable fluorescence, high photostability, and excellent water solubility. These properties make them ideal for optical pesticide sensors. The fluorescence of CQDs can be modulated upon interaction with pesticides, either through direct quenching or binding-mediated mechanisms. For example, a CQD-based "turn-off" sensor can detect the presence of pesticides like glyphosate by observing a measurable reduction in fluorescence intensity.

Fullerenes

Fullerenes like C_{60} have high electron affinity and redox activity, making them suitable for electrochemical biosensors targeting pesticides that act through redox reactions. Functionalized fullerenes can enhance the stability and sensitivity of enzymatic sensors for detecting pesticides that generate or consume reactive species.

Three-Dimensional Carbon Nanostructures

3D carbon materials such as mesoporous carbon and carbon nanofibers offer interconnected porosity and high surface areas, allowing the immobilization of multiple bioreceptors. These materials facilitate fast analyte diffusion and are being increasingly explored for developing portable pesticide biosensors, especially for real-time environmental monitoring.

Working Mechanisms of CBNs in Pesticide Biosensing

CBNs enhance pesticide detection through several critical mechanisms:

- Signal Amplification: Their high conductivity and large surface area enable strong signal amplification, essential for detecting low-concentration pesticide residues.
- Biomolecule Immobilization: Functional groups on CBNs allow stable and oriented attachment of enzymes, antibodies, and aptamers that specifically recognize pesticide molecules.
- Transduction Enhancement: CBNs improve electron and photon transfer efficiency, ensuring that biological interactions (e.g., enzyme inhibition or binding events) are quickly converted into measurable signals.

Types of CBN-Based Pesticide Biosensors

Electrochemical Biosensors

Electrochemical sensors are among the most developed platforms for pesticide detection. Graphene- or CNT-modified electrodes exhibit improved electron transfer and reduced background noise. AChE-based sensors can quantify organophosphates and carbamates by measuring current changes resulting from enzyme inhibition. CQDs have also been incorporated into electrochemical platforms to monitor redox-active pesticides.

Optical Biosensors

Optical biosensors detect pesticides through changes in fluorescence or absorbance. CQDs are used for "turn-on/turn-off" fluorescence detection, while GO-based FRET systems are commonly employed in aptamer or DNA probe-based sensors. These sensors are particularly useful for detecting neonicotinoids and glyphosate in food or water.

FET-Based Biosensors

Graphene and CNTs are frequently used as channels in field-effect transistor (FET) sensors due to their high charge-carrier mobility. Pesticide binding events on the functionalized surface of the nanomaterial cause changes in the local charge environment, which modulates the current across the FET. These systems have demonstrated real-time, label-free pesticide detection with ultralow detection limits.

Applications of Carbon-Based Nanomaterial Pesticide Sensors

The development of carbon-based nanomaterial (CBN) pesticide sensors has opened new frontiers in real-time, accurate, and on-site detection of toxic residues across various sectors. Due to their exceptional physicochemical properties—such as high surface area, excellent electrical conductivity, and tunable surface functionality—CBNs have found broad application in food safety, environmental monitoring, and precision agriculture.

Food Safety

Food contamination by pesticide residues poses a significant public health concern, particularly in fruits, vegetables, grains, and dairy products. Conventional detection techniques like gas chromatography and mass spectrometry, while accurate, require expensive equipment, complex sample preparation, and trained personnel. In contrast, CBN-based biosensors offer a more practical alternative for routine monitoring.

Electrochemical and colorimetric sensors incorporating materials such as graphene, graphene oxide (GO), carbon nanotubes (CNTs), and carbon quantum dots (CQDs) have demonstrated high sensitivity and rapid response times. For example, CQD-based fluorescent sensors can detect even trace amounts of pesticide residues in complex food matrices by exhibiting fluorescence quenching or enhancement upon interaction with specific analytes. Graphene-based field-effect transistor (FET) sensors and CNT-modified electrodes also enable label-free, low-cost detection with impressive selectivity. These portable devices are especially beneficial for on-site testing in food supply chains, retail markets, and inspection agencies, streamlining food safety assessment without the need for laboratory infrastructure.

Environmental Monitoring

The widespread and often unregulated use of pesticides has led to significant contamination of natural resources including soil, surface water, and groundwater. Monitoring environmental levels of pesticides is vital for ecological risk assessment and policy enforcement. Traditional testing methods often fall short due to their cost, logistical complexity, and the need for laboratory settings. CBN-based sensors offer a highly adaptable solution to these challenges. Their portability and robust performance make them suitable for in-field applications, even in remote or resource-limited settings. For instance, electrochemical sensors functionalized with graphene oxide or CNTs have been successfully used for detecting pesticides such as malathion, glyphosate, and paraquat in river water and agricultural runoff. Optical sensors based on CQDs can provide visible readouts for rapid screening. These technologies support real-time environmental monitoring and ensure regulatory compliance by enabling frequent, low-cost testing across multiple sites.

Agricultural Applications

In modern precision agriculture, real-time data is crucial for managing pesticide application effectively and sustainably. Overuse or misuse of pesticides not only harms the environment but also leads to increased resistance in pests. Integrating CBN-based sensors into smart farming systems can address these challenges. Wearable or embedded sensors utilizing materials like CNTs and graphene aerogels can continuously monitor pesticide residues on crop surfaces or in storage environments. These sensors can be incorporated into handheld devices, drones, or autonomous monitoring platforms to provide real-time data during harvesting, transportation, and storage. Their real-time feedback supports data-driven decisions for optimal pesticide application, reducing excess use and ensuring compliance with maximum residue limits (MRLs). Carbon-based nanomaterial pesticide sensors represent a transformative leap across food safety, environmental protection, and smart agriculture, offering scalable, sensitive, and sustainable solutions for residue monitoring.

CLASSIFICATION OF METAL-BASED NANOMATERIALS

Metal-based nanomaterials (MBNMs) encompass a diverse range of nanostructures that incorporate metallic elements in various chemical forms. These materials are central to nanotechnology due to their size-dependent properties and vast application potential. The classification of MBNMs is primarily based on their composition and structure. The major types include pure metal nanoparticles, metal oxide nanoparticles, metal chalcogenides (such as sulfides and selenides), and metal-organic frameworks (MOFs). Each class exhibits unique physicochemical characteristics and functionalities suited to specific technological and scientific applications, (Chavali & Nikolova, 2019; Cheng, Li, & Rao, 2021).

Metal Nanoparticles (MNPs)

Metal nanoparticles are composed of elemental metals and typically measure between 1–100 nanometers in size. Common examples include gold (Au), silver (Ag), platinum (Pt), copper (Cu), palladium (Pd), and nickel (Ni). These nanoparticles are especially valuable due to their high surface-area-to-volume ratio, quantum effects, and tunable surface properties. One of the most studied phenomena associated with noble metal nanoparticles, such as Au and Ag, is surface plasmon resonance (SPR). This refers to the collective oscillation of conduction electrons at the surface of the nanoparticles when excited by specific wavelengths of light, resulting in strong absorption and scattering. This property makes MNPs highly desirable in biomedical imaging, biosensing, and photothermal therapy, (Choudhury & Chowdhury, 2016; Chow, 2022).

Metal nanoparticles demonstrate remarkable catalytic efficiency in chemical reactions, including hydrogenation, oxidation, and environmental remediation processes. Their superior electrical conductivity and thermal stability also enable their integration into electronic circuits, conductive inks, and energy storage systems.

Metal Oxide Nanoparticles

Metal oxide nanoparticles are nanoscale materials composed of metal cations bonded to oxygen anions. Examples include titanium dioxide (TiO_2), zinc oxide (ZnO), iron oxide (Fe_2O_3 and Fe_3O_4), copper oxide (CuO), and cerium oxide (CeO_2). These materials are typically semiconducting and exhibit properties such as photocatalytic activity, magnetic behavior, and UV absorption. TiO_2 and ZnO are extensively used in photocatalysis, particularly for the degradation of organic pollutants in water and air, due to their high bandgap energies and stability under UV light. Iron oxides, especially magnetite (Fe_3O_4), exhibit magnetic properties that make them useful in magnetic resonance imaging (MRI), drug delivery, and magnetic separation technologies, (Cui & Yu, 2013; Das, Ray, & Raychaudhuri, 2013).

In the field of sensing, metal oxides like SnO_2 and WO_3 are employed in gas sensors owing to their ability to change electrical resistance in response to target gas molecules. Furthermore, metal oxides serve as active materials in lithium-ion batteries, solar cells, and anti-bacterial coatings, reinforcing their multifunctional significance.

Metal Sulfides and Selenides

Metal chalcogenides, particularly sulfides (e.g., CdS, ZnS, PbS) and selenides (e.g., CdSe, ZnSe), are another important class of metal-based nanomaterials. These compounds consist of metal atoms bonded with sulfur or selenium and are widely recognized for their semiconducting and photoluminescent properties. Due to their narrow and tunable bandgaps, metal sulfides and selenides are frequently employed in optoelectronic devices such as photodetectors, solar cells, and light-emitting diodes (LEDs). Cadmium-based nanocrystals, in particular, have been explored as quantum dots for applications in bioimaging, where their size-dependent fluorescence provides high-resolution cellular imaging. Despite their promising functionalities, concerns regarding the toxicity of cadmium and lead-based nanomaterials have prompted research into more environmentally benign alternatives like copper indium sulfide (CIS) and zinc sulfide (ZnS), (De, Mukhopadhyay, & Basak, 2020; Forrest & Thompson, 2007).

Metal-Organic Frameworks (MOFs)

Metal-organic frameworks are a relatively newer class of porous crystalline materials constructed by coordinating metal ions or metal clusters with organic linkers (ligands). MOFs are distinguished by their ultra-high surface areas, structural tunability, and well-defined pore sizes, which enable precise molecular recognition and storage. The metal centers in MOFs can range from transition metals like zinc, copper, and iron to lanthanides, and the organic ligands are typically carboxylates, imidazolates, or phosphonates. This hybrid architecture offers MOFs exceptional versatility, allowing for the incorporation of functional groups or catalytic sites directly into the framework. MOFs are being explored for a variety of applications, including gas storage (e.g., hydrogen, methane, CO_2), catalysis, drug delivery, and chemical sensing. Some MOFs also exhibit luminescence or electrical conductivity, expanding their utility into emerging areas such as electrochemical energy storage and optoelectronics, (Frens, 1972; Fu, Wu, & Zhang, 2021).

SYNTHESIS OF METAL-BASED NANOMATERIALS

The functionality and performance of metal-based nanomaterials (MBNMs) are largely governed by their size, shape, crystallinity, and surface properties all of which are dictated by the synthesis route and processing conditions. Broadly, the synthesis of MBNMs can be classified into two fundamental approaches: top-down and bottom-up. Top-down methods reduce bulk materials to the nanoscale, whereas bottom-up techniques build nanostructures atom-by-atom or molecule-by-molecule. Each strategy offers unique advantages and is selected based on the desired application, scalability, cost, and environmental impact, (Cheng, Li, & Rao, 2021; Gentil, Ruffino, & Grimaldi, 2016).

Physical Methods

Physical or top-down approaches utilize mechanical or thermal energy to convert bulk metals into nanoscale particles. One common technique is ball milling, where mechanical force grinds and fractures bulk metals into nanosized particles. While cost-effective and scalable, it often results in broad particle size distribution and surface defects. Laser ablation, another physical method, involves focusing high-energy laser pulses onto a solid metal target submerged in a liquid medium. This causes localized heating, vaporization, and condensation of the metal into nanoparticles. Evaporation-condensation techniques use controlled heating and cooling to vaporize metals and condense them into nanoparticles within an inert atmosphere. These methods are often solvent-free and produce relatively pure nanomaterials but may lack precise control over particle shape and uniformity, (Hassan & El Nemr, 2020; Hu, Zhang, & Liu, 2015).

Chemical Methods

Chemical or bottom-up synthesis routes are among the most versatile and widely adopted for fabricating MBNMs with controlled morphology and surface chemistry. In chemical reduction, metal ions from salts (e.g., silver nitrate or chloroauric acid) are reduced using agents such as sodium borohydride ($NaBH_4$) or trisodium citrate. This method enables the fine-tuning of particle size, shape, and stability by adjusting parameters like pH, temperature, and concentration. The sol-gel process involves the hydrolysis and condensation of metal alkoxides or inorganic salts to form a colloidal suspension or 'sol.' With time, this transforms into a gel-like network, which is then dried and calcined to obtain oxide nanoparticles with high purity and homogeneity. Co-precipitation entails the simultaneous precipitation of multiple metal ions in a controlled pH environment, facilitating the synthesis of doped or composite nanostructures. This technique is simple, cost-effective, and highly scalable, (Huston, DeBella, & DiCarmine, 2021; Jang & Jang, 2023).

Biological or Green Synthesis

In response to environmental and health concerns associated with conventional methods, green synthesis has emerged as an eco-friendly alternative. Using biological entities like plant extracts, bacteria, or fungi, metal ions are reduced and stabilized through natural biomolecules such as flavonoids, proteins, and polysaccharides. Green methods are non-toxic, sustainable, and often energy-efficient, though challenges remain in achieving reproducibility and uniformity, (Kalantari, Fahimi-Kashani, & Hormozi-Nezhada, 2021; Kelly, Coronado, & Liz, 2003).

Table 1. Methods to prepare MNPs and their features

Method	Type	Principle	Advantages	Limitations	Examples
Ball Milling	Top-down	Mechanical grinding of bulk metals	Simple, scalable	Broad size distribution, possible contamination	Fe, Cu, Al nanoparticles
Laser Ablation	Top-down	Laser-induced vaporization in liquid medium	Pure product, solvent-free	Expensive equipment, low yield	Au, Ag, Pt nanoparticles
Evaporation-Condensation	Top-down	Vaporization followed by condensation	Clean particles	High energy requirement	Zn, Cd nanoparticles
Chemical Reduction	Bottom-up	Reduction of metal salts using reducing agents	Control over size and shape	Toxic reagents, requires stabilizers	Au, Ag, Pt, Cu nanoparticles
Sol-Gel	Bottom-up	Hydrolysis and condensation of precursors	High purity, homogeneity	Requires calcination	TiO_2, SiO_2, ZnO nanoparticles
Co-precipitation	Bottom-up	Precipitation of metal ions under pH control	Economical, suitable for large scale	Poor size control, possible agglomeration	Fe_3O_4, MnO_2, ZnO nanoparticles
Green Synthesis	Bottom-up	Bio-reduction using plants/ microbes	Eco-friendly, biocompatible	Poor reproducibility, scalability challenges	Ag, Au, Fe nanoparticles

CHARACTERIZATION TECHNIQUES FOR METAL-BASED NANOPARTICLES

Characterization of MBNMs is vital to confirm successful synthesis and evaluate their physical, chemical, and functional attributes. Transmission Electron Microscopy (TEM) and Scanning Electron Microscopy (SEM) are widely used to visualize nanoparticle morphology, size distribution, and surface texture at high resolution, (Keyur, Vyas, & Modi, 2016; Kitagawa, 2014).

X-ray Diffraction (XRD) helps determine the crystalline structure, phase purity, and average crystallite size of the nanomaterials. Fourier-Transform Infrared Spectroscopy (FTIR) is employed to identify surface functional groups, especially after surface modifications or green synthesis.

UV-Visible Spectroscopy is instrumental in analyzing the optical behavior of metal nanoparticles, particularly in detecting surface plasmon resonance bands in noble metals like gold and silver. Dynamic Light Scattering (DLS) and Zeta Potential Analysis provide insights into the hydrodynamic size, dispersion quality, and colloidal stability of nanoparticle suspensions. These synthesis and characterization techniques form the foundation for designing application-specific metal-based nanomaterials with predictable performance and tailored functionalities, (Kongor, Vyas, & Jain, 2016).

Table 2. Characterization techniques for MBNMs

Technique	Measured Property	Information Provided	Common Use
Transmission Electron Microscopy (TEM)	Morphology, particle size	High-resolution internal structure and shape	Size and morphology analysis
Scanning Electron Microscopy (SEM)	Surface morphology	Surface topology and texture	Shape and surface features
X-ray Diffraction (XRD)	Crystallinity, phase structure	Crystal phase identification, grain size estimation	Phase confirmation
Fourier-Transform Infrared Spectroscopy (FTIR)	Functional groups	Identification of surface-bound molecules or coatings	Surface chemistry analysis
UV-Visible Spectroscopy	Optical properties	Absorption peak (e.g., SPR), band gap estimation	Optical behavior, especially Au/Ag
Dynamic Light Scattering (DLS)	Hydrodynamic diameter	Average particle size in suspension	Dispersion and stability analysis
Zeta Potential Analysis	Surface charge	Particle stability and tendency to aggregate	Colloidal stability

PROPERTIES OF METAL-BASED NANOMATERIALS

Metal-based nanomaterials (MBNMs) possess a wide array of physical and chemical properties that distinguish them from their bulk counterparts. These unique properties arise primarily due to quantum size effects, surface atom predominance, and enhanced surface energy. Understanding these characteristics is critical in optimizing their performance across various application domains, (Kuma, Pavithra, & Naushad, 2019; Li, Zhang, & Wang, 2023).

Key Properties of Metal-Based Nanomaterials

Optical Properties

One of the most notable features of noble metal nanoparticles, particularly gold (Au) and silver (Ag), is their surface plasmon resonance (SPR). SPR occurs when conduction electrons on the nanoparticle surface oscillate collectively in resonance with incident light. This leads to strong absorption and scattering at specific wavelengths, resulting in vibrant color changes. These optical responses are highly sensitive to particle size, shape, and dielectric environment, making MBNMs indispensable in biosensing, imaging, and optoelectronic applications, (Lue, 2007).

Catalytic Activity

MBNMs exhibit remarkable catalytic activity due to their high surface-to-volume ratio, which increases the number of active sites available for chemical interactions. This makes them particularly effective in facilitating redox reactions, including oxidation, reduction, and hydrogenation. Platinum (Pt), palladium (Pd), and gold (Au) nanoparticles, for instance, are frequently used as catalysts in industrial and environmental processes, including fuel cells and pollutant degradation, (Marczylo, Jacobs, & Grant, 2016).

Electrical Conductivity

Several metal nanoparticles, such as copper (Cu), silver (Ag), and gold (Au), display excellent electrical conductivity. Their small size and tunable electronic structures enable enhanced charge transport, positioning them as ideal components in sensors, conductive inks, and nanoelectronics circuits. Their integration into wearable electronics and printable technologies has opened new frontiers in flexible and miniaturized devices, (Li, Zhang, & Wang, 2023).

Magnetic Behaviour

Certain metal oxides, notably magnetite (Fe_3O_4) and maghemite (γ-Fe_2O_3), exhibit superparamagnetism at the nanoscale. Unlike ferromagnetic materials, superparamagnetic nanoparticles do not retain magnetization in the absence of an external field. This property is highly valuable for biomedical applications such as magnetic resonance imaging (MRI), magnetic hyperthermia for cancer treatment, and targeted drug delivery, where magnetic guidance ensures site-specific action, (Chavali & Nikolova, 2019).

Table 3. Comparison of MBNM types and common applications

Nanomaterial Type	Typical Composition	Key Properties	Common Applications
Metal Nanoparticles	Au, Ag, Cu, Pt	Plasmonic, catalytic, conductive	Sensors, catalysis, imaging, drug delivery
Metal Oxide Nanoparticles	TiO_2, ZnO, Fe_3O_4, CuO	Photocatalytic, magnetic, semiconducting	Water purification, batteries, gas sensing
Metal Sulfides/ Selenides	CdS, ZnS, PbS, CdSe	Semiconducting, optoelectronic	Solar cells, LEDs, quantum dots
Metal-Organic Frameworks	Cu-MOFs, Fe-MOFs, Zr-MOFs	Porous, high surface area, functionalizable	Gas storage, drug delivery, chemical separation

APPLICATIONS OF METAL-BASED NANOMATERIALS

The diverse and tenable properties of MBNMs have led to their deployment in various sectors, including healthcare, energy, environment, and electronics, (Li, Zhang, & Wang, 2023).

Biomedical Applications

Targeted Drug Delivery

Gold and iron oxide nanoparticles are engineered to deliver therapeutic agents specifically to diseased sites, minimizing side effects and enhancing efficacy, (Mauter & Elimelech, 2008; Mekuye & Abera, 2023).

Imaging and Diagnostics

MBNMs serve as contrast agents in imaging techniques like MRI (Fe_3O_4) and computed tomography (CT) (AuNPs), improving resolution and diagnostic precision.

Antimicrobial Action

Silver nanoparticles are well known for their broad-spectrum antimicrobial activity, capable of neutralizing bacteria, fungi, and even viruses, (Mie, 1908).

Cancer Therapy

Photothermal and photodynamic therapies utilize the SPR effect of Au nanoparticles to generate heat or reactive oxygen species for localized destruction of tumour cells with minimal impact on healthy tissues, (Mishra, Darjee, & Bhatt, 2015).

Environmental Applications

Water Treatment

MBNMs are employed in the removal of heavy metals, organic contaminants, and pathogens from water systems through adsorption, photocatalysis, and redox mechanisms.

Environmental Sensing

Metal oxide nanoparticles like ZnO, SnO_2, and TiO_2 are highly sensitive to environmental gases such as CO, NO_2, and H_2S, enabling real-time air quality monitoring and early detection of hazardous leaks, (Modan & Plăia um, 2020).

Energy Storage and Conversion

Batteries

Metal oxides such as MnO_2 and Co_3O_4 are used in lithium-ion and sodium-ion batteries due to their favourable redox potentials and cycling stability, (Morbioli, Mazzoni, & Furlan, 2017).

Fuel Cells

Noble metal nanoparticles, particularly platinum, are pivotal in enhancing the electrocatalytic activity of fuel cell electrodes, (Mourdikoudis, Pallares, & Thanh, 2018).

Photocatalysis

Semiconducting oxides like TiO_2 and ZnO are activated under light to degrade organic pollutants or split water molecules for hydrogen generation, (Narederan, Meyyanathan, & Babu, 2020).

Electronics and Optoelectronics

Flexible Electronics

Conductive metal nanoparticles are integrated into inks and films used in bendable circuits and wearable devices, (Nicolopoulou-Stamati, Maipas, & Kitampasi, 2016).

Photovoltaics

Plasmonic metal nanoparticles enhance light absorption in solar cells, improving energy conversion efficiency, (Pomal, Vyas, & Jain, 2021).

Quantum Dots

Metal sulphide and selenide nanoparticles, such as CdS and CdSe, function as quantum dots in LED displays, offering tenable emission and high brightness, (Pomerantseva, Dunn, & Meng, 2019).

TOXICITY AND ENVIRONMENTAL CONCERNS

Despite their promising capabilities, metal-based nanomaterials raise important questions about environmental safety and human health. Their nanoscale size allows for easy penetration of biological membranes, potentially leading to oxidative stress, DNA damage, cell apoptosis, or inflammatory responses. Silver, cadmium, and lead-based nanoparticles, in particular, have shown cytotoxic effects in various studies. The persistence and bioaccumulation of MBNMs in ecosystems is another pressing issue. Their uncontrolled release during manufacturing, usage, or disposal may harm aquatic organisms and disrupt ecological balance. Therefore, comprehensive toxicological evaluations—including in vitro, in vivo, and computational assessments—are crucial for regulatory approval and safe commercialization, (Ramanathan, Oluwafemi, & Kalarikkal, 2014; Rao & Biswas, 2009).

On this hand, the nanotechnology is the frequently used for their ultrafine particles size ranging 1-100 nanometers, (Rujiralai, Thongyoy, & Kaewkan, 2020). Within the list of nanoparticles, metal nanoparticles (MNPs) are investigated due to their electronic, optical and photo-thermal properties, (Sabela, Balme, & Bisetty, 2017). From machinery to medications, MNPs possess a wide range of applications. Specific features of nano-gold enclosing facile functionalization, size-related spectral characteristics, high molar extinction coefficients and unique Surface Plasmon Resonance (SPR), which are not observable at macroscopic or atomic scales, (Sengul & Asmatulu, 2020). The current analytical techniques to determine the pesticides are giving sensitive, selective, effective and satisfactorily results but these involves time consuming sample preparations, sophisticated instruments, and expert operators, (Shah, Zhu, & Li, 2013). Significantly, it is need of time to detect these chemicals on-site and rapidly. Recently, functionalized gold nanoparticles (AuNPs) has demonstrated the great success in the sensing of perilous pesticides, (Shamraiz, Riaz, & Badshah, 2016). This mini-review highlights recent research, covers the brief study of sensitive and selective detection of detrimental pesticides in different reaction conditions using functionalized AuNPs.

Figure 2. Chemical structures of targeted pesticides

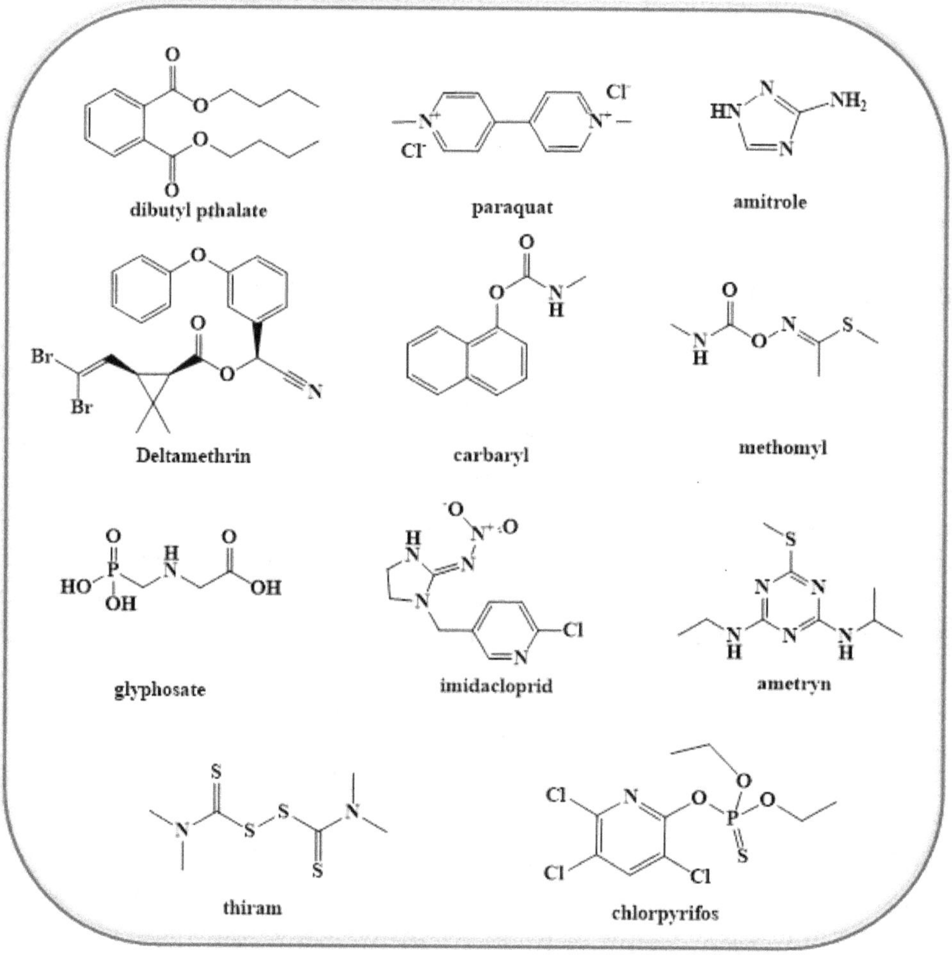

SYNTHESIS AND CHARACTERIZATION OF GOLD NANOPARTICLES:

In a broadways, nanoparticles can be synthesized by two ways; one is top-down approach and another is bottom-up approach as mentioned earlier in the chapter. In physical method, i.e. top-down approach, the bulk (gold) matter is converted to tine sized nano-form (Sharma, Kumar, & Kumar, 2019), while in chemical approach, viz. bottom-up technique, the ionic dimensions of particles of (gold) metal are transformed to bigger nano-structure, (Shuai, Wu, & Chen, 2023). For the large scale production top-down approach is inappropriate and slow whereas bottom-up approach is simple and economical, (Singh, Kumar, & Sharma, 2022).

The current research on chemical method of producing gold nanoparticles are indeed the improvised work of Turkevich who was firstly synthesized the AuNPs using aqueous solution of chloroauric acid ($HAuCl_4$) and tri-sodium citrate as reducing and capping agent, (Singh, Singh, & Srivastava, 2023). The Turkevich method was further modified by Frens (1972), other method was Burst-Schiffrin which involved sodium borohydride ($NaBH_4$) as reducing agent and tetra-octylammonium bromide as capping agent, (Tseng et al., 2020). In modern research numerous other reducing and capping agents are employed to achieve application oriented well stabilized AuNPs, (Tu, Tian, & Li, 2019). The synthesis process involve very simple laboratory procedure of redox reaction in which Au^{3+} or Au^{1+} ions gain the electrons from the reducing agents and converts to Au^0 form and following that nucleation and growth of MNPs occurs in which led to the final product i.e., functionalized AuNPs, (Turkevish, Stevenson, & Hillier, 1951).

Figure 3. Graphical illustration of synthesis of functionalized gold nanoparticles

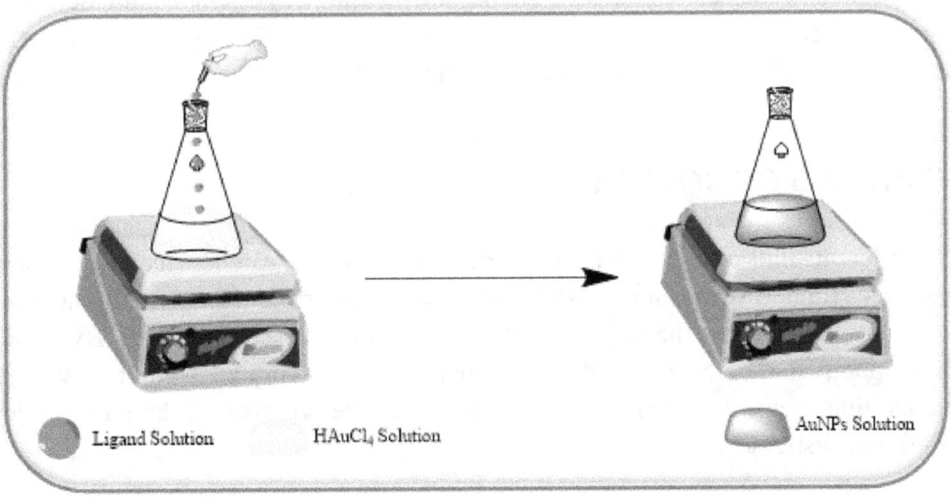

The bare gold nanoparticles, do not meet the criterion that could be satisfied in the colorimetric event of any analytes, but the modified or functionalized gold nanoparticles offers the aggregation of AuNPs in the presence of analyte that furnish the visual variation of solution and avail surface transformation. This is the case to be considered when nano-chemists develop these kind of sensors.

Various parameters regarding functionalized AuNPs, including size, shape, stability, morphology are to be optimized and depends upon certain important factors such as pH of solution, temperature, potential of reductant, stabilizing power of capping agent, and reaction kinetics etc., (Umapathi, Ghoreishian, & Ranjith, 2021). As produced AuNPs then to be characterized by various analytical tools in particular Transmission Electron Microscopy (TEM), Scanning Electron Microscopy (SEM), Dynamic Light Scattering (DLS), X-Ray Diffraction (XRD), Zeta potential, UV-Visible spectroscopy, Energy Dispersive X-Ray (EDX), and others, (Vajtai, 2013; Wang, Ma, & Wnag, 2020), from which the size, shape, size distribution, elemental analysis, SPR band, aggregation, surface charge are to be determined, (Yan, Li, & Su, 2018).

SURFACE PLASMON RESONANCE OF METAL NANOPARTICLES

Metal Nanoparticles manifests the signature property called Localized Surface Plasmon Resonance (LSPR) or in broad sense Surface Plasmon Resonance (SPR). The interaction of light and the metal nanoparticles, results into the collective and coherent oscillation of conduction electrons of MNPs is known as SPR, (Yan, Zhang, & Wang, 2021; Yang & Liu, 2015). The SPR and SPR frequency depends upon the various factors such as; metal, size, shape, surrounding environment and dielectric nature of surrounding which results in SPR shift in UV-Visible spectrophotometry and gives colorimetric changes, (Zhang, Wang, & Li, 2020).

FEATURES OF COLORIMETRIC ASSAYS

Different kinds of optical sensors are designed by the researchers mainly, Surface Enhanced Raman Scattering (SERS), fluorescence, colorimetric, chemiluminescent and more to recognize the target analytes, (Zhu, Wang, & Zhang, 2022). By virtue of simplicity and convenience the colorimetric assay is very powerful tool for the recognition event, in which AuNPs aggregates in the presence of the guest molecules, i.e. pesticides.

Figure 4. Graphical illustration of colorimetric assay of pesticide using functionalized gold nanoparticles

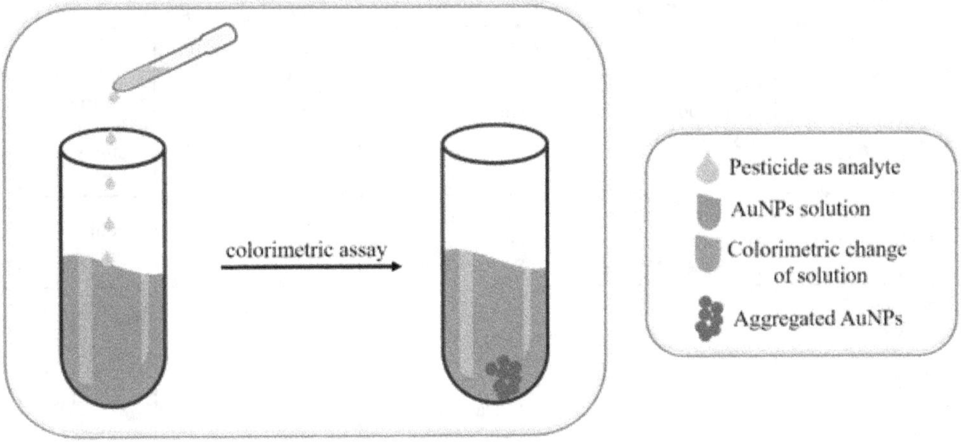

Without using any kind of advanced instruments and complex methods, merely a visual detection is possible. On top of that, more sophisticated colorimetric instruments have recently been designed by investigators; these devices are portable and rapidly detect perilous materials on site. These devices are based on papers, tubes, glass, plates, and other materials and are furnished with colorimetric reagents; instances of paper-based instruments are dipsticks and Enzyme base Immunosorbent Assay (ELISA) and two-dimensional microfluidic paper-based analytical devices (2D-μPADs). In the present paper similar kind of latest studies are tabulated (including important parameters) which magnifies the event of colorimetry based on functionalized gold nano-sensors (**Table 4**).

Table 4. Sensing of pesticides by different functionalized gold nanoparticles

Pesticide	Functionalized AuNPs used	Application on	Detection Limit	Linear Range	Visual Change	
Paraquat (PQ)	3-mercapto-1-propanesulfonate (AuNPs-3MPS)	Grape and Apple juice and Water	0.1 mg/L	0.001-0.5 ppm	red to blue-grey	
Deltamethrin (DEL)	2-mercapto-6-Nitrobenzothiazole gold nanoparticles (MNBT-AuNPs) μPADs	Apple, Mendarin orange, Spinach, Tomato and Cucumber	0.584 mg/L	0.001-0.5 mg/L	Red to purple	
Dibutyl phthalate (DBP)	arginine functionalized gold nanoparticles (ARG-AuNPs)	Baijiu samples	0.05 mg/L	0.0-2.8 mg/L	Red to blue	

continued on following page

Table 4. Continued

Pesticide	Functionalized AuNPs used	Application on	Detection Limit	Linear Range	Visual Change
Carbamate Pesticides a. Carbaryl b. Methomyl	L-cysteine-modified gold nanoparticles (Cys-AuNPs)	Chilis	a. 7.15 mg/L b. 3.49 mg/L	a. 10-114 mg/L b. 7.0 - 43 mg/L	a. Red to purple b. Red to blue-grey
Glyphosate (Gly)	cysteamine-modified gold nanoparticles (AuNPs-Cys)	Corn leaf, Apple peel, and Spinach leaf	0.026 mg/L	0.001-1000 mg/L	Red to blue/ purple
Triazole fungicides a. Amitrole (AMT), b. Prothioconazole (PTC), c. Ropiconazole (PPC), d. Tebuconazole (TBC)	Citrate-modified gold nanoparticles (AuNPs)	Wheat flour and Cucumber samples	0.35 mg/L	0.1 - 0.55 mg/L	a. red to Purple b. Red to blue c. Red to purple d. Red to purple
Ametryn	citrate-stabilized AuNP	Water	0.15 μg/L	0.0 - 80.0 μg/L	Wine red to grey
Imidacloprid (IMD)	Asparagine-modified gold nanoparticles (Asp-AuNPs)	Tap water, Drinking water and Apple	1.09 μM	10 - 1000 μM	Wine red to purple-blue.
Dithiocarbamate a. Ziram, b. Thiram c. Zineb	sulfydryl-mediated growth of gold nanobipyramids (AuNBPs)	Apple and Black tea	a. 17 nM b. 17 nM c. 18 nM	-	Multicol-our sensor
Chlorpyrifos (CPP)	γ-Cyclodextrin capped gold nanoparticles (γ-CD-AuNPs)	Aqueous Solution, Apple, and Tomato	10 ppb	-	Red to grey

CONCLUSION

This review underscores the emerging significance of functionalized gold nanoparticles in the development of effective, sensitive, and easy-to-use colorimetric sensors for pesticide detection. By leveraging the unique surface plasmon resonance (SPR) behavior of AuNPs, researchers have been able to create sensors that allow for rapid visual detection through distinct color changes upon interaction with pesticide molecules. These platforms offer remarkable advantages such as portability, affordability, and user-friendliness, which are critical for field-based and point-of-care applications. However, challenges remain, particularly in achieving reproducible synthesis, enhancing long-term stability, and improving the sensitivity of sensors in complex real-world samples. Future efforts should aim at refining functionalization strategies, integrating AuNP-based sensors with smartphone or microfluidic devices, and developing standardized protocols for large-scale deployment. With continued innovation, AuNP-based colorimetric assays hold great promise for transforming pesticide monitoring and enhancing global food and environmental safety.

REFERENCES

Alavi, M., & Rai, M. (2019). Recent advances in antibacterial applications of metal nanoparticles (MNPs) and metal nanocomposites (MNCs) against multidrug-resistant (MDR) bacteria. *Expert Review of Anti-Infective Therapy*, *17*(6), 419–428. DOI: 10.1080/14787210.2019.1614914 PMID: 31046483

Amaraneni, S. R. (2018). Potential impact of pesticides on environment and human health. *Chemistry International*, *40*(3), 46–48. DOI: 10.1515/ci-2018-0329

Arshad, F., Hassan, I. U., & Naikoo, G. A. (2023). Characterization of nanomaterials. In *Nanotechnology in herbal medicine* (pp. 37–61). Elsevier. DOI: 10.1016/B978-0-323-99527-6.00005-7

Bala, R., Sharma, R. K., & Wangoo, N. (2018). Ultra-sensitive detection of malathion using quantum dots-polymer based fluorescence aptasensor. *Biosensors & Bioelectronics*, *104*, 45–49. DOI: 10.1016/j.bios.2017.12.034 PMID: 29306032

Beyene, H. D., Werkneh, A. A., Bezabh, H. K., & Ambaye, T. G. (2017). Synthesis paradigm and applications of silver nanoparticles (AgNPs), a review. *Sustainable Materials and Technologies*, *13*, 18–23. DOI: 10.1016/j.susmat.2017.08.001

Bhagyaraj, S., Oluwafemi, O. S., & Kalarikkal, N. (Eds.). (2018). *Characterization of nanomaterials: Advances and key technologies*. Woodhead Publishing.

Bhatt, K. D., Vyas, D. J., & Modi, K. M. (2014). Highly stable water dispersible calix[4]pyrrole octa-hydrazide protected gold nanoparticles as colorimetric and fluorometric chemosensors for selective signaling of Co(II) ions. *Spectrochimica Acta. Part A: Molecular and Biomolecular Spectroscopy*, *121*, 94–100. DOI: 10.1016/j.saa.2013.10.076 PMID: 24231744

Chadha, R., Sharma, R., & Maiti, N. (2022). γ-Cyclodextrin capped silver and gold nanoparticles as colorimetric and Raman sensor for detecting traces of pesticide "Chlorpyrifos" in fruits and vegetables. *Colloids and Surfaces. A, Physicochemical and Engineering Aspects*, *641*, 128558. DOI: 10.1016/j.colsurfa.2022.128558

Chandrakala, V., Aruna, V., & Angajala, G. (2022). Review on metal nanoparticles as nanocarriers: Current challenges and perspectives in drug delivery systems. *Emergent Materials*, *5*(6), 1593–1615. DOI: 10.1007/s42247-021-00335-x PMID: 35005431

Chavali, M. S., & Nikolova, M. P. (2019). Metal oxide nanoparticles and their applications in nanotechnology. *SN Applied Sciences*, *1*(6), 607. DOI: 10.1007/s42452-019-0592-3

Cheng, Z., Li, M., & Rao, J. (2021). Nanomaterials for cancer therapy: Current progress and perspectives. *Journal of Hematology & Oncology, 14*(1), 85. DOI: 10.1186/s13045-021-01096-0 PMID: 34059100

Choudhury, S. A., & Chowdhury, M. H. (2016). Use of plasmonic metal nanoparticles to increase the light absorption efficiency of thin-film solar cells. In *2016 IEEE International Conference on Sustainable Energy Technologies (ICSET)* (pp. 268–272). IEEE. DOI: 10.1109/ICSET.2016.7811781

Chow, J. C. (Ed.). (2022). *Application of nanomaterials in biomedical imaging and cancer therapy*. MDPI., DOI: 10.3390/books978-3-0365-3514-2

Cui, C.-H., & Yu, S.-H. (2013). Engineering interface and surface of noble metal nanoparticle nanotubes toward enhanced catalytic activity for fuel cell applications. *Accounts of Chemical Research, 46*(7), 1427–1437. DOI: 10.1021/ar300254b PMID: 23425040

Das, S., Ray, S., & Raychaudhuri, A. (2013). Nanomaterials for biomedical applications. *Frontiers in Life Science, 7*(3–4), 90–98. DOI: 10.1080/21553769.2013.869510

De, J., Mukhopadhyay, B., & Basak, P. (2020). Luminescent conductive columnar π-gelators for Fe(II) sensing and bio-imaging applications. *The Journal of Physical Chemistry B, 124*(45), 10257–10265. DOI: 10.1021/acs.jpcb.0c07052 PMID: 33136408

Forrest, S. R., & Thompson, M. E. (2007). Introduction: Organic electronics and optoelectronics. *Chemical Reviews, 107*(4), 923–925. DOI: 10.1021/cr0501590

Frens, G. (1972). Particle size and sol stability in metal colloids. *Kolloid-Zeitschrift und Zeitschrift für Polymere, 250*(7), 736–741. DOI: 10.1007/BF01498565

Fu, B., Wu, W., & Zhang, Z. (2021). Recent progress on metal-based nanomaterials: Fabrications, optical properties, and applications in ultrafast photonics. *Advanced Functional Materials, 31*(49), 2107363. DOI: 10.1002/adfm.202107363

Gentile, A., Ruffino, F., & Grimaldi, M. G. (2016). Complex-morphology metal-based nanostructures: Fabrication, characterization, and applications. *Nanomaterials (Basel, Switzerland), 6*(6), 110. DOI: 10.3390/nano6060110 PMID: 28335236

Hassaan, M. A., & El Nemr, A. (2020). Pesticides pollution: Classifications, human health impact, extraction and treatment techniques. *Egyptian Journal of Aquatic Research, 46*(3), 207–220. DOI: 10.1016/j.ejar.2020.08.007

Hu, H., Zhang, J., & Liu, Y. (2015). Metal-free graphene-based catalyst—Insight into the catalytic activity: A short review. *Applied Catalysis A, General*, *492*, 1–9. DOI: 10.1016/j.apcata.2014.11.041

Huston, M., DeBella, M., & DiCarmine, M. (2021). Green synthesis of nanomaterials. *Nanomaterials (Basel, Switzerland)*, *11*(8), 2130. DOI: 10.3390/nano11082130 PMID: 34443960

Jang, E., & Jang, H. (2023). Quantum dot light-emitting diodes. *Chemical Reviews*, *123*(8), 4663–4692. DOI: 10.1021/acs.chemrev.2c00695 PMID: 36795794

Kalantari, K., Fahimi-Kashani, N., & Hormozi-Nezhada, M. R. (2021). Development of a colorimetric sensor array based on monometallic and bimetallic nanoparticles for discrimination of triazole fungicides. *Analytical and Bioanalytical Chemistry*, *413*(7), 1907–1917. PMID: 33855603

Kelly, K. L., Coronado, E., & Liz, L. M. (2003). The optical properties of metal nanoparticles: The influence of size, shape, and dielectric environment. *The Journal of Physical Chemistry B*, *107*(3), 668–677. DOI: 10.1021/jp026731y

Keyur, D. B., Vyas, D. J., & Modi, K. M. (2016). Emerging trends based on calix protected metal nanoparticles. In *Advanced Materials* (Vol. 1, pp. 145–148). TechConnect Briefs.

Kitagawa, S. (2014). Metal–organic frameworks (MOFs). *Chemical Society Reviews*, *43*(16), 5415–5418. DOI: 10.1039/C4CS90059F PMID: 25011480

Kongor, A. R., Vyas, V. K., & Jain, V. K. (2016). Calix-based nanoparticles: A review. *Topics in Current Chemistry*, *374*(3), 28. DOI: 10.1007/s41061-016-0029-z PMID: 27573268

Li, C., Zhang, Y., & Wang, J. (2023). Recent progress in metal–organic frameworks (MOFs) for electrocatalysis. *Industrial Chemistry & Materials*, *1*(1), 9–38. DOI: 10.1039/D2IM00063F

Lue, J. T. (2007). Physical properties of nanomaterials. In *Encyclopedia of nanoscience and nanotechnology* (Vol. 10, pp. 1–46). American Scientific Publishers.

Marczylo, E. L., Jacobs, M. N., & Gant, T. W. (2016). Environmentally induced epigenetic toxicity: Potential public health concerns. *Critical Reviews in Toxicology*, *2016*(8), 676–700. DOI: 10.1080/10408444.2016.1175417 PMID: 27278298

Mauter, M. S., & Elimelech, M. (2008). Environmental applications of carbon-based nanomaterials. *Environmental Science & Technology*, *42*(16), 5843–5859. DOI: 10.1021/es8006904 PMID: 18767635

Mekuye, B., & Abera, B. (2023). Nanomaterials: An overview of synthesis, classification, characterization, and applications. *Nano Select*, *4*(8), 486–501. DOI: 10.1002/nano.202300038

Mie, G. (1908). Beiträge zur Optik trüber Medien, speziell kolloidaler Metalllösungen. *Annalen der Physik*, *330*(3), 377–445. DOI: 10.1002/andp.19083300302

Mishra, D. R., Darjee, S. M., Bhatt, K. D., Modi, K. M., & Jain, V. K. (2015). Calix protected gold nanobeacon as turn-off fluorescent sensor for phenylalanine. *Journal of Inclusion Phenomena and Macrocyclic Chemistry*, *82*(3–4), 425–436. DOI: 10.1007/s10847-015-0509-8

Modan, E. M., & Plăia u, A. G. (2020). Advantages and disadvantages of chemical methods in the elaboration of nanomaterials. *The Annals of "Dunarea de Jos" University of Galati. Fascicle IX. Metallurgy and Materials Science*, *43*(1), 53–60. DOI: 10.35219/mms.2020.1.06

Morbioli, G. G., Mazzoni, T. S., & Furlan, A. P. (2017). Technical aspects and challenges of colorimetric detection with microfluidic paper-based analytical devices (μPADs)—A review. *Analytica Chimica Acta*, *970*, 1–22. DOI: 10.1016/j.aca.2017.03.037 PMID: 28433054

Mourdikoudis, S., Pallares, R. M., & Thanh, N. T. K. (2018). Characterization techniques for nanoparticles: Comparison and complementarity upon studying nanoparticle properties. *Nanoscale*, *10*(27), 12871–12934. DOI: 10.1039/C8NR02278J PMID: 29926865

Narenderan, S. T., Meyyanathan, S. N., & Babu, B. (2020). Review of pesticide residue analysis in fruits and vegetables. Pre-treatment, extraction and detection techniques. *Food Research International*, *133*, 109141. DOI: 10.1016/j.foodres.2020.109141 PMID: 32466907

Nicolopoulou-Stamati, P., Maipas, S., Kotampasi, C., Stamatis, P., & Hens, L. (2016). Chemical pesticides and human health: The urgent need for a new concept in agriculture. *Frontiers in Public Health*, *4*, 148. DOI: 10.3389/fpubh.2016.00148 PMID: 27486573

Pomal, N. C., Vyas, V. K., & Jain, V. K. (2021). Functionalized silver nanoparticles as colorimetric and fluorimetric sensor for environmentally toxic mercury ions: An overview. *Journal of Fluorescence*, *31*(3), 635–649. DOI: 10.1007/s10895-021-02699-z PMID: 33609215

Pomerantseva, E., Dunn, B., & Meng, F. (2019). Energy storage: The future enabled by nanomaterials. *Science, 366*(6468), eaan8285. Advance online publication. DOI: 10.1126/science.aan8285 PMID: 31753970

Ramanathan, R., Oluwafemi, O. S., & Kalarikkal, N. (2014). Green synthesis of nanomaterials using biological routes. In *Nanomaterials for environmental protection* (pp. 329–348). Wiley. DOI: 10.1002/9781118845530.ch20

Rao, C., & Biswas, K. (2009). Characterization of nanomaterials by physical methods. *Annual Review of Analytical Chemistry (Palo Alto, Calif.), 2*(1), 435–462. DOI: 10.1146/annurev-anchem-060908-155236 PMID: 20636070

Rujiralai, T., Thongyoy, S., & Kaewkan, P. (2020). Hydrolyzed product mediated aggregation of l-cysteine-modified gold nanoparticles as a colorimetric probe for carbamate residues in chilis. *Analytical Letters, 53*(4), 574–588. DOI: 10.1080/00032719.2019.1659804

Sabela, M., Balme, S., & Bisetty, K. (2017). A review of gold and silver nanoparticle-based colorimetric sensing assays. *Advanced Engineering Materials, 19*(12), 1700270. DOI: 10.1002/adem.201700270

Sengul, A. B., & Asmatulu, E. (2020). Toxicity of metal and metal oxide nanoparticles: A review. *Environmental Chemistry Letters, 18*(5), 1659–1683. DOI: 10.1007/s10311-020-01033-6

Shah, P., Zhu, X., & Li, C.-Z. (2013). Development of paper-based analytical kit for point-of-care testing. *Expert Review of Molecular Diagnostics, 13*(1), 83–91. DOI: 10.1586/erm.12.130 PMID: 23256705

Shamraiz, U., Riaz, B., & Badshah, A. (2016). Functional metal sulfides and selenides for the removal of hazardous dyes from water. *Journal of Photochemistry and Photobiology. B, Biology, 159*, 33–41. DOI: 10.1016/j.jphotobiol.2016.03.013 PMID: 27010842

Sharma, A., Kumar, V., & Kumar, P. (2019). Worldwide pesticide usage and its impacts on ecosystem. *SN Applied Sciences, 1*(11), 1446. DOI: 10.1007/s42452-019-1485-1

Shuai, H., Wu, Y., & Chen, J. (2023). Recent advances of transition metal sulfides/selenides cathodes for aqueous zinc-ion batteries. *Advanced Energy Materials, 13*(4), 2202992. DOI: 10.1002/aenm.202202992

Singh, R., Kumar, V., & Sharma, P. (2022). Colorimetric assay for visual determination of imidacloprid in water and fruit samples using asparagine modified gold nanoparticles. *Journal of the Iranian Chemical Society, 19*(2), 599–607. DOI: 10.1007/s13738-021-02334-5

Singh, R. P., Singh, P., & Srivastava, A. (Eds.). (2023). *Heavy metal toxicity: Environmental concerns, remediation and opportunities*. Springer. DOI: 10.1007/978-981-99-0397-9

Singh, S. K., Kulkarni, P. P., & Dash, D. (2013). Biomedical applications of nanomaterials: An overview. In *Bio-nanotechnology: A revolution in food, biomedical and health sciences* (pp. 1–32). Wiley.DOI: 10.1002/9781118451915.ch1

Tseng, W.-B. H., Chen, C.-H., Chiu, T.-C., & Tseng, W.-L. (2020). Functionalized gold nanoparticles for sensing of pesticides: A review. *Yao Wu Shi Pin Fen Xi, 28*(4), 517–534. DOI: 10.38212/2224-6614.1092 PMID: 35696144

Tu, Q., Tian, Y., & Li, T. (2019). In situ colorimetric detection of glyphosate on plant tissues using cysteamine-modified gold nanoparticles. *Analyst, 144*(6), 2017–2025. DOI: 10.1039/C8AN02473A PMID: 30702090

Turkevich, J., Stevenson, P. C., & Hillier, J. (1951). A study of the nucleation and growth processes in the synthesis of colloidal gold. *Discussions of the Faraday Society, 11*, 55–75. DOI: 10.1039/df9511100055

Umapathi, R., Ghoreishian, S. M., & Ranjith, K. S. (2021). Colorimetric based on-site sensing strategies for the rapid detection of pesticides in agricultural foods: New horizons, perspectives, and challenges. *Coordination Chemistry Reviews, 446*, 214061. DOI: 10.1016/j.ccr.2021.214061

Vajtai, R. (Ed.). (2013). *Science and engineering of nanomaterials*. In *Springer handbook of nanomaterials* (pp. 1–36). Springer. DOI: 10.1007/978-3-642-20595-8_1

Wang, Z., Ma, Q., & Wang, J. (2020). Multicolor visual screening of total dithiocarbamate pesticides in foods based on sulfydryl-mediated growth of gold nanobipyramids. *Analytica Chimica Acta, 1139*, 59–67. DOI: 10.1016/j.aca.2020.09.032 PMID: 33190710

Yan, X., Li, H., & Su, X. (2018). Review of optical sensors for pesticides. *Trends in Analytical Chemistry, 103*, 1–20. DOI: 10.1016/j.trac.2018.03.004

Yan, Y., Zhang, Y., & Wang, J. (2021). Colorimetric assay based on arginine-functionalized gold nanoparticles for the detection of dibutyl phthalate in Baijiu samples. *Analytical Methods, 13*(43), 5179–5186. DOI: 10.1039/D1AY01464A PMID: 34672311

Yang, J., & Liu, H. (2015). *Metal-based composite nanomaterials*. Springer. DOI: 10.1007/978-3-319-12220-5

Zhang, Y., Wang, J., & Li, M. (2020). Colorimetric detection of paraquat in aqueous and fruit juice samples based on functionalized gold nanoparticles. *Journal of Food Composition and Analysis*, *92*, 103574. DOI: 10.1016/j.jfca.2020.103574

Zhu, J., Wang, J., & Zhang, X. (2022). Colorimetric measurement of deltamethrin pesticide using a paper sensor based on aggregation of gold nanoparticles. *Coatings*, *12*(1), 63. DOI: 10.3390/coatings12010063

Chapter 7
Nanomedicines and Its Applications Beyond

Popat Mohite
https://orcid.org/0000-0002-4536-4444
AET's St. John Institute of Pharmacy and Research, Palghar, India

Roshan Dave
https://orcid.org/0009-0002-8889-9139
AET's St. John Institute of Pharmacy and Research, Palghar, India

Shubham Munde
AET's St. John Institute of Pharmacy and Research, Palghar, India

Anil Raosaheb Pawar
MES's College of Pharmacy, India

Sudarshan Singh
https://orcid.org/0000-0002-7929-3322
Department of Pharmaceutical Sciences, Faculty of Pharmacy, Chiang Mai University, Thailand

Bhupendra Prajapati
https://orcid.org/0000-0001-8242-4541
Shree S.K. Patel College of Pharmaceutical Education and Research, Ganpat University, India

Aarati Budhar
AET's St. John Institute of Pharmacy and Research, Palghar, India

ABSTRACT

This chapter delves into the diverse applications of nanotechnology, focusing on its role in nanomedicines and its potential to revolutionize healthcare. It covers nanocarriers in drug delivery, diagnostic applications, and therapeutic approaches, highlighting nanotechnology's impact on regenerative medicine, as well as its industrial and environmental applications. The chapter also discusses challenges and considerations, such as biocompatibility, toxicity, regulatory frameworks, and ethical implications, providing a comprehensive perspective on nanomedicine complexities.

DOI: 10.4018/979-8-3693-2774-6.ch007

INTRODUCTION

Nanomedicine encompasses the utilization of nanomaterials in medicine, spanning from diagnostic to therapeutic uses. While there is no universally agreed-upon definition among scientific and international regulatory entities, a construct is generally considered a nanomedicine if it possesses at least one dimension within the nanoscale range (measured in nanometres, up to 1000 nm) and demonstrates properties influenced by these dimensions, (Abbasi et al., 2014). Nanomedicine constitutes a broad field that encompasses various elements, including nanoparticles that serve as biological mimics (e.g., functionalized carbon nanotubes), "nanomachines" (such as those constructed from interchangeable DNA components and DNA scaffolds like octahedron and stick cube), nanofibers, and polymeric nanostructures utilized as biomaterials (e.g., molecular self-assembly and nanofibers composed of peptides and peptide-amphiphiles for tissue engineering, shape-memory polymers acting as molecular switches, nanoporous membranes). It also involves devices based on nanoscale microfabrication (e.g., silicon microchips for drug release, micromachined hollow needles, and two-dimensional needle arrays created from single-crystal silicon), sensors, and laboratory diagnostics, (Abdollahiyan et al., 2021). The historical development of nanomedicine can be broadly categorized into three stages, as illustrated in Figure 1. The initial stage, spanning 30 years from the discovery of the structure of liposomes in 1964 to the approval of the first nanotherapeutic liposomal drug doxorubicin (marketed under brand names such as Doxil, Adriamycin, and Lipodox) by the Food and Drug Administration (FDA) in 1995, marked significant progress. The second stage, from 1995 to 2007, witnessed certain nanotherapeutics' clinical validation and commercialization. The third stage, from 2008 to the present, has experienced notable advancements, particularly in the rapid expansion of new types of nanotherapeutics, with a focus on intelligent nanotherapeutics, (Abdollahiyan et al., 2021). The evolution of nanomedicine development is shown in Figure 1.

Figure 1. Evolution of nanomedicine development

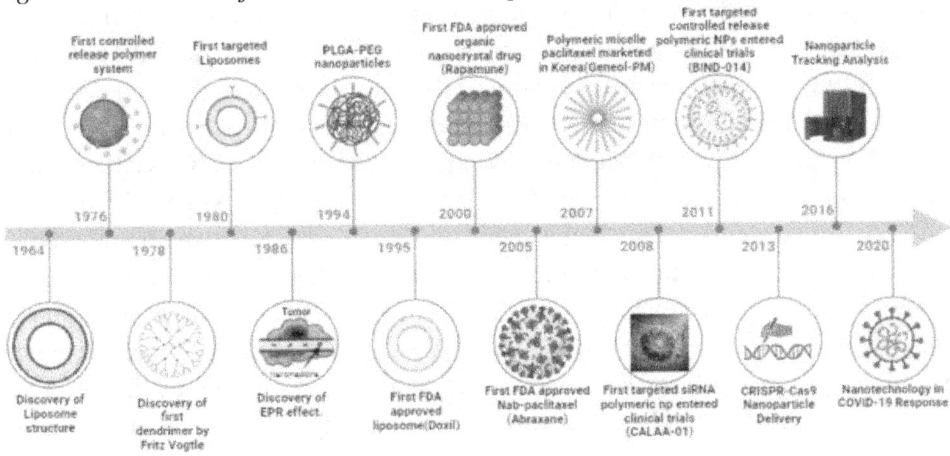

Nanomedicine encompasses a diverse range of scientific and technological endeavours, including disease treatment and diagnosis, disease prevention, pain management, Enhancement of human health, nanotechnology for addressing traumatic injuries, and various therapeutic interventions for diseases. This field adopts an interdisciplinary approach, integrating biotechnology, nanomaterials, biomedical robotics, and genetic engineering to advance medical applications. Scaling down medical technologies to the nano level enhances efficiency, responsiveness, and functionality in biological and chemical processes involved in medical material production. Consequently, ongoing research continually unveils new applications for nanomedicine, offering promising prospects for the future, (Aghababai & Jabbari, 2022). Research on nanoparticles (NPs) has typically focused on their intended purpose, such as solving a specific problem or serving as a tool for investigation or therapy. However, there often needs to be a more comprehensive examination regarding the potential interactions of NPs with their biological surroundings beyond their primary function. NPs offer versatility for modification with various molecules to achieve specific biological effects. Still, the non-payload portion of NPs, which constitutes a significant portion of the particle, also interacts with biology in substantial yet often overlooked ways. Understanding these interactions is crucial for predicting nanomedicine's safety and overall impact. The distinctive properties of three-dimensional structures with numerous nanoscale components make applications in nanomedicine so appealing.

Meanwhile, because of their complexity, nanomedicine products need to be carefully designed and engineered, their physicochemical properties must be strictly characterized, and their scale-up and manufacturing processes must be repeatable to produce a consistent product with stable physicochemical properties, biological behaviours, and pharmacological profiles, (Ahmad et al., 2014). Systemically administered nanomedicines often accumulate because of macrophage confinement in organs such as the liver, kidneys, spleen, and lungs. Certain particles, like polymeric filo micelles, circulate for days because of decreased macrophage uptake, but particles smaller than 100 nm, particularly neutral, PEG-stabilized ones, have longer circulation times. However, patients may experience adverse reactions if these nanomedicines build up and release cytotoxic substances, (Aikins, Xu, & Moon, 2020).

Nano drugs, incorporating dendrimers, liposomes, and polymeric nanoparticles, improve pharmacokinetic parameters by providing targeted delivery, enhanced bioavailability, and prolonged action. Smaller nanoparticles exhibit better uptake, whereas macrophages rapidly clear larger ones. Hydrophilic polymers or biodegradable copolymers for coating reduce opsonization and extend circulation time. Surface modifications enhance transport and targeting. These nano drugs can be administered through various routes, with pulmonary delivery being auspicious due to extended residence times, thereby boosting therapeutic potential and expediting drug development, (Anselmo & Mitragotri, 2016).

Enhancing our knowledge of nanobiology can lead to the development of improved NPs to address current limitations. Additionally, unexpected effects of NPs in biological systems might be harnessed for innovative therapeutic approaches in nanomedicine, (Barathur et al., 2002). The chapter's objective typically involves providing a comprehensive overview of the field, including its principles, applications, recent advancements, and prospects. The chapter explores various aspects of nanomedicine, such as nanomaterials, drug delivery systems, diagnostic techniques, therapeutic approaches, and their implications in healthcare. Additionally, it may discuss the challenges and opportunities in nanomedicine research, including regulatory considerations, ethical concerns, and commercialization aspects. Overall, the chapter seeks to offer valuable insights and contribute to the understanding and advancement of nanomedicines in improving healthcare outcomes.

NANOCARRIERS IN DRUG DELIVERY

Liposomes and Micelles

Liposomes are vesicles formed by the self-assembly of (phosphor)lipids, which can arrange themselves into either a single bilayer (uni-lamellar) or multiple concentric bilayers (multilamellar). These vesicles enclose a central aqueous compartment, (Baruah & Dutta, 2009). In 1964, Liposomes were initially observed by British haematologist Dr. Alec D. Bangham at the Babraham Institute in Cambridge. Their discovery occurred during experiments with the institute's recently acquired electron microscope, where Bangham and R. W. Horne added a negative stain to dried phospholipids, (Boulaiz et al., 2011). So far, liposomes have been extensively studied in various pharmaceutical investigations as carriers for delivering drugs and remain a highly active area of research. Liposomes are recognized as effective drug delivery systems because of their flexible structure and their characteristics of being biocompatible, biodegradable, non-toxic, and non-immunogenic, (Cai et al., 2019). Liposomes are typically created by hydrating a dehydrated layer of phospholipids above the lipid's primary phase transition temperature. Liposome sizes vary between 20 nanometers and several hundred nanometres in diameter, with the phospholipid bilayer membrane measuring around 4 to 7 nanometres in thickness, (Cai et al., 2019). Liposomes are categorized according to their structural characteristics, with classifications including Unilamellar vesicles, Oligolamellar vesicles, and Multilamellar vesicles. The classification of liposomes is summarized in Table 1.

Table 1. Types of liposome vesicles and their sizes

Sr. No.	Type	Number of bilayers	Subtypes	Size range
1	Unilamellar Vesicles (ULV)	1	Small Unilamellar Vesicles (SUV)	20-40 nm
			Medium Unilamellar Vesicles (MUV)	40-80 nm
			Large Unilamellar Vesicles (LUV)	100-1000 nm
2	Oligo Lamellar Vesicles (OLV)	2 – 10		100-1000 nm
3	Multi Lamellar Vesicles (MLV)	More than 10		>500 nm

Micelles, which also encompass polymeric micelles, present another encouraging category of pharmaceutical carriers. They are colloidal dispersions characterized by particle sizes ranging from 5 to 100 nanometres, (Castro, Costa, & Campos, 2022). The molecules of their surfactant can group through cationic, anionic, zwitterionic, or non-ionic interactions. When in water, the non-polar hydrocarbon chain, referred to as the "tail," tends to cluster in the core. At the same time, the hydrophilic or "water-loving" "head" forms the outer layer, creating a spherical micelle structure. These micelles can originate from compounds such as fatty acids, salts (soaps), phospholipids, or similar substances, (Chen, Wu, & Lu, 2020). Polymeric micelles have a unique core-shell structure, making them practical for encapsulating hydrophobic drugs. Their hydrophilic shell prevents biological interactions and enhances stability. They can carry multiple drugs efficiently and accumulate in target areas like tumours. Despite their promising applications, challenges remain, including improving drug loading efficiency, stability post-injection, and facilitating cellular membrane transport. These hurdles must be addressed to fully harness the potential of polymeric micelles as drug carriers, (Choudhary et al., 2023).

Polymeric Nanoparticles

Polymeric nanoparticles (PNPs) are created using biocompatible and biodegradable polymers, typically ranging in size from 10 to 1000 nm. Within these nanoparticles, the drug can be dissolved, entrapped, encapsulated, or attached to a matrix, (Corradetti & Ferrari, 2016). Polymeric nanocarriers are characterized by their unique physicochemical compositions, such as solid polymeric nanoparticles, polymeric micelles, polymer conjugates, dendrimers, polymersomes, polyplexes, and lipomers. Over recent years, polymeric nanoparticles have garnered significant interest and are deemed crucial in drug delivery applications. These nanoparticles are acknowledged as colloidal systems with sizes from 1 to 1000 nm, (Crucho & Barros, 2017). Polymeric nanoparticles (PNPs) have garnered significant attention recently due to their distinct characteristics stemming from their small size. Various authors have highlighted their potential for diverse applications, including diagnostics and drug delivery. PNPs offer advantages as carriers, including controlled release, the capability for theranostics (therapy and imaging combined), protection of drug molecules, and targeted delivery, thus facilitating enhancements in the therapeutic index, (Cui & Mumper, 2001). Biodegradable synthetic polymers are utilized for the controlled and targeted delivery of drugs, with notable examples being polylactic acid (PLA), polyglycolic acid (PGA), and lactic-co-glycolic acid (PLGA) derived polymers, which possess biodegradable and biocompatible properties. However, there is increasing attention towards protein and polysaccharide biopolymers such as chitosan, alginate, cellulose, hyaluronic acid, starch, and collagen due to their biodegradability, biocompatibility, low immunogenicity, and antibacterial properties, (Daraee et al., 2016). Various methods can be utilized for producing polymeric nanoparticles (NPs) depending on the type of drug to be loaded and the requirements for a specific administration route. Two primary approaches are typically employed: dispersing preformed polymers or polymerizing monomers. Table 2 outlines the most frequently used techniques, (Fang & Chen, 2013).

Table 2. Polymeric nanoparticle production methods

Polymeric Nanoparticles	Production Methods
Nanospheres	Solvent evaporation
	Emulsification/solvent diffusion
	Nanoprecipitation
	Emulsification/reverse salting-out
Nanocapsules	Nanoprecipitation

Dendrimers and Nanogels

Dendrimers are nano-size molecules exhibiting radial symmetry, characterized by a well-defined, uniform, and consistent structure comprising tree-like arms or branches. Fritz Vogtle discovered these hyperbranched molecules in 1978, while Donald Tomalia and colleagues made similar findings in the early 1980s, coinciding with independent research by George R. Newkome, (Gao et al., 2013). Tomalia referred to the polymers they created as dendrimers. These dendrimers exhibited a central, hollow core and tendrils that branched outward precisely and predictably, (Ghaderi, Ramesh, & Seifalian, 2011). Dendrimers have found extensive applications in supramolecular chemistry, particularly in host-guest reactions and self-assembly processes. Their distinctive features render them promising candidates for numerous applications, (Gao et al., 2013). Due to their unique characteristics, the chemical makeup of dendrimers surpasses that of other polymers, such as linear, branched, or crosslinked ones. These include uniform structures, which are highly desirable for applications in nanomedicine, precise control over growth at the molecular level, versatile surface functionality, internal spaces capable of accommodating small molecules, improved solubility, and numerous other advantages. These features make dendrimers applicable in various fields. One prominent application is nanomedicine, where they are effective carriers for bioactive molecules, imaging agents, and gene transfection, (Gorain et al., 2017).

Nanogels

Nanogels are hydrophilic polymers crosslinked in a three-dimensional structure, allowing them to absorb significant amounts of water or bodily fluids. Typically, they have a size of less than 200 nm, (Guan et al., 2023). Nanogels are composed of synthetic polymers or biopolymers, which are physically and chemically crosslinked. Within these nanogels, the pores are filled with either micromolecules or macromolecules, (Guerra et al., 2018). Usually, they exhibit a spherical shape, although different configurations may arise depending on the manufacturing techniques employed. For instance, micromoulding techniques and photolithography offer means to regulate the size and shape of nano gels by adjusting surface energy and chemical interactions among the polymers, (Guerrini et al., 2008). The synthesis of nanogels can be categorized into two main methods based on their structures and constituent building blocks: chemical crosslinking and physical self-assembly. Nanogels formed through chemical crosslinking typically demonstrate enhanced stability than those formed through physical crosslinking due to the covalent bonding between functional groups on polymer chains.

In contrast, physically crosslinked nanogels rely on reversible connections primarily driven by noncovalent interactions, including hydrogen bonding, Van der Waals forces, hydrophobic interactions, host-guest interactions, and electrostatic interactions. While the bonds in physical self-assembly are comparatively weaker than those in chemical crosslinking, the process is more flexible and convenient as it does not necessitate complex reactions, (Hameedat et al., 2023). Nanogels are versatile carriers in various drug delivery applications, including anti-psoriatic, anti-inflammatory, transdermal, ocular, anticancer, and protein delivery. Their stimuli-responsive nature enables targeted drug delivery, with applications in passive and active anticancer targeting and controlled release at inflammatory sites or specific locations in the eye. Nanogels can be tailored during synthesis to control drug release, and surface modifications enhance topical delivery to target cells. Nanogels offer enhanced permeability and site-specific delivery, making them valuable in treating skin diseases like psoriasis while minimizing systemic toxicities, (Hanafy, El-Kemary, & Leporatti, 2018).

Quantum Dots and Nanocrystals

Quantum dots (QDs), alternatively termed nanoscale semiconductor crystals, were initially elucidated by Ekimov and Onushenko in a glass matrix as far back as 1981, with their first biological imaging application documented in 1998, (Hare et al., 2017). They have become essential instruments in biomedical investigations, particularly for multifaceted, precise, and enduring fluorescence imaging and detection, (He et al., 2023). Multiple research teams have concentrated on integrating diagnostic and therapeutic functionalities into a single nanoparticle-based substance. Quantum dots (QDs) are promising for theranostic applications, serving as the primary nanocarrier or incorporated within a more intricate structure, such as fluorescent markers, (Jin & Ye, 2007). Drugs can be introduced into QD nanocarriers through methods such as dissolution, dispersion, adsorption, coupling, and similar techniques, (Kang et al., 2020). Various synthesis techniques are used to create Quantum Dots (QDs), with organic and aqueous methods employed to produce luminescent nanocrystals. Traditionally, high-quality nanocrystals are synthesized in organic solvents, which limits their direct use in biological systems due to their hydrophobic nature. However, advances have led to developing water-dispersible QDs through ligand exchange with hydrophilic components, although this process is complex. Alternatively, direct synthesis in an aqueous environment is feasible for biological applications. Cadmium and tellurium are critical precursors for QD assembly in nanotechnology, forming 2 to 10 nm in diameter cores. QDs exhibit unique optical properties at such small sizes, making them attractive for various applications, including biological imaging, cancer detection, and targeted drug delivery. However, caution is advised due to the potential health risks associated with the cadmium content in QDs, especially in their ionic state, with limited research on the adverse effects of nanoparticle exposure, (Khin et al., 2012).

Nanocrystals

The emergence of nanocrystals in response to limitations in current targeted therapy delivery methods highlights a significant advancement. These nanocrystals, ranging in size from tens to hundreds of nanometers, serve as drug carriers. In certain instances, pure drug crystals can be physically stabilized through the use of surfactants and polymers, (Kumar et al., 2023). Regarding drug loading and stability, they differ from carrier-based nanoparticles (such as liposomes, micelles, and polymeric nanoparticles). Nanocrystals offer an advantage in drug loading, typically achieving rates between 50% and 90% (w/w) due to the absence of carrier chemicals. Carrier-based systems often face physical instability problems, primarily avoided using crystalline formulations, (Lanza et al., 1998). Drug nanocrystals offer a promising approach for improving the oral delivery of poorly soluble drugs. By reformulating such drugs into nanocrystal dosage forms, it becomes possible to enhance their bioavailability, efficacy, and safety. This technology has shown rapid industrial applicability, with products reaching the market within less than a decade of invention, and the number of products continues to grow, (Lanza et al., 1996). Drug nanocrystals enhance pharmacological outcomes by targeting specific sites such as the brain, tumours, and macrophages. However, there are hurdles like bioavailability, stability, solubility, safety, and toxicity. To tackle these obstacles, nanocrystals with different functionalities and modifications have been devised to enable targeted drug delivery, (Liu & Zhang, 2011).

DIAGNOSTIC APPLICATIONS

The cellular and molecular aspects of the body, including its physiology, pathological conditions, and metabolic levels, can now be unveiled by utilizing targeted nanomaterials. These nanomaterials are nanocontrast agents or nanoprobes for various imaging devices, (Liu, Chen, & Zhang, 2022). Nanomaterials used as contrast agents visualize internal body structures, helping clinicians differentiate between healthy and diseased tissues for precise treatment recommendations. Engineered nanoparticles can exhibit diverse contrast properties, employed in modalities like computed tomography (CT), magnetic resonance imaging (MRI), radioactivity imaging such as positron emission tomography (PET) or single photon emission computed tomography (SPECT), fluorescence imaging, and photoacoustic imaging, (Lu, Li, & Wu, 2016). Nanomaterials exhibit substantial specific surface areas, improved catalytic abilities, and flexibility for surface modification using diverse functional groups. The extensive specific surface area facilitates efficient immobilization of biomolecules on the transducer surface, thereby boosting sensor sensitivity. Wunderbaldinger and colleagues illustrated the systemic injection of nanoparticles for targeted localization in the liver, spleen, and lymph nodes, (Luo & Saltzman, 2000). The research conducted by Lanza et al. showcased the precise delivery of drugs to specific locations and the real-time imaging of individual nanospheres in areas such as thrombi and revascularization during tumour formation, (Malik, Muhammad, & Waheed, 2023; Manimaran et al., 2023; Matea et al., 2017). Sosnovik, Nahrendorf, and Weissleder (2008) focus on using magnetic nanoparticle (MNP) enhanced MRI for cardiovascular imaging. They have demonstrated the effectiveness of these agents in imaging macrophage infiltration, apoptosis, and VCAM-1 expression in small animal models. They aim to improve sensitivity by utilizing agents with higher relativities and optimizing pharmacokinetics. The versatility of imaging techniques (T1, T2, T2*, SSFP, off-resonance) ensures probe specificity, and future MNP iterations are anticipated to enable multi-modality imaging (MRI, fluorescence, PET), (Matos, Correia, & Rasteiro, 2017). Additional uses involve the application of gold nanoparticles in DNA/RNA diagnostics and employing DNA microarrays for analyzing genotypes and evaluating drug responses, (Mauri et al., 2021).

THERAPEUTIC APPROACHES

In the last ten years, a crucial advancement in nanotechnology applications has been the development of nanovesicles nanosized compounds utilized as therapeutic tools. These nanovesicles are designed to selectively gather at specific locations in the body, enhancing the effectiveness of pharmacotherapy to reduce toxicity.

The primary goal is to improve therapeutic outcomes, leading to the creation of nanodrugs and Nano diagnostics, (Mehata et al., 2023). By utilizing nanocarriers to transport therapeutic medications and directing their delivery to specific areas, it becomes feasible to enhance the concentration of therapeutic drugs in the targeted region while minimizing potential side effects and harm to non-targeted areas, thereby improving overall patient health, (Mironov, Kasyanov, & Markwald, 2008).

Hyperthermia and Photothermal Therapy

Hyperthermia involves raising the temperature of a specific area in the body affected by cancer or other illnesses to eliminate malignant cells. This therapeutic approach, typically executed at temperatures ranging from 40°C to 44°C, has attracted attention for its low toxicity, noninvasive nature, simplicity, and potential effectiveness in treating tumours locally, (Moghimi, Hunter, & Murray, 2005). Temperature Treatment (HTT) can reduce cell viability by causing protein denaturation and activating and deactivating various downstream pathways. Temperatures between 41 and 47 °C trigger apoptosis in tissue cells, activating specific caspases, leading to the proteolytic dismantling of cells. If temperatures exceed 50 °C, cell death is more likely to occur via necrosis, a faster and disorganized form of cell death that does not require sequential protein activation, (Nagavarma et al., 2012). Recently, there has been growing interest in the research on photo hyperthermia or photothermal therapy as a treatment for cancer. Photothermal therapy (PTT) is a novel and effective tumour treatment within nanomedicine. This method utilizes photothermal agents (PTAs) to convert light energy into heat energy, effectively eliminating tumour cells when exposed to external light sources, particularly near-infrared light (NIR), (Nichols & Bae, 2014). In photothermal therapy, patients are administered small-molecule photoactive agents. When irradiating the target area, these agents get excited and undergo internal conversion, converting light energy to heat. In tissues with insufficient blood and oxygen, the generated hyperthermia causes irreversible cell damage at 42°C and 46°C over several minutes. Higher temperatures reduce the required treatment time. This method has proven successful in ablating tumours and improving drug delivery efficiency by enhancing blood flow and tumour vessel permeability, (Onoue, Yamada, & Chan, 2014). The commonly employed photothermal agents for photo hyperthermia include gold nanomaterials like gold nanoparticles, gold nanorods, and gold nanoshells, as well as magnetic nanoparticles and carbon nanotubes, (Pelaz et al., 2017). Pitsillides et al. (2003) pioneered enhanced photothermal therapy by targeting cells using superparamagnetic iron oxide-doped latex microspheres and gold nanoparticles. They conjugated these particles with antibodies specific to T-lymphocytes, mixed them with the lymphocytes, and selectively attached particles to the cells. The targeted cells were irradiated with light to destroy them, (Pitsillides et al., 2003). Zharov, Letfullin, and Galitovskaya (2005) propose a novel method for using lasers and nanoparticles to kill cancer cells called the "microbubbles-overlapping mode". They explored the use of laser irradiation to absorb nanoparticles inside cancer cells. Nanoparticle absorption of laser energy leads to localized heating and potential microbubble formation around them. The authors propose that overlapping these microbubbles can significantly enhance laser treatment's damage range and effectiveness, (Rani, Chauhan, & Kaur, 2014). Vardarajan et al. (2011) investigated the potential of magnetic carbon nanoparticles (MCNPs) for photothermal therapy of cancer cells. Synthesized MCNPs with specific properties for efficient light absorption and magnetic guidance. Introduced MCNPs into cancer cells and applied an external magnetic field to localize the MCNPs at the desired tumour site. Then, they irradiated the localized MCNPs with a near-infrared laser to generate heat and kill cancer cells, (Riehemann et al., 2009).

Nanoparticle-Mediated Gene Therapy

Gene therapy involves introducing and activating a gene construct in cells by directly injecting naked DNA plasmids. However, the uptake of DNA by cells in different target tissues is limited (1-3%), resulting in reduced production of the encoded protein. Additionally, serum nucleases break down much of the naked DNA, making this method inefficient for in vivo transport of genetic material into cells, (Rios-Doria et al., 2015). Research is currently concentrating on two primary categories of gene delivery systems: one employing viral carriers and the other using nonviral carriers. Concerns about safety regarding the utilization of viral vectors have prompted exploration into alternative gene carriers. Nonviral nanoparticles (NPs) present a potential alternative, offering a superior approach due to their lower immunogenicity and cytotoxicity compared to viral vectors.

In contrast to gene transfer through viral vectors, using NPs does not entail introducing additional genes into the transduced cells, (Robles- García et al., 2016). Polymeric nanocarriers typically include polycations that frequently feature ionizable amino groups. These amine groups engage with negatively charged nucleic acids that possess phosphate groups through attractive coulomb forces, facilitating self-assembly and the creation of compact nanoscale structures known as polyplexes. However, it is essential to note that cationic polymers may exhibit toxicity due to their impact on cell membranes, (Rouhani, Samih, & Kalantari, 2012). Moreover, carriers like peptides, dendrimers, and magnetic nanoparticles have been employed for gene delivery, (Roy et al., 2021). Cui and Mumper (2001) explored chitosan-based nanoparticles with plasmid DNA for genetic immunization. Two types were tested: pDNA-condensed chitosan nanoparticles and pDNA-coated cationic chitosan/CMC nanoparticles. Results show stable formulations, with topical application resulting in luciferase expression in mouse skin after 24 hours and significant antigen-specific IgG levels at 28 days, (Ruan et al., 2020). Present nonviral techniques formulated to address challenges in gene delivery frequently lead to modest improvements in transfection efficiency. Luo and Saltzman (2000) introduce an innovative method, positing that the restriction in transfection efficiency arises from a primary physical obstacle—namely, low DNA concentration at the cell surface. By utilizing dense silica nanoparticles to amplify DNA–vector complexes at the cell surface, the scientists attained an up to 8.5-fold Enhancement in transfection efficiency compared to standard reagents, (Saad et al., 2022). Wilson, Stephens, and Jones (2023). developed a placenta-specific nanoparticle gene therapy, enhancing human insulin-like growth factor 1 (hIGF1) expression for in-utero FGR treatment. Using guinea pigs, they examined the impact of FGR on hepatic pathways and found maternal nutrient restriction reduced liver weight in both female and male foetuses. The hIGF1 therapy effectively reversed these effects, providing insights into sex-specific adaptations and demonstrating the potential for placental treatment to normalize disrupted fetal development, (Sainz et al., 2015).

Targeted Drug Delivery Strategies

Recent progress in nanomedicine-driven delivery platforms, like nanoparticles, liposomes, and self-nano emulsifying drug delivery systems (SNEDDS), has led to advanced systems for delivering active biomolecules. Nanomedicines as drug delivery carriers are gaining attention, especially for targeted treatment of specific cells. These nanosized carriers provide benefits such as biocompatibility, biodegradability, low toxicity, improved drug delivery and targeting efficiency, enhanced solubility, bioavailability, and bioactivities, (Sanhai et al., 2008). Guan et al. (2023) created a pH-responsive nanocarrier by combining mesoporous polydopamine nanoparticles (MPDA NPs) with polyacrylic acid (PAA) polymerization, forming a "core-shell" structure. This nanocarrier efficiently loaded the anti-inflammatory drug Sulfasalazine (SAP). The resulting nanomedicine effectively targeted the inflamed colon in mice, alleviating colitis symptoms by reducing pro-inflammatory factors, enhancing the mucosal barrier, and providing anti-inflammatory and antioxidative effects, (Sharma et al., 2016). Targeted delivery can be achieved actively or passively. In active targeting, the therapeutic agent is connected to a tissue-specific ligand. Passive targeting involves placing the therapeutic agent in a macromolecule or nanoparticle that naturally reaches the target organ. Nanoparticles and drugs can also passively target tumours through the EPR effect. Alternatively, catheters can infuse nanoparticles into target organs or tissues, such as delivering drug-loaded nanoparticles to sites of vascular restenosis for sustained drug release on the arterial wall, (Shi & Lammers,

2019). Hameedat et al. (2023) developed effective peptide delivery using targeted zein nanoparticles (ZNPs) for pulmonary administration, addressing degradation and absorption issues. Insulin, used as a model, showed 72% association efficiency. Targeted ZNPs demonstrated controlled insulin release, confirming safety and enhanced permeability in vitro. PEGylated ZNPs were effective carriers, solving pulmonary administration challenges, (Singh & Lillard, 2009).

Immunotherapy and Vaccines

Recent advancements in cancer biology and knowledge of anticancer immunity have significantly contributed to the progress of immunotherapy. This is especially evident in identifying key immunosuppressive pathways, (Singh & Verma, 2015). The cancer vaccine is a highly appealing strategy for triggering robust priming of cytotoxic T lymphocytes (CTLs) and initiating the development of long-term memory T cells. Accurate targeting of tumour antigens and immunostimulant adjuvants to lymphoid organs enhances the functionality of antigen-presenting cells (APCs), leading to effective adaptive immunity that helps prevent tumour growth and relapse, (Sivaram et al., 2015). Improving cancer immunotherapy can be achieved by designing nanoparticles, enabling (1) accurate transport of cargo materials to targeted organs, tissues, and cells, (2) delivering antigens and adjuvants concurrently to antigen-presenting cells (APCs) for robust immune activation, (3) delivering therapeutics to cancer cells in a noninvasive and responsive manner, and (4) establishing a secure and biocompatible platform for combined immunotherapy, (Sosnovik, Nahrendorf, & Weissleder, 2008). A study by Rios-Doria et al. (2015) explored how doxorubicin and liposomal doxorubicin (Doxil) enhance the effectiveness of cancer immunotherapies. Both substances synergized with checkpoint blockers in mouse models, mainly showing effectiveness with Doxil in the presence of a functional immune system. Combining Doxil with immunotherapies demonstrated strong antitumor effects, more complete responders, and increased survival rates. Doxil also influenced immune cell populations, suggesting its potential to activate an antitumor T-cell response, (Stater et al., 2021). Another study was done by Tang et al. (2018) to develop a unique approach using nano gels to enhance Adoptive Cell Therapy (ACT) with antigen-specific T cells, focusing on improving T cell function in solid tumours. These nanogels release protein drugs selectively upon T cell receptor activation, responding to antigen recognition and limiting drug release to tumour sites. Incorporating an interleukin-15 super-agonist into the nanogels led to a 16-fold expansion of T cells in tumours, allowing higher cytokine doses without toxicity. This enhanced therapeutic approach significantly improved tumour clearance by both mouse and human CAR-T cell therapy *in vivo*, (Stater et al., 2021). Utilizing nanocarriers for vaccine administration not only facilitates the delivery of vaccines to tissues and cells but also enhances antigens' stability and immunogenic properties, (Tang et al., 2018).

NANOTECHNOLOGY IN REGENERATIVE MEDICINE

Nanotechnology introduces innovative platforms that could facilitate the development of novel tissue-engineering methods. Despite focusing on objects ranging from 1 to 100 nm, nanotechnology enables the construction of even large tissue-engineered blood vessels through nanostructured and nanopatterned biomaterials, (Tinkle et al., 2014). 3D nanostructured scaffolds can replicate the extracellular matrix (ECM) structure, allowing precise control over cell arrangement and the distribution of bioactive nanomaterials crucial for cell adhesion, proliferation, and differentiation. Nanofibrous scaffolds, resembling extracellular proteins, are particularly significant in this regard. Moreover, there's growing interest in using nanocomposite-based scaffolds and carbon nanotubes for constructing 3D scaffolds.

Additionally, researchers are exploring nanotopographic modifications of the microenvironment by incorporating grooves and ridges due to their similarity to the ECM structure, (Torchilin, 2005). Nanotechnology offers a promising pathway for enhancing tissue regeneration by guiding cell differentiation through molecular signalling pathways. Nanopatterned scaffolds and controlled architecture at the nanoscale have shown superior regenerative responses. Incorporating growth factors and proteins within nanoparticles regulates cell behaviour for programmed proliferation and differentiation. Advanced polymer-based multi-delivery systems enable precise control over delivery kinetics. Integrating nanotechnology and tissue engineering holds great potential for translational regenerative medicine, (van Rijt & Habibovic, 2017). Stem cells are a group of cells capable of multiplying, renewing themselves, and maturing into different types of specialized cells, (Vardarajan et al., 2011).

Currently, nanocombinatorics are under development for analyzing stem cells, which exhibit various chemical and physical structures varying in size. This method utilizes these structures to identify the conditions for stem cells to differentiate into osteocytes. Similarly, at the nanoscale level, a treatment for nerve damage involves using silver nanotubes injected into the brains of mice suffering from strokes. These nanotubes can transport sufficient oxygen and nutrients to the affected area to sustain cell viability. This approach could be combined with drugs to mitigate or repair damage caused by strokes and other types of nervous system damage, (Verma, Domb, & Kumar, 2011). In specific conditions like inflammation or injury, resident and circulating cells are recruited into the niche, creating gradients of bioactive factors that influence stem cell behaviour. Spatial information from the extracellular matrix controls stem cell shape, size, and interactions. At the nanoscale level, mechanical and physical cues affect stem cell behaviour. Proper interaction with the niche is crucial for tissue-specific progenitor cell maintenance and preventing malignant transformation. Nanostructured materials resembling ECM are developed to promote stem cell adhesion, proliferation, and differentiation, (Wang et al., 2009). Integrating nanotechnology drug delivery with stem cell-based therapy holds great promise in effectively regulating neuro-regeneration. This combination, known as nano-stem cell-based treatment, presents a crucial opportunity for managing various neurological disorders. Successful applications of neural stem cell (NSC) or neural progenitor cell (NPC)--based therapy has been observed in conditions such as ischemic stroke, spinal cord injury (SCI), multiple sclerosis, Parkinson's disease, and others, (Wang et al., 2022). It is widely recognized that mesenchymal stem cells (MSCs) naturally possess tumor-tropic and inflammatory migratory characteristics, providing distinct advantages compared to other cell types. Leveraging these properties, MSCs demonstrate considerable potential for use in tissue regeneration, antitumor treatments, and studies related to inflammation, (Wang et al., 2021). Regenerative medicine provides a shift from managing conditions to actively restoring health by utilizing the body's innate ability to regenerate. This approach aims to deliver the potential for restoration rather than prolonging decline. Using nanotechnology to enhance existing tissue and organ regeneration methods has garnered growing interest. Specifically, nanoparticles present intriguing characteristics that hold promise for advancing the field of regenerative medicine, (Wei, Mehtala, & Patri, 2012). The primary uses of different nanomaterials in regenerative medicine and related biomedical applications encompass:

- Serving as scaffolds to support cell growth
- Acting as delivery vehicles for drugs, growth factors, and genes
- Facilitating cellular modification, isolation, and tracking
- Functioning as nanodevices, including biosensors, (Wei et al., 2023).

In recent times, there's been a growing focus on stem cell therapies within the realm of regenerative medicine, especially in addressing central nervous system (CNS) repair. Stem-cell-mediated regenerative medicine enables the restoration of damaged neuronal tissues by directly replacing damaged cells in the targeted tissues or indirectly stimulating the paracrine signalling network at the injured sites to activate endogenous stem cells, (Wickline & Lanza, 2003).

BEYOND MEDICINE: INDUSTRIAL AND ENVIRONMENTAL APPLICATIONS

Nanotechnology in Agriculture

Nanoparticles have gained recent interest in agriculture for their potential applications in disease treatment, quick pathogen detection through nano-based kits, and enhancing production, conservation, intelligent delivery systems, and biosensors, (Wilson, Stephens, & Jones, 2023). The administration of pesticides or medications can be either preventative or given once the disease-causing organism has proliferated and symptoms are noticeable in the plant. Nanotechnology enhances their effectiveness and acceptance by improving performance, safety, and patient adherence and ultimately lowering healthcare expenses, (Wu, Wang, & Li, 2020). Using a green chemistry approach, Kumar et al. (2023) developed an eco-friendly alternative to chemical fungicides in modern agriculture. Guar gum nanoemulsions (NEs) with the fungicide mancozeb were effective against various fungi, showing superior antifungal effects on tomatoes and potatoes. Compared to commercial alternatives, the slower release of mancozeb from NEs suggests potential environmental benefits, offering a promising solution to address pollution concerns associated with traditional pesticides and safeguard vegetable crops, (Xie, 2023). Rouhani, Samih, and Kalantari (2012) investigated the insecticidal activity of silver nanoparticles (Ag) against the oleander aphid (*Aphis nerii*), a common pest affecting ornamental plants. Ag and Ag-Zn nanoparticles were synthesized and tested at different concentrations on A. nerii, with imidacloprid used for comparison as a conventional insecticide. The results revealed that Ag nanoparticles exhibited the highest insect mortality effect in the experiments, (Yin et al., 2020). Integrating nanoparticles into fertilizers offers numerous advantages, including hastening nutrient absorption, minimizing nutrient runoff, and enhancing crops' quality and yield, (Zhang et al., 2018). In a work done by Saad et al. (2022), shrimp waste, considered harmful to the environment, was repurposed to create chitosan nanoparticles, serving as an eco-friendly nano-nitrogen fertilizer. The study applied chitosan nanoparticles on wheat crops and compared them with traditional mineral nitrogen fertilizers, assessing their impact on agronomic traits. Results indicated significant improvements in chlorophyll content, plant height, 1000-grain weight, and yield-contributing characteristics, with the Misr-1 wheat cultivar performing better, (Zhang et al., 2020). Extensive research highlights the diverse applications of nanotechnology, demonstrating significant promise in the field of agriculture. Despite financial obstacles, nanotechnology can transform agriculture, alleviate environmental harm, and enhance crop productivity. Successfully addressing production scale and cost reduction challenges is essential for its widespread adoption. Before gaining broad acceptance, it is imperative to tackle safety concerns, including the possible transportation of toxins by nanomaterials, (Zhang et al., 2018).

Environmental Remediation Using Nanomaterials

Nanomaterials show promise in efficiently removing pollutants and biological contaminants in environmental remediation. Their shapes, like nanoparticles and fibres, act as effective adsorbents and catalysts. Combined with polymers, they detect and remove gases, contaminated chemicals, organic pollutants, and biological substances. Nanomaterials outperform conventional methods due to their high surface area and reactivity, (Zhang et al., 2018). Nanotechnology in the environment focuses on two main aspects: monitoring pollutants using nanotechnology and eliminating environmental pollution through nanotechnology applications, (Zhao et al., 2018). Monitoring the environment involves detecting the presence of contaminants in both the air and wastewater. Various sensors are employed for this task, each operating on distinct principles, (Zhao & Zhu, 2016). In a study done by Guerrini et al. (2018), surface-enhanced Raman scattering (SERS) is employed to detect the pesticide endosulfan at deficient concentrations (10-6 M). The detection is made possible by utilizing silver nanoparticles (Ag NPs) customized with viologen lucigenin, attaining a detection limit of 20 parts per billion, (Zhao et al., 2023). Wang et al. (2009) developed a novel molecularly imprinted polymer (MIP)-based optosensor for room-temperature phosphorescence (RTP) detection. They achieved this by attaching the MIP layer onto the surface of manganese-doped zinc sulfide quantum dots (Mn-doped ZnS QDs) through a surface molecular imprinting process. They tested it by finding small amounts of a chemical called pentachlorophenol (PCP) in water. This sensor could help detect and monitor pollutants in water, (Zharov, Letfullin, & Galitovskaya, 2005). In environmental remediation, various methods like surface assimilation, absorption, synthetic reactions, catalysis using light, and filtering are employed to eliminate pollutants, (Zielińska et al., 2020). Developing new nanomaterials for environmental remediation requires addressing crucial challenges such as capturing specific targets, ensuring cost-effectiveness, facilitating synthesis, embracing green chemistry practices, prioritizing non-toxicity, promoting biodegradability, enabling recyclability, and exploring the potential for recovery after use through regeneration. Carbon Nanotubes, known for their exceptional properties, were investigated by Matos, Correia, and Rasteiro (2017) for their ability to immobilize heavy metal ions, such as lead, copper, nickel, and zinc, commonly found in contaminated soils. The researchers prepared an aqueous CNT suspension to prevent accumulation and loss of properties. Ruan et al. (2020) introduced nanofibrous membranes for air filtration, addressing indoor and outdoor air quality issues associated with PM2.5 emissions. They electrospun polyacrylonitrile: TiO2 and develop polyacrylonitrile-co-polyacrylate: TiO2 composite nanofiber membranes by controlling nanofiber diameter and membrane thickness.

Nanoparticles in Energy Storage and Production

Nanoparticles, made from materials like lithium iron phosphate or titanium dioxide, are revolutionizing battery technology. Their small size provides a large surface area and improved electron mobility, leading to faster charging, higher capacity, and enhanced overall battery efficiency. This innovation extends battery lifespan, enables lightweight energy storage solutions, and finds applications in various devices, from electronics to electric vehicles. Copper nanoparticles through laser ablation in acetone and methanol, confirming their metallic nature. These nanoparticles were integrated between transparent electrodes of aluminium-doped zinc oxide and zirconium-doped indium oxide for potential use in photovoltaic applications. Singh and Lillard (2009) created a cost-effective electrode for microbial fuel cells (MFCs) using carbon micro-nanofibers (ACFs/CNFs) embedded with nickel (Ni) nanoparticles. The electrode, made through chemical vapour deposition with Ni NPs as catalysts and benzene as the carbon source, allowed efficient electron transfer from Escherichia coli to the MFC anode. This transition metal-CNFs-based electrode is a promising and economical substitute for the typically expensive noble metal-based electrodes in microbial fuel cells. Ongoing research in nanoparticle-based energy solutions is expected to impact the future of energy storage and conservation significantly and positively. By addressing limitations and leveraging the strengths of nanoparticles, there is optimism for a more sustainable and resilient energy landscape, reducing reliance on fossil fuels and mitigating climate change impacts.

CHALLENGES AND CONSIDERATIONS

Nanotherapeutics offer distinct clinical benefits compared to traditional small molecular drugs. Over the last thirty years, extensive research in nanomedicine has resulted in the global commercialization of numerous nanomedical products, along with a pipeline of candidates in translation. Before reaching the targeted disease sites, nano-formulation is crucial in aiding drugs in traversing various biological barriers. For oral nanoparticles, stability in the gastrointestinal tract and the ability to penetrate intestinal barriers are crucial. Intravenous administration is often used for large molecular drugs, but overcoming barriers like the blood-brain barrier (BBB) and penetrating solid tumours is challenging, (Ahmad et al., 2014). The issue becomes more complex due to various barriers, including enzymatic breakdown, absorption by scavenging phagocytic cells, irregular blood flow and hydrostatic pressure at targeted locations, and molecular and ionic efflux pumps that eject drugs from target cells. Gaining comprehensive knowledge of the fundamental science behind nanoparticle transport, considering these barriers, will enhance our understanding of their basic biology, provide better control over potential negative effects, and ultimately contribute to the development of superior therapeutic agents. Physical and chemical characteristics such as size, shape, surface area, and larger-scale aggregation can modify biodistribution and interactions with cells and biomolecules, introducing complexities to safety considerations. If not excreted, nanoparticles can accumulate in cells and tissues, posing an unknown risk of long-term diseases. Nanomaterials can enter the human body through accidental exposure, especially through the lungs, leading to rapid translocation to vital organs. On a cellular level, nanoparticles can serve as gene vectors, and certain types, like carbon black nanoparticles, have been linked to interference with cell signalling. In recent years, there has been a surge in stem cell therapies for CNS repair in regenerative medicine. Nanomedicines hold promise in biomedical applications, but the lack of standardized protocols hampers their clinical impact. Regulatory efforts are underway, yet challenges persist due to nanomedicines' unique properties. Collaboration between regulatory agencies is improving, but divergent perspectives exist. Manufacturing processes pose scalability challenges, addressed through quality-by-design concepts. International cooperation aims to harmonize regulatory approaches. Sensitive assays are needed to assess nanomedicine safety and efficacy alongside alternative imaging techniques. In Europe, early-stage scientific counselling facilitates nanomedicine development. Pharmacoeconomic studies are essential before commercialization to demonstrate added value compared to established treatments. Nanomedicines are increasingly utilized in clinical practice, with various sophisticated options entering trials. The European market includes nanoparticles, liposomes, and other nanocomplexes. Regulatory frameworks, particularly for follow-on nanomedicines, are evolving, drawing from biological frameworks. Reflection papers by the EMA address complexities in nanosystems, offering guidance for pharmaceutical development and early clinical studies. These documents emphasize factors crucial for understanding toxicity and pharmacokinetics. Effective large-scale synthesis of nanotherapeutics is vital for successful clinical use. Challenges arise in bringing nanotherapeutics to market due to issues in scaling up synthesis and ensuring reproducibility. The complexity of nanotherapeutics complicates large-scale drug production, leading to variations in properties like drug composition and size. The associated high costs and regulatory hurdles, especially when similar products exist, can impede the success of novel nanotherapeutics. Many successful nanotherapeutics are modified formulations of existing drugs, minimizing financial risks. The elevated costs of nanomedical systems can hinder their development and adoption as mainstream treatments, emphasizing the need for early consideration of synthesis challenges and cost-benefit analysis in nano drug development. The gap between preclinical studies in animals and human clinical trials is a significant challenge in translating nanoparticles to clinical use. For instance, while the Enhanced Permeability and Retention (EPR) effect is observed in small animal models, similar evidence is lacking for humans. The complexity is heightened by species-barrier issues and the diverse nature of human diseases. However, ongoing efforts to correlate data between humans and small animals are increasing, especially with the number of clinical trials. Despite challenges, a united effort within nanomedicine to embrace new clinically focused approaches will aid the next generation of nanomedicines in overcoming obstacles faced in current clinical trials. Investing in the scientific foundations of nanomedicine could significantly contribute to delivering effective nanomedicine therapies to patients.

CONCLUSION

In conclusion, nanomedicine stands at the forefront of medical innovation, offering many promising applications across diagnostics, therapeutics, regenerative medicine, and beyond. Through the ingenious utilization of nanocarriers like liposomes, polymeric nanoparticles, and dendrimers, along with cutting-edge techniques such as hyperthermia and nanoparticle-mediated gene therapy, nanomedicine has revolutionized drug delivery and therapeutic approaches. Moreover, its potential extends into diagnostic precision, regenerative medicine, and even industrial and environmental sectors. However, as with any burgeoning field, challenges abound, from ensuring biocompatibility and addressing toxicity concerns to navigating regulatory and ethical landscapes. Nonetheless, with ongoing research advancements and interdisciplinary collaborations, the future of nanomedicine holds boundless promise, poised to reshape healthcare and industry alike, offering a glimpse into a more efficient, effective, and sustainable future.

REFERENCES

Abbasi, E., Aval, S. F., Akbarzadeh, A., Milani, M., Nasrabadi, H. T., Joo, S. W., Hanifehpour, Y., Nejati-Koshki, K., & Pashaei-Asl, R. (2014). Dendrimers: Synthesis, applications, and properties. *Nanoscale Research Letters*, *9*(1), 247. DOI: 10.1186/1556-276X-9-247 PMID: 24994950

Abdollahiyan, P., Oroojalian, F., Hejazi, M., de la Guardia, M., & Mokhtarzadeh, A. (2021). Nanotechnology, and scaffold implantation for the effective repair of injured organs: An overview on hard tissue engineering. *Journal of Controlled Release*, *333*, 391–417. DOI: 10.1016/j.jconrel.2021.04.003 PMID: 33823222

Aghababai Beni, A., & Jabbari, H. (2022). Nanomaterials for environmental applications. *Results in Engineering*, *15*, 100467. DOI: 10.1016/j.rineng.2022.100467

Ahmad, Z., Shah, A., Siddiq, M., & Kraatz, H.-B. (2014). Polymeric micelles as drug delivery vehicles. *RSC Advances*, *4*(33), 17028–17038. DOI: 10.1039/C3RA47370H

Aikins, M. E., Xu, C., & Moon, J. J. (2020). Engineered nanoparticles for cancer vaccination and immunotherapy. *Accounts of Chemical Research*, *53*(10), 2094–2105. DOI: 10.1021/acs.accounts.0c00456 PMID: 33017150

Anselmo, A. C., & Mitragotri, S. (2016). Nanoparticles in the clinic. *Bioengineering & Translational Medicine*, *1*(1), 10–29. DOI: 10.1002/btm2.10003 PMID: 29313004

Barathur, R., Bookout, J., Sreevatsan, S., Gordon, J., Werner, M., Thor, G., & Worthington, M. (2002). New disc-based technologies for diagnostic and research applications. *Psychiatric Genetics*, *12*(4), 193–206. DOI: 10.1097/00041444-200212000-00002 PMID: 12454524

Baruah, S., & Dutta, J. (2009). Nanotechnology applications in pollution sensing and degradation in agriculture: A review. *Environmental Chemistry Letters*, *7*(3), 191–204. DOI: 10.1007/s10311-009-0228-8

Boulaiz, H., Alvarez, P. J., Ramirez, A., Marchal, J. A., Prados, J., Rodríguez-Serrano, F., Perán, M., Melguizo, C., & Aranega, A. (2011). Nanomedicine: Application areas and development prospects. *International Journal of Molecular Sciences*, *12*(5), 3303–3321. DOI: 10.3390/ijms12053303 PMID: 21686186

Cai, J., Wang, H., Wang, D., & Li, Y. (2019). Improving cancer vaccine efficiency by nanomedicine. *Advanced Biosystems*, *3*(5), 1800287. DOI: 10.1002/adbi.201800287 PMID: 32627400

Castro, K. C., Costa, J. M., & Campos, M. G. N. (2022). Drug-loaded polymeric nanoparticles: A review. *International Journal of Polymeric Materials, 71*(1), 1–13. DOI: 10.1080/00914037.2020.1798436

Chen, Z., Wu, W., & Lu, Y. (2020). What is the future for nanocrystal-based drug-delivery systems? *Therapeutic Delivery, 11*(4), 225–229. DOI: 10.4155/tde-2020-0016 PMID: 32157960

Choudhary, S. K., Mali, P. C., Jangir, R. N., Meena, S., & Tiwari, R. (2023). Applications of nanoparticles in agriculture improvement. *International Journal of Life Sciences (Amravati), 9*(4), 3223–3228. DOI: 10.53730/ijls.v9n4.14970

Corradetti, B., & Ferrari, M. (2016). Nanotechnology for mesenchymal stem cell therapies. *Journal of Controlled Release, 240*, 242–250. DOI: 10.1016/j.jconrel.2015.12.042 PMID: 26732556

Crucho, C. I. C., & Barros, M. T. (2017). Polymeric nanoparticles: A study on the preparation variables and characterization methods. *Materials Science and Engineering C, 80*, 771–784. DOI: 10.1016/j.msec.2017.06.004 PMID: 28866227

Cui, Z., & Mumper, R. J. (2001). Chitosan-based nanoparticles for topical genetic immunization. *Journal of Controlled Release, 75*(3), 409–419. DOI: 10.1016/S0168-3659(01)00407-2 PMID: 11489327

Daraee, H., Etemadi, A., Kouhi, M., Alimirzalu, S., & Akbarzadeh, A. (2016). Application of liposomes in medicine and drug delivery. *Artificial Cells, Nanomedicine, and Biotechnology, 44*(1), 381–391. DOI: 10.3109/21691401.2014.953633 PMID: 25222036

Fang, J., & Chen, Y.-C. (2013). Nanomaterials for photohyperthermia: A review. *Current Pharmaceutical Design, 19*(37), 6622–6634. DOI: 10.2174/1381612811319370006 PMID: 23621537

Gao, L., Liu, G., Ma, J., Wang, X., Zhou, L., Li, X., & Wang, F. (2013). Application of drug nanocrystal technologies on oral drug delivery of poorly soluble drugs. *Pharmaceutical Research, 30*(2), 307–324. DOI: 10.1007/s11095-012-0889-z PMID: 23073665

Ghaderi, S., Ramesh, B., & Seifalian, A. M. (2011). Fluorescence nanoparticles "quantum dots" as drug delivery system and their toxicity: A review. *Journal of Drug Targeting, 19*(7), 475–486. DOI: 10.3109/1061186X.2010.526227 PMID: 20964619

Gorain, B., Tekade, M., Kesharwani, P., Iyer, A. K., Kalia, K., & Tekade, R. K. (2017). The use of nanoscaffolds and dendrimers in tissue engineering. *Drug Discovery Today, 22*(4), 652–664. DOI: 10.1016/j.drudis.2016.12.007 PMID: 28219742

Guan, H., Xu, Z., Du, G., Liu, Q., Tan, Q., Chen, Y., Chen, S., Wu, J., Wang, F., Zhang, J., Sun, L., & Xiao, W. (2023). A mesoporous polydopamine-derived nanomedicine for targeted and synergistic treatment of inflammatory bowel disease by pH-responsive drug release and ROS scavenging. *Materials Today. Bio*, *19*, 100610. DOI: 10.1016/j.mtbio.2023.100610 PMID: 37009068

Guerra, F. D., Attia, M. F., Whitehead, D. C., & Alexis, F. (2018). Nanotechnology for environmental remediation: Materials and applications. *Molecules (Basel, Switzerland)*, *23*(7), 1760. DOI: 10.3390/molecules23071760 PMID: 30021974

Guerrini, L., Aliaga, A. E., Cárcamo, J., Gómez-Jeria, J. S., Sanchez-Cortes, S., Campos-Vallette, M. M., & García-Ramos, J. V. (2008). Functionalization of Ag nanoparticles with the bis-acridinium lucigenin as a chemical assembler in the detection of persistent organic pollutants by surface-enhanced Raman scattering. *Analytica Chimica Acta*, *624*(2), 286–293. DOI: 10.1016/j.aca.2008.06.038 PMID: 18706335

Hameedat, F., Pinto, S., Marques, J., Dias, S., & Sarmento, B. (2023). Functionalized zein nanoparticles targeting neonatal Fc receptor to enhance lung absorption of peptides. *Drug Delivery and Translational Research*, *13*(6), 1699–1715. DOI: 10.1007/s13346-022-01286-4 PMID: 36587110

Hanafy, N. A. N., El-Kemary, M., & Leporatti, S. (2018). Micelles structure development as a strategy to improve smart cancer therapy. *Cancers (Basel)*, *10*(7), 238. DOI: 10.3390/cancers10070238 PMID: 30037052

Hare, J. I., Lammers, T., Ashford, M. B., Puri, S., Storm, G., & Barry, S. T. (2017). Challenges and strategies in anticancer nanomedicine development: An industry perspective. *Advanced Drug Delivery Reviews*, *108*, 25–38. DOI: 10.1016/j.addr.2016.04.025 PMID: 27137110

He, X., Zhang, S., Tian, Y., Cheng, W., & Jing, H. (2023). Research progress of nanomedicine-based mild photothermal therapy in tumor. *International Journal of Nanomedicine*, *18*, 1433–1468. DOI: 10.2147/IJN.S405020 PMID: 36992822

Jin, S., & Ye, K. (2007). Nanoparticle-mediated drug delivery and gene therapy. *Biotechnology Progress*, *23*(1), 32–41. DOI: 10.1021/bp060348j PMID: 17269667

Kang, J. K., Kim, J. C., Shin, Y., Han, S. M., Won, W. R., Her, J., Park, J. Y., & Oh, K. T. (2020). Principles and applications of nanomaterial-based hyperthermia in cancer therapy. *Archives of Pharmacal Research*, *43*(1), 46–57. DOI: 10.1007/s12272-020-01206-5 PMID: 31993968

Khin, M. M., Nair, A. S., Babu, V. J., Murugan, R., & Ramakrishna, S. (2012). A review on nanomaterials for environmental remediation. *Energy & Environmental Science, 5*(8), 8075–8109. DOI: 10.1039/c2ee21818f

Kumar, R., Nehra, M., Kumar, D., Saharan, B. S., Chawla, P., Sadh, P. K., Manuja, A., & Duhan, J. S. (2023). Evaluation of cytotoxicity, release behavior and phytopathogens control by mancozeb-loaded guar gum nanoemulsions for sustainable agriculture. *Journal of Xenobiotics, 13*(2), 270–283. DOI: 10.3390/jox13020020 PMID: 37367496

Lanza, G. M., Trousil, R. L., Wallace, K. D., Rose, J. H., Hall, C. S., Scott, M. J., Miller, J. G., Eisenberg, P. R., Gaffney, P. J., & Wickline, S. A. (1998). In vitro characterization of a novel, tissue-targeted ultrasonic contrast system with acoustic microscopy. *The Journal of the Acoustical Society of America, 104*(6), 3665–3672. DOI: 10.1121/1.423948 PMID: 9857523

Lanza, G. M., Wallace, K. D., Scott, M. J., Cacheris, W. P., Abendschein, D. R., Christy, D. H., Sharkey, A. M., Miller, J. G., Gaffney, P. J., & Wickline, S. A. (1996). A novel site-targeted ultrasonic contrast agent with broad biomedical application. *Circulation, 94*(12), 3334–3340. DOI: 10.1161/01.CIR.94.12.3334 PMID: 8989148

Liu, C., & Zhang, N. (2011). Nanoparticles in gene therapy: Principles, prospects, and challenges. In Villaverde, A. (Ed.), *Progress in molecular biology and translational science* (Vol. 104, pp. 509–562). Academic Press., DOI: 10.1016/B978-0-12-416020-0.00013-9

Liu, P., Chen, G., & Zhang, J. (2022). A review of liposomes as a drug delivery system: Current status of approved products, regulatory environments, and future perspectives. *Molecules (Basel, Switzerland), 27*(4), 1372. DOI: 10.3390/molecules27041372 PMID: 35209162

Lu, Y., Li, Y., & Wu, W. (2016). Injected nanocrystals for targeted drug delivery. *Acta Pharmaceutica Sinica. B, 6*(2), 106–113. DOI: 10.1016/j.apsb.2015.11.005 PMID: 27006893

Luo, D., & Saltzman, W. M. (2000). Enhancement of transfection by physical concentration of DNA at the cell surface. *Nature Biotechnology, 18*(8), 893–895. DOI: 10.1038/78523 PMID: 10932162

Malik, S., Muhammad, K., & Waheed, Y. (2023). Emerging applications of nanotechnology in healthcare and medicine. *Molecules (Basel, Switzerland), 28*(18), 6624. DOI: 10.3390/molecules28186624 PMID: 37764400

Manimaran, V., Nivetha, R. P., Tamilanban, T., Narayanan, J., Vetriselvan, S., Fuloria, N. K., Chinni, S. V., Sekar, M., Fuloria, S., Wong, L. S., Biswas, A., Ramachawolran, G., & Selvaraj, S. (2023). Nanogels are novel drug nanocarriers for CNS drug delivery. *Frontiers in Molecular Biosciences*, *10*, 1232109. DOI: 10.3389/fmolb.2023.1232109 PMID: 37621994

Matea, C. T., Mocan, T., Tabaran, F., Pop, T., Mosteanu, O., Puia, C., Iancu, C., & Mocan, L. (2017). Quantum dots in imaging, drug delivery and sensor applications. *International Journal of Nanomedicine*, *12*, 5421–5431. DOI: 10.2147/IJN.S138624 PMID: 28814860

Matos, M. P. S. R., Correia, A. A. S., & Rasteiro, M. G. (2017). Application of carbon nanotubes to immobilize heavy metals in contaminated soils. *Journal of Nanoparticle Research*, *19*(4), 126. DOI: 10.1007/s11051-017-3830-x

Mauri, E., Giannitelli, S. M., Trombetta, M., & Rainer, A. (2021). Synthesis of nanogels: Current trends and future outlook. *Gels (Basel, Switzerland)*, *7*(1), 36. DOI: 10.3390/gels7020036 PMID: 33805279

Mehata, A. K., Setia, A., Vikas, V., Malik, A. K., Hassani, R., Dailah, H. G., Alhazmi, H. A., Albarraq, A. A., Mohan, S., & Muthu, M. S. (2023). Vitamin E TPGS-based nanomedicine, nanotheranostics, and targeted drug delivery: Past, present, and future. *Pharmaceutics*, *15*(2), 722. DOI: 10.3390/pharmaceutics15030722 PMID: 36986583

Mironov, V., Kasyanov, V., & Markwald, R. R. (2008). Nanotechnology in vascular tissue engineering: From nanoscaffolding towards rapid vessel biofabrication. *Trends in Biotechnology*, *26*(6), 338–344. DOI: 10.1016/j.tibtech.2008.03.001 PMID: 18423666

Moghimi, S. M., Hunter, A. C., & Murray, J. C. (2005). Nanomedicine: Current status and prospects. *The FASEB Journal*, *19*(3), 311–330. DOI: 10.1096/fj.04-2747rev PMID: 15746175

Nagavarma, B., Yadav, H. K., Ayaz, A., Vasudha, L., & Shivakumar, H. (2012). Different techniques for preparation of polymeric nanoparticles—A review. *Asian Journal of Pharmaceutical and Clinical Research*, *5*(3), 16–23.

Nichols, J. W., & Bae, Y. H. (2014). EPR: Evidence and fallacy. *Journal of Controlled Release*, *190*, 451–464. DOI: 10.1016/j.jconrel.2014.03.057 PMID: 24794900

Onoue, S., Yamada, S., & Chan, H.-K. (2014). Nanodrugs: Pharmacokinetics and safety. *International Journal of Nanomedicine*, *9*, 1025–1037. DOI: 10.2147/IJN.S38378 PMID: 24591825

Pelaz, B., Alexiou, C., Alvarez-Puebla, R. A., Alves, F., Andrews, A. M., Ashraf, S., Balogh, L. P., Ballerini, L., Bestetti, A., Brendel, C., Bosi, S., Carril, M., Chan, W. C. W., Chen, C., Chen, X., Chen, X., Cheng, Z., Cui, D., Du, J., & Parak, W. J. (2017). Diverse applications of nanomedicine. *ACS Nano*, *11*(3), 2313–2381. DOI: 10.1021/acsnano.6b06040 PMID: 28290206

Pitsillides, C. M., Joe, E. K., Wei, X., Anderson, R. R., & Lin, C. P. (2003). Selective cell targeting with light-absorbing microparticles and nanoparticles. *Biophysical Journal*, *84*(6), 4023–4032. DOI: 10.1016/S0006-3495(03)75128-5 PMID: 12770906

Rani, K., Chauhan, C., & Kaur, H. (2014). Potential and automotive applications of nanomaterials in combating cancer and stem cell therapy: An informative overview. *Journal of Nanotechnology and Smart Materials*, *1*(1), 1–6.

Riehemann, K., Schneider, S. W., Luger, T. A., Godin, B., Ferrari, M., & Fuchs, H. (2009). Nanomedicine—Challenge and perspectives. *Angewandte Chemie International Edition*, *48*(5), 872–897. DOI: 10.1002/anie.200802585 PMID: 19142939

Rios-Doria, J., Durham, N., Wetzel, L., Rothstein, R., Chesebrough, J., Holoweckyj, N., Zhao, W., Leow, C. C., & Hollingsworth, R. (2015). Doxil synergizes with cancer immunotherapies to enhance antitumor responses in syngeneic mouse models. *Neoplasia : an International Journal for Oncology Research*, *17*(8), 661–670. DOI: 10.1016/j.neo.2015.08.004 PMID: 26408258

Robles-García, M. A., Rodríguez-Félix, F., Marquez-Rios, E., Aguilar, J. A., Barrera-Rodríguez, A., Aguilar, J., Ruiz-Cruz, S., & Del-Toro-Sánchez, C. L. (2016). Applications of nanotechnology in the agriculture, food, and pharmaceuticals. *Journal of Nanoscience and Nanotechnology*, *16*(8), 8188–8207. DOI: 10.1166/jnn.2016.12925

Rouhani, M., Samih, M. A., & Kalantari, S. (2012). Insecticide effect of silver and zinc nanoparticles against *Aphis nerii* Boyer de Fonscolombe (Hemiptera: Aphididae). *Chilean Journal of Agricultural Research*, *72*(4), 590–594. DOI: 10.4067/S0718-58392012000400020

Roy, A., Sharma, A., Yadav, S., Jule, L. T., & Krishnaraj, R. (2021). Nanomaterials for remediation of environmental pollutants. *Bioinorganic Chemistry and Applications*, *1764647*, 1–16. Advance online publication. DOI: 10.1155/2021/1764647 PMID: 34992641

Ruan, D., Qin, L., Chen, R., Xu, G., Su, Z., Cheng, J., Xie, S., Cheng, F., & Ko, F. (2020). Transparent PAN:TiO2 and PAN-co-PMA:TiO2 nanofiber composite membranes with high efficiency in particulate matter pollutants filtration. *Nanoscale Research Letters*, *15*(1), 7. DOI: 10.1186/s11671-019-3225-2 PMID: 31933114

Saad, A. M., Alabdali, A. Y. M., Ebaid, M., Salama, E., El-Saadony, M. T., Selim, S., Safhi, F. A., ALshamrani, S. M., Abdalla, H., Mahdi, A. H. A., & El-Saadony, F. M. A. (2022). Impact of green chitosan nanoparticles fabricated from shrimp processing waste as a source of nano nitrogen fertilizers on the yield quantity and quality of wheat (*Triticum aestivum* L.) cultivars. *Molecules (Basel, Switzerland), 27*(17), 5640. DOI: 10.3390/molecules27175640 PMID: 36080408

Sainz, V., Conniot, J., Matos, A. I., Peres, C., Zupan i , E., Moura, L., Silva, L. C., Florindo, H. F., & Gaspar, R. S. (2015). Regulatory aspects on nanomedicines. *Biochemical and Biophysical Research Communications, 468*(3), 504–510. DOI: 10.1016/j.bbrc.2015.08.023 PMID: 26260323

Sanhai, W. R., Sakamoto, J. H., Canady, R., & Ferrari, M. (2008). Seven challenges for nanomedicine. *Nature Nanotechnology, 3*(5), 242–244. DOI: 10.1038/nnano.2008.114 PMID: 18654511

Sharma, A., Garg, T., Aman, A., Panchal, K., Sharma, R., Kumar, S., & Markandeywar, T. (2016). Nanogel—An advanced drug delivery tool: Current and future. *Artificial Cells, Nanomedicine, and Biotechnology, 44*(1), 165–177. DOI: 10.3109/21691401.2014.930745 PMID: 25053442

Shi, Y., & Lammers, T. (2019). Combining nanomedicine and immunotherapy. *Accounts of Chemical Research, 52*(6), 1543–1554. DOI: 10.1021/acs.accounts.9b00148 PMID: 31120725

Singh, R., & Lillard, J. W.Jr. (2009). Nanoparticle-based targeted drug delivery. *Experimental and Molecular Pathology, 86*(3), 215–223. DOI: 10.1016/j.yexmp.2008.12.004 PMID: 19186176

Singh, S., & Verma, N. (2015). Fabrication of Ni nanoparticles-dispersed carbon micro-nanofibers as the electrodes of a microbial fuel cell for bio-energy production. *International Journal of Hydrogen Energy, 40*(2), 1145–1153. DOI: 10.1016/j.ijhydene.2014.11.073

Sivaram, A. J., Rajitha, P., Maya, S., Jayakumar, R., & Sabitha, M. (2015). Nanogels for delivery, imaging and therapy. *Wiley Interdisciplinary Reviews. Nanomedicine and Nanobiotechnology, 7*(4), 509–533. DOI: 10.1002/wnan.1328 PMID: 25581024

Sosnovik, D. E., Nahrendorf, M., & Weissleder, R. (2008). Magnetic nanoparticles for MR imaging: Agents, techniques and cardiovascular applications. *Basic Research in Cardiology, 103*(2), 122–130. DOI: 10.1007/s00395-008-0710-7 PMID: 18324368

Srilatha, B. (2011). Nanotechnology in agriculture. *Journal of Nanomedicine & Nanotechnology, 2*(7), 123. DOI: 10.4172/2157-7439.1000123

Stater, E. P., Sonay, A. Y., Hart, C., & Grimm, J. (2021). The ancillary effects of nanoparticles and their implications for nanomedicine. *Nature Nanotechnology*, *16*(11), 1180–1194. DOI: 10.1038/s41565-021-01017-9 PMID: 34759355

Tang, L., Zheng, Y., Melo, M. B., Mabardi, L., Castaño, A. P., Xie, Y.-Q., Li, N., Kudchodkar, S. B., Wong, H. C., Jeng, E. K., Maus, M. V., & Irvine, D. J. (2018). Enhancing T cell therapy through TCR-signaling-responsive nanoparticle drug delivery. *Nature Biotechnology*, *36*(8), 707–716. DOI: 10.1038/nbt.4181 PMID: 29985479

Tinkle, S., McNeil, S. E., Mühlebach, S., Bawa, R., Borchard, G., Barenholz, Y., Tamarkin, L., & Desai, N. (2014). Nanomedicines: Addressing the scientific and regulatory gap. *Annals of the New York Academy of Sciences*, *1313*(1), 35–56. DOI: 10.1111/nyas.12403 PMID: 24673240

Torchilin, V. P. (2005). Fluorescence microscopy to follow the targeting of liposomes and micelles to cells and their intracellular fate. *Advanced Drug Delivery Reviews*, *57*(1), 95–109. DOI: 10.1016/j.addr.2004.06.002 PMID: 15518923

van Rijt, S., & Habibovic, P. (2017). Enhancing regenerative approaches with nanoparticles. *Journal of the Royal Society, Interface*, *14*(129), 20170093. DOI: 10.1098/rsif.2017.0093 PMID: 28404870

Vardarajan, V., Gu, L., Kanneganti, A., Mohanty, S., & Koymen, A. (2011). Photothermal therapy of cancer cells using magnetic carbon nanoparticles. In *SPIE Proceedings* (Vol. 7897). DOI: 10.1117/12.875971

Verma, S., Domb, A. J., & Kumar, N. (2011). Nanomaterials for regenerative medicine. *Nanomedicine (London)*, *6*(1), 157–181. DOI: 10.2217/nnm.10.146 PMID: 21182426

Wang, H.-F., He, Y., Ji, T.-R., & Yan, X.-P. (2009). Surface molecular imprinting on Mn-doped ZnS quantum dots for room-temperature phosphorescence optosensing of pentachlorophenol in water. *Analytical Chemistry*, *81*(4), 1615–1621. DOI: 10.1021/ac802375a PMID: 19170523

Wang, J., Li, B., Qiu, L., Qiao, X., & Yang, H. (2022). Dendrimer-based drug delivery systems: History, challenges, and latest developments. *Journal of Biological Engineering*, *16*(1), 18. DOI: 10.1186/s13036-022-00298-5 PMID: 35879774

Wang, M., Xin, Y., Cao, H., Li, W., Hua, Y., Webster, T. J., Zhang, C., Tang, W., & Liu, Z. (2021). Recent advances in mesenchymal stem cell membrane-coated nanoparticles for enhanced drug delivery. *Biomaterials Science*, *9*(4), 1088–1103. DOI: 10.1039/D0BM01164A PMID: 33332490

Wei, A., Mehtala, J. G., & Patri, A. K. (2012). Challenges and opportunities in the advancement of nanomedicines. *Journal of Controlled Release*, *164*(2), 236–246. DOI: 10.1016/j.jconrel.2012.10.007 PMID: 23064314

Wei, M., Yang, Z., Li, S., & Le, W. (2023). Nanotherapeutic and stem cell therapeutic strategies in neurodegenerative diseases: A promising therapeutic approach. *International Journal of Nanomedicine*, *18*, 611–626. DOI: 10.2147/IJN.S395010 PMID: 36760756

Wickline, S. A., & Lanza, G. M. (2003). Nanotechnology for molecular imaging and targeted therapy. *Circulation*, *107*(8), 1092–1095. DOI: 10.1161/01.CIR.0000059651.17045.77 PMID: 12615782

Wilson, R. L., Stephens, K. K., & Jones, H. N. (2023). Placental nanoparticle gene therapy normalizes gene expression changes in the fetal liver associated with fetal growth restriction in a fetal sex-specific manner. *Journal of Developmental Origins of Health and Disease*, *14*(3), 325–332. DOI: 10.1017/S2040174423000016 PMID: 36794386

Wu, L.-P., Wang, D., & Li, Z. (2020). Grand challenges in nanomedicine. *Materials Science and Engineering C*, *106*, 110302. DOI: 10.1016/j.msec.2019.110302 PMID: 31753337

Xie, X. (2023). Application of nanomedicine in diagnostic technology. *Highlights in Science. Engineering and Technology*, *40*, 125–131. DOI: 10.54097/hset.v40i.6577

Yin, Y., Hu, B., Yuan, X., Cai, L., Gao, H., & Yang, Q. (2020). Nanogel: A versatile nano-delivery system for biomedical applications. *Pharmaceutics*, *12*(3), 290. DOI: 10.3390/pharmaceutics12030290 PMID: 32210184

Zhang, B., Yan, W., Zhu, Y., Yang, W., Le, W., Chen, B., Zhu, R., & Cheng, L. (2018). Nanomaterials in neural-stem-cell-mediated regenerative medicine: Imaging and treatment of neurological diseases. *Advanced Materials*, *30*(17), 1705694. DOI: 10.1002/adma.201705694 PMID: 29543350

Zhang, C., Yan, L., Wang, X., Zhu, S., Chen, C., Gu, Z., & Zhao, Y. (2020). Progress, challenges, and future of nanomedicine. *Nano Today*, *35*, 101008. DOI: 10.1016/j.nantod.2020.101008

Zhang, G., Khan, A. A., Wu, H., Chen, L., Gu, Y., & Gu, N. (2018). The application of nanomaterials in stem cell therapy for some neurological diseases. *Current Drug Targets*, *19*(3), 279–298. DOI: 10.2174/1389450118666170328115801 PMID: 28356028

Zhao, J., Liu, Y., Wang, L., Zhou, Y., Du, J., & Wang, Y. (2018). Functional and modified nanocrystals technology for target drug delivery. *Journal of Nanoscience and Nanotechnology*, *18*(8), 5207–5221. DOI: 10.1166/jnn.2018.15421 PMID: 29458570

Zhao, M.-X., & Zhu, B.-J. (2016). The research and applications of quantum dots as nanocarriers for targeted drug delivery and cancer therapy. *Nanoscale Research Letters*, *11*(1), 207. DOI: 10.1186/s11671-016-1394-9 PMID: 27090658

Zhao, Q., Cheng, N., Sun, X., Yan, L., & Li, W. (2023). The application of nanomedicine in clinical settings. *Frontiers in Bioengineering and Biotechnology*, *11*, 1219054. DOI: 10.3389/fbioe.2023.1219054 PMID: 37441195

Zharov, V. P., Letfullin, R. R., & Galitovskaya, E. N. (2005). Microbubbles-overlapping mode for laser killing of cancer cells with absorbing nanoparticle clusters. *Journal of Physics. D, Applied Physics*, *38*(15), 2571–2581. DOI: 10.1088/0022-3727/38/15/007

Zielińska, A., Carreiró, F., Oliveira, A. M., Neves, A., Pires, B., Venkatesh, D. N., Durazzo, A., Lucarini, M., Eder, P., Silva, A. M., Santini, A., & Souto, E. B. (2020). Polymeric nanoparticles: Production, characterization, toxicology and ecotoxicology. *Molecules (Basel, Switzerland)*, *25*(16), 3731. DOI: 10.3390/molecules25163731 PMID: 32824172

Compilation of References

Abbasi, E., Aval, S. F., Akbarzadeh, A., Milani, M., Nasrabadi, H. T., Joo, S. W., Hanifehpour, Y., Nejati-Koshki, K., & Pashaei-Asl, R. (2014). Dendrimers: Synthesis, applications, and properties. *Nanoscale Research Letters*, *9*(1), 247. DOI: 10.1186/1556-276X-9-247 PMID: 24994950

Abdelsattar, A. S., Dawoud, A., & Helal, M. A. (2021). Interaction of nanoparticles with biological macromolecules: A review of molecular docking studies. *Nanotoxicology*, *15*(1), 66–95. DOI: 10.1080/17435390.2020.1842537 PMID: 33283572

Abdel-Sattar, N. E. A., El-Naggar, A. M., & Abdel-Mottaleb, M. S. A. (2017). Novel Thiazole Derivatives of Medicinal Potential: Synthesis and Modeling. *Journal of Chemistry*, *4102796*, 1–11. Advance online publication. DOI: 10.1155/2017/4102796

Abdollahiyan, P., Oroojalian, F., Hejazi, M., de la Guardia, M., & Mokhtarzadeh, A. (2021). Nanotechnology, and scaffold implantation for the effective repair of injured organs: An overview on hard tissue engineering. *Journal of Controlled Release*, *333*, 391–417. DOI: 10.1016/j.jconrel.2021.04.003 PMID: 33823222

Abhilasha, A., Sreenivasulu, A., & Manimozhi, T. (2022). The Model of Smart Sensing Device For Sensitive Nanoclusters Modification in Sensing Properties. 2022 2nd International Conference on Advance Computing and Innovative Technologies in Engineering (ICACITE), 1043-1047, DOI: 10.1109/ICACITE53722.2022.9823846

Abo-zeid, Y., Ismail, N. S. M., McLean, G. R., & Hamdy, N. M. (2020). A molecular docking study repurposes FDA-approved iron oxide nanoparticles to treat and control COVID-19 infection. *European Journal of Pharmaceutical Sciences*, *153*, 105465. DOI: 10.1016/j.ejps.2020.105465 PMID: 32668312

Agarwal, S., & Mehrotra, R. (2016). An overview of molecular docking. *JSM Chemistry*, *4*(2), 1024–1028.

Aghababai Beni, A., & Jabbari, H. (2022). Nanomaterials for environmental applications. *Results in Engineering*, *15*, 100467. DOI: 10.1016/j.rineng.2022.100467

Agu, P. C., Afiukwa, C. A., Orji, O. U., Ezeh, E. M., Ofoke, I. H., Ogbu, C. O., Ugwuja, E. I., & Aja, P. M. (2023). Molecular docking as a tool for the discovery of molecular targets of nutraceuticals in diseases management. *Scientific Reports*, *13*(1), 13398. DOI: 10.1038/s41598-023-40160-2 PMID: 37592012

Ahirwar, J., Ahirwar, D., Lanjhiyana, S., Jha, A. K., Dewangan, D., & Badwaik, H. (2018). Analgesic and Anti-inflammatory Potential of Merged Pharmacophore Containing 1,2,4-triazoles and Substituted Benzyl Groups via Thio Linkage. *Journal of Heterocyclic Chemistry*, *55*(9), 2130–2141. DOI: 10.1002/jhet.3258

Ahmadivand, A., Gerislioglu, B., Ahuja, R., & Mishra, Y. K. (2020). Terahertz plasmonics: The rise of toroidal metadevices towards immunobiosensings. *Materials Today*, *32*, 108–130. DOI: 10.1016/j.mattod.2019.08.002

Ahmad, Z., Shah, A., Siddiq, M., & Kraatz, H.-B. (2014). Polymeric micelles as drug delivery vehicles. *RSC Advances*, *4*(33), 17028–17038. DOI: 10.1039/C3RA47370H

Ahn, J., Miura, Y., Yamada, N., Chida, T., Liu, X., Kim, A., & Yasunaga, M. (2015). Antibody fragment-conjugated polymeric micelles incorporating platinum drugs for targeted therapy of pancreatic cancer. *Biomaterials*, *39*, 23–30. DOI: 10.1016/j.biomaterials.2014.10.069 PMID: 25477168

Aikins, M. E., Xu, C., & Moon, J. J. (2020). Engineered nanoparticles for cancer vaccination and immunotherapy. *Accounts of Chemical Research*, *53*(10), 2094–2105. DOI: 10.1021/acs.accounts.0c00456 PMID: 33017150

Ailincai, D., Morariu, S., Rosca, I., Sandu, A. I., & Marin, L. (2023). Drug delivery based on a supramolecular chemistry approach by using chitosan hydrogels. *International Journal of Biological Macromolecules*, *248*, 125800. DOI: 10.1016/j.ijbiomac.2023.125800 PMID: 37442500

Aissani, N., Urgeghe, P. P., Oplos, C., Saba, M., Tocco, G., Petretto, G. L., Eloh, K., Menkissoglu-Spiroudi, U., Ntalli, N., & Caboni, P. (2015). Nematicidal Activity of the Volatilome of Eruca sativa on Meloidogyne incognita. *Journal of Agricultural and Food Chemistry*, *63*(27), 6120–6125. DOI: 10.1021/acs.jafc.5b02425 PMID: 26082278

Akash, B. A., Kanagaraj, S., Sundaravadivelu, S., Varalakshmi, G. S., Manikantan, V., Pillai, A. S., Alexander, A., & Enoch, I. V. M. V. (2024). Cytotoxicity of 5-fluorouracil- loaded, β-cyclodextrin-tethered metallic copper nanoparticles on triple-negative breast cancer cells. *Journal of Molecular Structure*, *1295*, 136530. DOI: 10.1016/j.molstruc.2023.136530

Akbarzadeh, I., Saremi Poor, A., Khodarahmi, M., Abdihaji, M., Moammeri, A., Jafari, S., Salehi Moghaddam, Z., Seif, M., Moghtaderi, M., Lalami, Z. A., Heydari, M., Adelnia, H., & Farasati Far, B. (2022). Gingerol/letrozole-loaded mesoporous silica nanoparticles for breast cancer therapy: In-silico and in-vitro studies. *Microporous and Mesoporous Materials*, *337*, 111919. DOI: 10.1016/j.micromeso.2022.111919

Alam, A., Karmakar, R., Rengan, A. K., & Khandelwal, M. (2023). Nanofiber-based systems for stimuli-responsive and dual drug delivery: Present scenario and the way forward. *ACS Biomaterials Science & Engineering*, *9*(6), 3160–3184. DOI: 10.1021/acsbiomaterials.3c00363 PMID: 37218599

Alavi, M., & Rai, M. (2019). Recent advances in antibacterial applications of metal nanoparticles (MNPs) and metal nanocomposites (MNCs) against multidrug-resistant (MDR) bacteria. *Expert Review of Anti-Infective Therapy*, *17*(6), 419–428. DOI: 10.1080/14787210.2019.1614914 PMID: 31046483

Alfaifi, G. H., Farghaly, T. A., & Magda, H. A. (2023). Indenyl-thiazole and indenyl-formazan derivatives: Synthesis, anticancer screening studies, molecular-docking, and pharmacokinetic/ molin-spiration properties. *PLoS One*, *18*(3), e0274459. DOI: 10.1371/journal.pone.0274459 PMID: 36857383

Aliabadi, A., Vakili-Azghandi, M., Abnous, K., Taghdisi, S. M., Babaei, M., Ramezani, M., & Alibolandi, M. (2024). Amphiphilic polylactic acid-b-poly(N-(3-aminopropyl) methacrylamide) copolymers: Self-assembly to polymeric micelles for gene delivery. *Journal of Drug Delivery Science and Technology*, *91*, 105236. DOI: 10.1016/j.jddst.2023.105236

Ali, I., Jamal, A., Saqib, Z., Ali, M. I., & Masood, A. B. (2024). Biosurfactants: introduction and classification. In Aslam, R., Aslam, J., & Hussain, C. M. (Eds.), *Industrial Applications of Biosurfactants and Microorganisms* (pp. 1–23). Academic Press. DOI: 10.1016/B978-0-443-13288-9.00021-8

Alom, N.-E., Wu, F., & Li, W. (2017). One-Pot Strategy for Thiazoline Synthesis from Alkenes and Thioamides. *Organic Letters*, *19*(4), 930–933. DOI: 10.1021/acs.orglett.7b00079 PMID: 28177643

Alsharif, Z. A., & Alam, M. A. (2017). Modular synthesis of thiazoline and thiazole derivatives by using a cascade protocol. *RSC Advances*, *7*(52), 32647–32651. DOI: 10.1039/C7RA05993K PMID: 29170713

Altınoğlu, E. I., Russin, T. J., Kaiser, J. M., Barth, B. M., Eklund, B. C., Kester, M., & Adair, J. H. (2008). Near-Infrared Emitting Fluorophore-Doped Calcium Phosphate Nanoparticles for In Vivo Imaging of Human Breast Cancer. *ACS Nano*, *2*(10), 2075–2084. DOI: 10.1021/nn800448r PMID: 19206454

Amaraneni, S. R. (2018). Potential impact of pesticides on environment and human health. *Chemistry International, 40*(3), 46–48. DOI: 10.1515/ci-2018-0329

Amin, M. C. I. M., Ahmad, N., Pandey, M., Abeer, M. M., & Mohamad, N. (2015). Recent advances in the role of supramolecular hydrogels in drug delivery. *Expert Opinion on Drug Delivery, 12*(7), 1149–1161. DOI: 10.1517/17425247.2015.997707 PMID: 25547588

Anselmo, A. C., & Mitragotri, S. (2016). Nanoparticles in the clinic. *Bioengineering & Translational Medicine, 1*(1), 10–29. DOI: 10.1002/btm2.10003 PMID: 29313004

Anselmo, A. C., & Mitragotri, S. (2019). Nanoparticles in the clinic: An update. *Bioengineering & Translational Medicine, 4*(3), e10143. DOI: 10.1002/btm2.10143 PMID: 31572799

Antuma, L. J., Braitmaier, S. H., Garamus, V. M., Hinrichs, J., Boom, R. M., & Keppler, J. K. (2024). Engineering artificial casein micelles for future food: Preparation rate and coagulation properties. *Journal of Food Engineering, 366*, 111868. DOI: 10.1016/j.jfoodeng.2023.111868

Aoyagi, M., Kamoi, T., Kato, M., Sasako, H., Tsuge, N., & Imai, S. (2011). Structure and Bioactivity of Thiosulfinates Resulting from Suppression of Lachrymatory Factor Synthase in Onion. *Journal of Agricultural and Food Chemistry, 59*(20), 10893–10900. DOI: 10.1021/jf202446q PMID: 21905712

Arora, K., Singh, A., Javaid, A., & Mudavath, S. L. (2022). Promising Functional Supramolecules in Antiviral Drugs. In Goel, N., & Kumar, N. (Eds.), *Pharmaceutical Applications of Supramolecules* (pp. 135–155). Springer International Publishing. DOI: 10.1007/978-3-031-21900-9_6

Arruda, T. R., Silva, R. R. A., Marques, C. S., Moraes, A. R. F. e., Bernardes, P. C., de Oliveira, T. V., de Oliveira, S. O., Muranyi, P., & Soares, N. F. F. (2024). β-cyclodextrin versus hydroxypropyl-β-cyclodextrin: Is inclusion complexation a suitable alternative to improve the properties of hop-derived β-acids? *Food Hydrocolloids, 149*, 109622. DOI: 10.1016/j.foodhyd.2023.109622

Arshad, F., Hassan, I. U., & Naikoo, G. A. (2023). Characterization of nanomaterials. In *Nanotechnology in herbal medicine* (pp. 37–61). Elsevier. DOI: 10.1016/B978-0-323-99527-6.00005-7

Asadi, N., Pazoki-Toroudi, H., Del Bakhshayesh, A. R., Akbarzadeh, A., Davaran, S., & Annabi, N. (2021). Multifunctional hydrogels for wound healing: Special focus on biomacromolecular based hydrogels. *International Journal of Biological Macromolecules, 170*, 728–750. DOI: 10.1016/j.ijbiomac.2020.12.202 PMID: 33387543

Attia, M. F., Ogunnaike, E. A., Pitz, M., Elbaz, N. M., Panda, D. K., Alexander-Bryant, A., Saha, S., Whitehead, D. C., & Kabanov, A. (2024). Enhancing drug delivery with supramolecular amphiphilic macrocycle nanoparticles: Selective targeting of CDK4/6 inhibitor palbociclib to melanoma. *Biomaterials Science*, *xx*(3), xx–xx. DOI: 10.1039/D3BM01888A PMID: 38099834

Ayesha, I., Imran, M., Haider, A., Shahzadi, I., Moeen, S., Ul-Hamid, A., Nabgan, W., Shahzadi, A., Alshahrani, T., & Ikram, M. (2023). Polyvinylpyrrolidone and chitosan-coated magnetite (Fe3O4) nanoparticles for catalytic and antimicrobial activity with molecular docking analysis. *Journal of Environmental Chemical Engineering*, *11*(3), 110088. DOI: 10.1016/j.jece.2023.110088

Bai, M.-Y., Zhou, Q., Zhang, J., Li, T., Cheng, J., Liu, Q., Xu, W.-R., & Zhang, Y.-C. (2022). Antioxidant and antibacterial properties of essential oils-loaded β-cyclodextrin-epichlorohydrin oligomer and chitosan composite films. *Colloids and Surfaces. B, Biointerfaces*, *215*, 112504. DOI: 10.1016/j.colsurfb.2022.112504 PMID: 35453062

Baker, S. N., & Baker, G. A. (2010). Luminescent Carbon Nanodots: Emergent Nanolights. *Angewandte Chemie International Edition*, *49*(38), 6726–6744. DOI: 10.1002/anie.200906623 PMID: 20687055

Bala, R., Sharma, R. K., & Wangoo, N. (2018). Ultra-sensitive detection of malathion using quantum dots-polymer based fluorescence aptasensor. *Biosensors & Bioelectronics*, *104*, 45–49. DOI: 10.1016/j.bios.2017.12.034 PMID: 29306032

Bandara, C. S., Gamage, N., Gunaratne, J. C., Waruni, K., Nelumdeniya, R. M., & Udayana, R. (2023). Modeling and simulation in drug delivery. In Das, S., Thomas, S., & Das, P. P. (Eds.), *Novel Platforms for Drug Delivery Applications* (pp. 491–536). Woodhead Publishing. DOI: 10.1016/B978-0-323-91376-8.00004-5

Bansal, P., Kumar, R., Singh, J., & Dhanda, S. (2021). In silico molecular docking of SARS-CoV-2 surface proteins with microbial non-ribosomal peptides: Identification of potential drugs. *Journal of Proteins and Proteomics*, *12*(3), 177–184. DOI: 10.1007/s42485-021-00072-z PMID: 34456530

Barathur, R., Bookout, J., Sreevatsan, S., Gordon, J., Werner, M., Thor, G., & Worthington, M. (2002). New disc-based technologies for diagnostic and research applications. *Psychiatric Genetics*, *12*(4), 193–206. DOI: 10.1097/00041444-200212000-00002 PMID: 12454524

Baruah, S., & Dutta, J. (2009). Nanotechnology applications in pollution sensing and degradation in agriculture: A review. *Environmental Chemistry Letters*, *7*(3), 191–204. DOI: 10.1007/s10311-009-0228-8

Bashandy, M. S. (2016). Synthesis and Characterization of Thiazole Derivatives and Evaluating their Performance as Additives in Polyurethane Coating. *Egyptian Journal of Chemistry*, *59*(1), 115–129. DOI: 10.21608/ejchem.2016.1375

Baughman, R. H., Zakhidov, A. A., & de Heer, W. A. (2002). Carbon Nanotubes—The Route toward. *Applied Sciences (Basel, Switzerland)*, *297*(5582), 787–792. DOI: 10.1126/science.1060928 PMID: 12161643

Bertilsson, L., Dahl, M.-L., Dalén, P., & Al-Shurbaji, A. (2002). Molecular genetics of CYP2D6: Clinical relevance with focus on psychotropic drugs. *British Journal of Clinical Pharmacology*, *53*(2), 111–122. DOI: 10.1046/j.0306-5251.2001.01548.x PMID: 11851634

Beyene, H. D., Werkneh, A. A., Bezabh, H. K., & Ambaye, T. G. (2017). Synthesis paradigm and applications of silver nanoparticles (AgNPs), a review. *Sustainable Materials and Technologies*, *13*, 18–23. DOI: 10.1016/j.susmat.2017.08.001

Bhagyaraj, S., Oluwafemi, O. S., & Kalarikkal, N. (Eds.). (2018). *Characterization of nanomaterials: Advances and key technologies*. Woodhead Publishing.

Bhatt, K. D., Vyas, D. J., & Modi, K. M. (2014). Highly stable water dispersible calix[4]pyrrole octa-hydrazide protected gold nanoparticles as colorimetric and fluorometric chemosensors for selective signaling of Co(II) ions. *Spectrochimica Acta. Part A: Molecular and Biomolecular Spectroscopy*, *121*, 94–100. DOI: 10.1016/j.saa.2013.10.076 PMID: 24231744

Bitsch, R. (2003). BIOTIN I Properties and Determination. In Caballero, B. (Ed.), *Encyclopedia of Food Sciences and Nutrition* (2nd ed., pp. 506–516). Academic Press. DOI: 10.1016/B0-12-227055-X/00101-2

Block, E., Dethier, B., Bechand, B., Cotelesage, J. J. H., George, G. N., Goto, K., Pickering, I. J., Mendoza Rengifo, E., Sheridan, R., Sneeden, E. Y., & Vogt, L. (2018). Ajothiolanes: 3,4-Dimethylthiolane Natural Products from Garlic (Allium sativum). *Journal of Agricultural and Food Chemistry*, *66*(39), 10193–10204. DOI: 10.1021/acs.jafc.8b03638 PMID: 30196701

Borah, P., Hazarika, S., Chettri, A., Sharma, D., Deka, S., Venugopala, K. N., & Deb, P. K. (2023). Heterocyclic compounds as antimicrobial agents. In Bagchi, D., Das, A., & Downs, B. W. (Eds.), *Viral, Parasitic, Bacterial, and Fungal Infections* (pp. 781–804). Academic Press. DOI: 10.1016/B978-0-323-85730-7.00068-0

Borthakur, U., & Saikia, A. K. (2019). Bismuth(III)-Triflate-Catalyzed Highly Diastereoselective Synthesis of Substituted Tetrahydrothiophene via Tandem Isomerization, Michael and Aldol Reactions. *ChemistrySelect*, *4*(37), 11136–11139. DOI: 10.1002/slct.201902871

Boulaiz, H., Alvarez, P. J., Ramirez, A., Marchal, J. A., Prados, J., Rodríguez-Serrano, F., Perán, M., Melguizo, C., & Aranega, A. (2011). Nanomedicine: Application areas and development prospects. *International Journal of Molecular Sciences*, *12*(5), 3303–3321. DOI: 10.3390/ijms12053303 PMID: 21686186

Braegelman, A. S., & Webber, M. J. (2019). Integrating Stimuli-Responsive Properties in Host-Guest Supramolecular Drug Delivery Systems. *Theranostics*, *9*(11), 3017–3040. DOI: 10.7150/thno.31913 PMID: 31244940

Braga, S. S., El-Saleh, F., Oliveira, C., Barbosa, J. S., & Paz, F. A. A. (2023). Co-milling of β-cyclodextrin with simvastatin: Solid-state studies and dissolution profile. *Journal of Drug Delivery Science and Technology*, *89*, 105077. DOI: 10.1016/j.jddst.2023.105077

Buddhadev, S. S. C., Garala, K. S. S., Rahamathulla, M., Ahmed, M. M., Farhana, S. A., & Pasha, I. (2024). Quality by design aided self-nano emulsifying drug delivery systems development for the oral delivery of Benidipine: Improvement of biopharmaceutical performance. *Drug Delivery*, *31*(1), 2288801. DOI: 10.1080/10717544.2023.2288801 PMID: 38073402

Burkeev, M., Fazylov, S., Bakirova, R., Iskineyeva, A., Sarsenbekova, A., Tazhbaev, E., & Davrenbekov, S. (2021). Thermal decomposition of β-cyclodextrin and its inclusion complex with vitamin E. *Mendeleev Communications*, *31*(1), 76–78. DOI: 10.1016/j.mencom.2021.01.023

Cai, J., Wang, H., Wang, D., & Li, Y. (2019). Improving cancer vaccine efficiency by nanomedicine. *Advanced Biosystems*, *3*(5), 1800287. DOI: 10.1002/adbi.201800287 PMID: 32627400

Cai, Y., Qi, J., Lu, Y., He, H., & Wu, W. (2022). The in vivo fate of polymeric micelles. *Advanced Drug Delivery Reviews*, *188*, 114463. DOI: 10.1016/j.addr.2022.114463 PMID: 35905947

Castro, K. C., Costa, J. M., & Campos, M. G. N. (2022). Drug-loaded polymeric nanoparticles: A review. *International Journal of Polymeric Materials*, *71*(1), 1–13. DOI: 10.1080/00914037.2020.1798436

Chadha, R., Sharma, R., & Maiti, N. (2022). γ-Cyclodextrin capped silver and gold nanoparticles as colorimetric and Raman sensor for detecting traces of pesticide "Chlorpyrifos" in fruits and vegetables. *Colloids and Surfaces. A, Physicochemical and Engineering Aspects*, *641*, 128558. DOI: 10.1016/j.colsurfa.2022.128558

Chakraborty, S., & Newkome, G. R. (2018). Terpyridine-based metallosupramolecular constructs: Tailored monomers to precise 2D-motifs and 3D-metallocages. *Chemical Society Reviews*, *47*(11), 3991–4016. DOI: 10.1039/C8CS00030A PMID: 29594272

Chandrakala, V., Aruna, V., & Angajala, G. (2022). Review on metal nanoparticles as nanocarriers: Current challenges and perspectives in drug delivery systems. *Emergent Materials*, *5*(6), 1593–1615. DOI: 10.1007/s42247-021-00335-x PMID: 35005431

Changsan, N., Sawatdee, S., Suedee, R., Chunhachaichana, C., & Srichana, T. (2023). Aqueous cannabidiol β-cyclodextrin complexed polymeric micelle nasal spray to attenuate in vitro and ex vivo SARS-CoV-2-induced cytokine storms. *International Journal of Pharmaceutics*, *640*, 123035. DOI: 10.1016/j.ijpharm.2023.123035 PMID: 37182795

Chaudhary, M., & Tyagi, K. (2024). A review on molecular docking and its application. *International Journal of Advanced Research*, *12*(3), 1141–1153. DOI: 10.21474/IJAR01/18505

Chaudhuri, A., Ramesh, K., Kumar, D. N., Dehari, D., Singh, S., Kumar, D., & Agrawal, A. K. (2022). Polymeric micelles: A novel drug delivery system for the treatment of breast cancer. *Journal of Drug Delivery Science and Technology*, *77*, 103886. DOI: 10.1016/j.jddst.2022.103886

Chauhan, A., Anton, B., & Singh, M. K. (2020). 3 - Dendrimers for drug solubilization, dissolution and bioavailability. In Chauhan, A., & Kulhari, H. (Eds.), *Pharmaceutical Applications of Dendrimers* (pp. 59–92). Elsevier. DOI: 10.1016/B978-0-12-814527-2.00003-2

Chavali, M. S., & Nikolova, M. P. (2019). Metal oxide nanoparticles and their applications in nanotechnology. *SN Applied Sciences*, *1*(6), 607. DOI: 10.1007/s42452-019-0592-3

Chen, D., Bai, Y., Cheng, Q., Li, J., Tong, Z., Hou, J., Liu, T., Guo, Y., Tang, X., Yang, X., & Yang, X. (2022). Domino synthetic strategy for tetrahydrothiophene derivatives from 2-acetylfuran/2-acetylthiophene, benzaldehydes, and sulfur powder. *Arabian Journal of Chemistry*, *15*(9), 104097. DOI: 10.1016/j.arabjc.2022.104097

Cheng, X., Chen, H., Yang, F., Hong, J., Cheng, Y., & Hu, J. (2022). All-small-molecule supramolecular hydrogels assembled from guanosine 5′-monophosphate disodium salt and tobramycin for the treatment of bacterial keratitis. *Bioactive Materials, 16*, 293–300. DOI: 10.1016/j.bioactmat.2021.12.024 PMID: 35386321

Cheng, Z., Li, M., & Rao, J. (2021). Nanomaterials for cancer therapy: Current progress and perspectives. *Journal of Hematology & Oncology, 14*(1), 85. DOI: 10.1186/s13045-021-01096-0 PMID: 34059100

Chen, H., Chen, Y., Wu, H., Xu, J.-F., Sun, Z., & Zhang, X. (2018). Supramolecular polymeric chemotherapy based on cucurbit[7]uril-PEG copolymer. *Biomaterials, 178*, 697–705. DOI: 10.1016/j.biomaterials.2018.02.051 PMID: 29545011

Chen, R., & Riviere, J. E. (2017). Biological and environmental surface interactions of nanomaterials: Characterization, modeling, and prediction. *Wiley Interdisciplinary Reviews. Nanomedicine and Nanobiotechnology, 9*(3), e1440. DOI: 10.1002/wnan.1440 PMID: 27863136

Chen, W., & Liu, P. (2023). Design of GSH-triggered disintegrable dendritic polyurethane as bioreducible unimolecular micelles for tumor-specific chemotherapeutic drug delivery. *European Polymer Journal, 200*, 112521. DOI: 10.1016/j.eurpolymj.2023.112521

Chen, Z., Wu, W., & Lu, Y. (2020). What is the future for nanocrystal-based drug-delivery systems? *Therapeutic Delivery, 11*(4), 225–229. DOI: 10.4155/tde-2020-0016 PMID: 32157960

Chhabria, T. M., Patel, S., Modi, P., & Brahmkshatriya, S. P. (2016). Thiazole: A Review on Chemistry, Synthesis and Therapeutic Importance of its Derivatives. *Current Topics in Medicinal Chemistry, 16*(26), 2841–2862. DOI: 10.2174/1568026616666160506130731 PMID: 27150376

Chibber, S., & Ahmad, I. (2016). Molecular docking, a tool to determine the interaction of CuO and TiO2 nanoparticles with human serum albumin. *Biochemistry and Biophysics Reports, 6*, 63–67. DOI: 10.1016/j.bbrep.2016.03.004 PMID: 28955863

Cho, N., Cheong, T., Min, J., Wu, J., Lee, S., & Kim, D. (2011). JS yang, S. Kim, YK Kim and SY Seong. *Nature Nanotechnology, 6*(10), 675–682. DOI: 10.1038/nnano.2011.149 PMID: 21909083

Choudhary, S. K., Mali, P. C., Jangir, R. N., Meena, S., & Tiwari, R. (2023). Applications of nanoparticles in agriculture improvement. *International Journal of Life Sciences (Amravati), 9*(4), 3223–3228. DOI: 10.53730/ijls.v9n4.14970

Choudhury, S. A., & Chowdhury, M. H. (2016). Use of plasmonic metal nanoparticles to increase the light absorption efficiency of thin-film solar cells. In *2016 IEEE International Conference on Sustainable Energy Technologies (ICSET)* (pp. 268–272). IEEE. DOI: 10.1109/ICSET.2016.7811781

Chouhan, D., Thatikonda, N., Nilebäck, L., Widhe, M., Hedhammar, M., & Mandal, B. B. (2018). Recombinant Spider Silk Functionalized Silkworm Silk Matrices as Potential Bioactive Wound Dressings and Skin Grafts. *ACS Applied Materials & Interfaces, 10*(28), 23560–23572. DOI: 10.1021/acsami.8b05853 PMID: 29940099

Chowdhury, A. R., & Goswami, S. (2023). Study of drug delivery using purely organic macrocyclic containers—Cucurbit[7]uril and pillararene. *ACS Omega, 8*(50), 47340–47366. DOI: 10.1021/acsomega.3c05465 PMID: 38144095

Chow, E. K., & Ho, D. (2013). Cancer Nanomedicine: From Drug Delivery to Imaging. *Science Translational Medicine, 5*(216), 216rv4. Advance online publication. DOI: 10.1126/scitranslmed.3005872 PMID: 24353161

Chow, J. C. (Ed.). (2022). *Application of nanomaterials in biomedical imaging and cancer therapy*. MDPI., DOI: 10.3390/books978-3-0365-3514-2

Colombo, M., Carregal-Romero, S., Casula, M. F., Gutiérrez, L., Morales, M. P., Böhm, I. B., Heverhagen, J. T., Prosperi, D., & Parak, W. J. (2012). Biological Applications of Magnetic Nanoparticles. *Chemical Society Reviews, 41*(11), 4306–4334. DOI: 10.1039/c2cs15337h PMID: 22481569

Corradetti, B., & Ferrari, M. (2016). Nanotechnology for mesenchymal stem cell therapies. *Journal of Controlled Release, 240*, 242–250. DOI: 10.1016/j.jconrel.2015.12.042 PMID: 26732556

Côrte-Real, L., Robalo, M. P., Marques, F., Nogueira, G., Avecilla, F., Silva, T. J., & Valente, A. (2015). The key role of coligands in novel ruthenium (II)-cyclopentadienyl bipyridine derivatives: Ranging from non-cytotoxic to highly cytotoxic compounds. *Journal of Inorganic Biochemistry, 150*, 148–159. DOI: 10.1016/j.jinorgbio.2015.06.015 PMID: 26150132

Crucho, C. I. C., & Barros, M. T. (2017). Polymeric nanoparticles: A study on the preparation variables and characterization methods. *Materials Science and Engineering C, 80*, 771–784. DOI: 10.1016/j.msec.2017.06.004 PMID: 28866227

Csige, K., Szabó, J. P., Kálmán-Szabó, I., Dénes, N. S., Szikra, D., Képes, Z., Opposits, G., Méhes, G., Kertész, I., Fenyvesi, F., Trencsényi, G., & Hajdu, I. (2022). In vivo investigation of Gallium-68 and Bismuth-205/206 labeled beta cyclodextrin for targeted alpha therapy of prostaglandin E2 receptor-expressing tumors in mice. *International Journal of Pharmaceutics*, *625*, 122132. DOI: 10.1016/j.ijpharm.2022.122132 PMID: 36028082

Cui, C.-H., & Yu, S.-H. (2013). Engineering interface and surface of noble metal nanoparticle nanotubes toward enhanced catalytic activity for fuel cell applications. *Accounts of Chemical Research*, *46*(7), 1427–1437. DOI: 10.1021/ar300254b PMID: 23425040

Cui, Q., Hernandez, R., Mason, S. E., Frauenheim, T., Pedersen, J. A., & Geiger, F. (2016). Sustainable nanotechnology: Opportunities and challenges for theoretical/computational studies. *The Journal of Physical Chemistry B*, *120*(31), 7297–7306. DOI: 10.1021/acs.jpcb.6b03976 PMID: 27388532

Cui, Z., & Mumper, R. J. (2001). Chitosan-based nanoparticles for topical genetic immunization. *Journal of Controlled Release*, *75*(3), 409–419. DOI: 10.1016/S0168-3659(01)00407-2 PMID: 11489327

Dahiya, R., Dahiya, S., Fuloria, N. K., Kumar, S., Mourya, R., Chennupati, S. V., Jankie, S., Gautam, H., Singh, S., Karan, S. K., Maharaj, S., Fuloria, S., Shrivastava, J., Agarwal, A., Singh, S., Kishor, A., Jadon, G., & Sharma, A. (2020). Natural Bioactive Thiazole-Based Peptides from Marine Resources: Structural and Pharmacological Aspects. *Marine Drugs*, *18*(6), 329. Advance online publication. DOI: 10.3390/md18060329 PMID: 32599909

Dar, A., & Mir, S. (2017). Molecular docking: Approaches, types, applications, and basic challenges. *Journal of Analytical & Bioanalytical Techniques*, *8*(2), 356. DOI: 10.4172/2155-9872.1000356

Daraee, H., Etemadi, A., Kouhi, M., Alimirzalu, S., & Akbarzadeh, A. (2016). Application of liposomes in medicine and drug delivery. *Artificial Cells, Nanomedicine, and Biotechnology*, *44*(1), 381–391. DOI: 10.3109/21691401.2014.953633 PMID: 25222036

Das, A., Dey, S., Chakraborty, S., Barman, A., Naresh Yadav, R., Gazi, R., Jana, M., & Firoj Hossain, M. (2021). Metal-Free One-Pot Synthesis of 2-(2-Hydrazinyl) Thiazole Derivatives Using Graphene Oxide in a Green Solvent and Molecular Docking Studies. *ChemistrySelect*, *6*(36), 9552–9558. DOI: 10.1002/slct.202102642

Das, S., Ray, S., & Raychaudhuri, A. (2013). Nanomaterials for biomedical applications. *Frontiers in Life Science*, *7*(3–4), 90–98. DOI: 10.1080/21553769.2013.869510

De, A., Sarkar, S., & Majee, A. (2021). Recent advances on heterocyclic compounds with antiviral properties. *Chemistry of Heterocyclic Compounds, 57*(4), 410–416. DOI: 10.1007/s10593-021-02917-3 PMID: 33994556

De, J., Mukhopadhyay, B., & Basak, P. (2020). Luminescent conductive columnar π-gelators for Fe(II) sensing and bio-imaging applications. *The Journal of Physical Chemistry B, 124*(45), 10257–10265. DOI: 10.1021/acs.jpcb.0c07052 PMID: 33136408

Deng, Z., Chen, G., Shi, Y., Lin, Y., Ou, J., Zhu, H., Wu, J., Li, G., & Lv, L. (2022). Curcumin and its nanoformulations: Defining triple-negative breast cancer targets through network pharmacology, molecular docking, and experimental verification. *Frontiers in Pharmacology, 13*, 1001046. DOI: 10.3389/fphar.2022.920514 PMID: 36003508

DeRosa, F., Kibbe, M. R., Najjar, S. F., Citro, M. L., Keefer, L. K., & Hrabie, J. A. (2007). Nitric oxide-releasing fabrics and other acrylonitrile-based diazeniumdiolates. *Journal of the American Chemical Society, 129*(13), 3786–3787. DOI: 10.1021/ja0686864 PMID: 17343382

Dewan, M., Adhikari, A., Jana, R., & Chattopadhyay, D. (2023). Development, evaluation and recent progress of ocular in situ gelling drug delivery vehicle based on poloxamer 407. *Journal of Drug Delivery Science and Technology, 88*, 104885. DOI: 10.1016/j.jddst.2023.104885

Dhandapani, G., Wachtel, E., & Patchornik, G. (2022). Detergent micelle conjugates containing amino acid monomers allow purification of human IgG near neutral pH. *Journal of Chromatography. B, Analytical Technologies in the Biomedical and Life Sciences, 1206*, 123358. DOI: 10.1016/j.jchromb.2022.123358 PMID: 35780745

Dietzel, P. D., Panella, B., Hirscher, M., Blom, R., & Fjellvåg, H. (2006). Hydrogen adsorption in a nickel based coordination polymer with open metal sites in the cylindrical cavities of the desolvated framework. *Chemical Communications*, (9), 959–961. DOI: 10.1039/b515434k PMID: 16491175

Dolai, J., Mandal, K., & Jana, N. R. (2021). Nanoparticle size effects in biomedical applications. *ACS Applied Nano Materials, 4*(7), 6471–6496. DOI: 10.1021/acsanm.1c00987

Dong, R., Zhou, Y., Huang, X., Zhu, X., Lu, Y., & Shen, J. (2015). Functional supramolecular polymers for biomedical applications. *Advanced Materials, 27*(3), 498–526. DOI: 10.1002/adma.201402975 PMID: 25393728

Dragutan, I., Dragutan, V., Simionescu, B. C., Demonceau, A., & Fischer, H. (2015). Recent advances in metathesis-derived polymers containing transition metals in the side chain. *Beilstein Journal of Organic Chemistry, 11*(1), 2747–2762. DOI: 10.3762/bjoc.11.296 PMID: 26877797

Dua, R., Shrivastava, S., Sonwane, S., & Srivastava, S. (2011). Pharmacological significance of synthetic heterocycles scaffold: a review. Academic Press.

Du, L., Wang, X., Cui, G., & Xu, B. (2021). Design, synthesis and biological evaluation of novel thiazole-based derivatives as human Pin1 inhibitors. *Bioorganic & Medicinal Chemistry, 29*, 115878. DOI: 10.1016/j.bmc.2020.115878 PMID: 33246256

Dutt, S., Kumar, R., & Siril, P. F. (2015). Green synthesis of a palladium–polyaniline nanocomposite for green Suzuki–Miyaura coupling reactions. *RSC Advances, 5*(43), 33786–33791. DOI: 10.1039/C5RA05007C

Dzhardimalieva, G. I., Rabinskiy, L. N., Kydralieva, K. A., & Uflyand, I. E. (2019). Recent advances in metallopolymer-based drug delivery systems. *RSC Advances, 9*(63), 37009–37051. DOI: 10.1039/C9RA06678K PMID: 35539076

Ebrahimi, A., Hamishehkar, H., & Amjadi, S. (2023). Development of gelatin-coated nanoliposomes loaded with β-cyclodextrin/vitamin D3 inclusion complex for nutritional therapy. *Food Chemistry, 424*, 136346. DOI: 10.1016/j.foodchem.2023.136346 PMID: 37201470

Edilino de Lima, F., Giordano Viegas, R., & Vital de Oliveira, O. (2023). In silico study of the encapsulation of α-tocopherol and α-tocotrienol vitamins E into cucurbit[7]uril. *Chemical Physics Letters, 826*, 140647. DOI: 10.1016/j.cplett.2023.140647

Edris, A., Abdelrahman, M., Osman, W., Sherif, A. E., Ashour, A., Garelnabi, E. A. E., Ibrahim, S. R. M., Bafail, R., Samman, W. A., Ghazawi, K. F., Mohamed, G. A., & Alzain, A. A. (2023). Design of novel letrozole analogues targeting aromatase for breast cancer: Molecular docking, molecular dynamics, and theoretical studies on gold nanoparticles. *Metabolites, 13*(5), 583. DOI: 10.3390/metabo13050583 PMID: 37233624

El-Aasr, M., Fujiwara, Y., Takeya, M., Ono, M., Nakano, D., Okawa, M., Kinjo, J., Ikeda, T., Miyashita, H., Yoshimitsu, H., & Nohara, T. (2011). Garlicnin A from the Fraction Regulating Macrophage Activation of Allium sativum. *Chemical & Pharmaceutical Bulletin, 59*(11), 1340–1343. DOI: 10.1248/cpb.59.1340 PMID: 22041068

El-Helw, E. A. E., & El-Badawy, A. A. (2020). Synthesis of chromenone, pyrimidinone, thiazoline, and quinolone derivatives as prospective antitumor agents. *Journal of Heterocyclic Chemistry*, *57*(6), 2354–2364. DOI: 10.1002/jhet.3948

El-Helw, E. A. E., Sallam, H. A., & Elgubbi, A. S. (2019). Antioxidant activity of some N-heterocycles derived from 2-(1-(2-oxo-2H-chromen-3-yl)ethylidene) hydrazinecarbothioamide. *Synthetic Communications*, *49*(20), 2651–2661. DOI: 10.1080/00397911.2019.1638938

El-Sayed, A. F., Aboulthana, W. M., Sherief, M. A., El-Bassyouni, G. T., & Mousa, S. M. (2024). Synthesis, structural, molecular docking, and in vitro biological activities of Cu-doped ZnO nanomaterials. *Scientific Reports*, *14*(1), 9027. DOI: 10.1038/s41598-024-59088-2 PMID: 38641640

Eskin, N. A. M., Ho, C.-T., & Shahidi, F. (2013). Browning Reactions in Foods. In Eskin, N. A. M., & Shahidi, F. (Eds.), *Biochemistry of Foods* (3rd ed., pp. 245–289). Academic Press. DOI: 10.1016/B978-0-08-091809-9.00006-6

Evans, N., Grygorash, R., Williams, P., Kyle, A., Kantner, T., Pathak, R., Sheng, X. B., Simoes, F., Makwana, H., Resende, R., de Juan, E., Jenkins, A., Morris, D., Michelet, A., Jewitt, F., Rudge, F., Camper, N., Manin, A., McDowell, W., & Bird, M. (2022). Incorporation of hydrophilic macrocycles into drug-linker reagents produces antibody–drug conjugates with enhanced in vivo performance. *Frontiers in Pharmacology*, *13*, 764540. DOI: 10.3389/fphar.2022.764540 PMID: 35784686

Ezike, T. C., Okpala, U. S., Onoja, U. L., Nwike, C. P., Ezeako, E. C., Okpara, O. J., Okoroafor, C. C., Eze, S. C., Kalu, O. L., Odoh, E. C., Nwadike, U. G., Ogbodo, J. O., Umeh, B. U., Ossai, E. C., & Nwanguma, B. C. (2023). Advances in drug delivery systems, challenges and future directions. *Heliyon*, *9*(6), e17488. DOI: 10.1016/j.heliyon.2023.e17488 PMID: 37416680

Fang, F., Zhang, Z., Zhang, P., Zhang, X., Ma, H., & Wei, Y. (2024). Fluorescence detection of amantadine based on competitive β-Cyclodextrin host-guest inclusion process. *Colloids and Surfaces. A, Physicochemical and Engineering Aspects*, *681*, 132801. DOI: 10.1016/j.colsurfa.2023.132801

Fang, J., & Chen, Y.-C. (2013). Nanomaterials for photohyperthermia: A review. *Current Pharmaceutical Design*, *19*(37), 6622–6634. DOI: 10.2174/1381612811319370006 PMID: 23621537

Feliciano, J. A., Leitgeb, A. J., Schrank, C. L., Allen, R. A., Minbiole, K. P. C., Wuest, W. M., & Carden, R. G. (2021). Trivalent sulfonium compounds (TSCs): Tetrahydrothiophene-based amphiphiles exhibit similar antimicrobial activity to analogous ammonium-based amphiphiles. *Bioorganic & Medicinal Chemistry Letters, 37*, 127809. DOI: 10.1016/j.bmcl.2021.127809 PMID: 33516911

Feng, M., Tang, B., Liang, S., & Jiang, X. (2015). Sulfur Containing Scaffolds in Drugs: Synthesis and Application in Medicinal Chemistry. *Current Topics in Medicinal Chemistry, 16*. Advance online publication. DOI: 10.2174/1568026615666150915111741 PMID: 26369815

Forero-Doria, O., Polo, E., Marican, A., Guzmán, L., Venegas, B., Vijayakumar, S., Wehinger, S., Guerrero, M., Gallego, J., & Durán-Lara, E. F. (2020). Supramolecular hydrogels based on cellulose for sustained release of therapeutic substances with antimicrobial and wound healing properties. *Carbohydrate Polymers, 242*, 116383. DOI: 10.1016/j.carbpol.2020.116383 PMID: 32564841

Forrest, S. R., & Thompson, M. E. (2007). Introduction: Organic electronics and optoelectronics. *Chemical Reviews, 107*(4), 923–925. DOI: 10.1021/cr0501590

Fratoddi, I. (2017). Hydrophobic and hydrophilic Au and Ag nanoparticles. Breakthroughs and perspectives. *Nanomaterials (Basel, Switzerland), 8*(1), 11. DOI: 10.3390/nano8010011 PMID: 29280980

Frens, G. (1972). Particle size and sol stability in metal colloids. *Kolloid-Zeitschrift und Zeitschrift für Polymere, 250*(7), 736–741. DOI: 10.1007/BF01498565

Fu, B., Wu, W., & Zhang, Z. (2021). Recent progress on metal-based nanomaterials: Fabrications, optical properties, and applications in ultrafast photonics. *Advanced Functional Materials, 31*(49), 2107363. DOI: 10.1002/adfm.202107363

Fuhrmann, G. (2023). Drug delivery as a sustainable avenue to future therapies. *Journal of Controlled Release, 354*, 746–754. DOI: 10.1016/j.jconrel.2023.01.045 PMID: 36690037

Gao, J., Karp, J. M., Langer, R., & Joshi, N. (2023). The future of drug delivery. *Chemistry of Materials, 35*(2), 359–363. DOI: 10.1021/acs.chemmater.2c03003 PMID: 37799624

Gao, L., Liu, G., Ma, J., Wang, X., Zhou, L., Li, X., & Wang, F. (2013). Application of drug nanocrystal technologies on oral drug delivery of poorly soluble drugs. *Pharmaceutical Research, 30*(2), 307–324. DOI: 10.1007/s11095-012-0889-z PMID: 23073665

Garcia Jimenez, D., Poongavanam, V., & Kihlberg, J. (2023). Macrocycles in drug discovery—Learning from the past for the future. *Journal of Medicinal Chemistry, 66*(8), 5377–5396. DOI: 10.1021/acs.jmedchem.3c00134 PMID: 37017513

Gautam, S., & Kaur, S. (2024). Revolutionizing biomedicine: Metal-organic frameworks combating multidrug resistance. *Materials Letters, 354*, 135306. DOI: 10.1016/j.matlet.2023.135306

Gentile, A., Ruffino, F., & Grimaldi, M. G. (2016). Complex-morphology metal-based nanostructures: Fabrication, characterization, and applications. *Nanomaterials (Basel, Switzerland), 6*(6), 110. DOI: 10.3390/nano6060110 PMID: 28335236

Ghaderi, S., Ramesh, B., & Seifalian, A. M. (2011). Fluorescence nanoparticles "quantum dots" as drug delivery system and their toxicity: A review. *Journal of Drug Targeting, 19*(7), 475–486. DOI: 10.3109/1061186X.2010.526227 PMID: 20964619

Ghasemi, J. B., Abdolmaleki, A., & Shiri, F. (2017). Molecular docking challenges and limitations. In *Pharmaceutical sciences: Breakthroughs in research and practice* (pp. 770–794). IGI Global. DOI: 10.4018/978-1-5225-1762-7.ch030

Ghitman, J., & Voicu, S. I. (2023). Controlled drug delivery mediated by cyclodextrin-based supramolecular self-assembled carriers: From design to clinical performances. *Carbohydrate Polymer Technologies and Applications, 5*, 100266. DOI: 10.1016/j.carpta.2022.100266

Gillies, R. J., Verduzco, D., & Gatenby, R. A. (2012). Evolutionary dynamics of carcinogenesis and why targeted therapy does not work. *Nature Reviews. Cancer, 12*(7), 487–493. DOI: 10.1038/nrc3298 PMID: 22695393

Gomha, S. M., Riyadh, S. M., Huwaimel, B., Zayed, M. E. M., & Abdellattif, M. H. (2022). Synthesis, Molecular Docking Study, and Cytotoxic Activity against MCF Cells of New Thiazole–Thiophene Scaffolds. *Molecules (Basel, Switzerland), 27*(14), 4639. Advance online publication. DOI: 10.3390/molecules27144639 PMID: 35889511

Gontero, D., Lessard-Viger, M., Brouard, D., Bracamonte, A. G., Boudreau, D., & Veglia, A. V. (2017). Smart multifunctional nanoparticles design as sensors and drug delivery systems based on supramolecular chemistry. *Microchemical Journal, 130*, 316–328. DOI: 10.1016/j.microc.2016.10.007

Gorain, B., Tekade, M., Kesharwani, P., Iyer, A. K., Kalia, K., & Tekade, R. K. (2017). The use of nanoscaffolds and dendrimers in tissue engineering. *Drug Discovery Today, 22*(4), 652–664. DOI: 10.1016/j.drudis.2016.12.007 PMID: 28219742

Guan, H., Xu, Z., Du, G., Liu, Q., Tan, Q., Chen, Y., Chen, S., Wu, J., Wang, F., Zhang, J., Sun, L., & Xiao, W. (2023). A mesoporous polydopamine-derived nanomedicine for targeted and synergistic treatment of inflammatory bowel disease by pH-responsive drug release and ROS scavenging. *Materials Today. Bio*, *19*, 100610. DOI: 10.1016/j.mtbio.2023.100610 PMID: 37009068

Guerra, F. D., Attia, M. F., Whitehead, D. C., & Alexis, F. (2018). Nanotechnology for environmental remediation: Materials and applications. *Molecules (Basel, Switzerland)*, *23*(7), 1760. DOI: 10.3390/molecules23071760 PMID: 30021974

Guerrini, L., Aliaga, A. E., Cárcamo, J., Gómez-Jeria, J. S., Sanchez-Cortes, S., Campos-Vallette, M. M., & García-Ramos, J. V. (2008). Functionalization of Ag nanoparticles with the bis-acridinium lucigenin as a chemical assembler in the detection of persistent organic pollutants by surface-enhanced Raman scattering. *Analytica Chimica Acta*, *624*(2), 286–293. DOI: 10.1016/j.aca.2008.06.038 PMID: 18706335

Guler, M. O., & Stupp, S. I. (2007). A self-assembled nanofiber catalyst for ester hydrolysis. *Journal of the American Chemical Society*, *129*(40), 12082–12083. DOI: 10.1021/ja075044n PMID: 17854188

Guo, P. (2010). The Emerging Field of RNA Nanotechnology. *Nature Nanotechnology*, *5*(12), 833–842. DOI: 10.1038/nnano.2010.231 PMID: 21102465

Halder, J., Mahanty, R., Rajwar, T. K., Rai, V. K., Kar, B., Ghosh, G., & Rath, G. (2023). Nanofibers of glycyrrhizin/hydroxypropyl-β-cyclodextrin inclusion complex: Enhanced solubility profile and anti-inflammatory effect of glycyrrhizin. *AAPS PharmSciTech*, *24*(7), 196. DOI: 10.1208/s12249-023-02662-0 PMID: 37783948

Hameedat, F., Pinto, S., Marques, J., Dias, S., & Sarmento, B. (2023). Functionalized zein nanoparticles targeting neonatal Fc receptor to enhance lung absorption of peptides. *Drug Delivery and Translational Research*, *13*(6), 1699–1715. DOI: 10.1007/s13346-022-01286-4 PMID: 36587110

Hanafy, N. A. N., El-Kemary, M., & Leporatti, S. (2018). Micelles structure development as a strategy to improve smart cancer therapy. *Cancers (Basel)*, *10*(7), 238. DOI: 10.3390/cancers10070238 PMID: 30037052

Han, G., Ghosh, P., De, M., & Rotello, V. M. (2007). Drug and gene delivery using gold nanoparticles. *NanoBiotechnology*, *3*(1), 40–45. DOI: 10.1007/s12030-007-0005-3

Harada, A., Takashima, Y., & Nakahata, M. (2014). Supramolecular polymeric materials via cyclodextrin-guest interactions. *Accounts of Chemical Research, 47*(7), 2128–2140. DOI: 10.1021/ar500109h PMID: 24911321

Hare, J. I., Lammers, T., Ashford, M. B., Puri, S., Storm, G., & Barry, S. T. (2017). Challenges and strategies in anticancer nanomedicine development: An industry perspective. *Advanced Drug Delivery Reviews, 108*, 25–38. DOI: 10.1016/j.addr.2016.04.025 PMID: 27137110

Hasanah, N., Manurung, R. V., Jenie, S. N. A., Indriyati, , Prastya, M. E., & Andreani, A. S. (2023). The effect of size control of Gold Nanoparticles Stabilized with α-cyclodextrin and β-cyclodextrin and their antibacterial activities. *Materials Chemistry and Physics, 302*, 127762. DOI: 10.1016/j.matchemphys.2023.127762

Hassaan, M. A., & El Nemr, A. (2020). Pesticides pollution: Classifications, human health impact, extraction and treatment techniques. *Egyptian Journal of Aquatic Research, 46*(3), 207–220. DOI: 10.1016/j.ejar.2020.08.007

Hazarika, B., & Singh, V. P. (2023). Macrocyclic supramolecular biomaterials in anticancer therapeutics. *Chinese Chemical Letters, 34*(11), 108220. DOI: 10.1016/j.cclet.2023.108220

He, C., Poon, C., Chan, C., Yamada, S. D., & Lin, W. (2016). Nanoscale coordination polymers codeliver chemotherapeutics and siRNAs to eradicate tumors of cisplatin-resistant ovarian cancer. *Journal of the American Chemical Society, 138*(18), 6010–6019. DOI: 10.1021/jacs.6b02486 PMID: 27088560

He, H.-H., Guan, Z.-J., Peng, Y., Liang, Y., Li, J., Zhang, L.-L., & Fang, Y. (2024). Engineering the interactions between metal-organic frameworks and modifying agents: Design, structures, and applications. *Coordination Chemistry Reviews, 499*, 215515. DOI: 10.1016/j.ccr.2023.215515

He, X., Zhang, S., Tian, Y., Cheng, W., & Jing, H. (2023). Research progress of nanomedicine-based mild photothermal therapy in tumor. *International Journal of Nanomedicine, 18*, 1433–1468. DOI: 10.2147/IJN.S405020 PMID: 36992822

Homberg, A. (2018). Synthesis and Applications of Chiral Polyether Macrocycles. éditeur non identifié, Hu, J., Zhang, G., & Liu, S. (2012). Enzyme-responsive polymeric assemblies, nanoparticles and hydrogels. *Chemical Society Reviews, 41*(18), 5933–5949. DOI: 10.1039/c2cs35103j

Huang, K.-X., Xie, M.-S., Sang, J.-W., Qu, G.-R., & Guo, H.-M. (2020). Asymmetric Synthesis of 3-Amine-tetrahydrothiophenes with a Quaternary Stereocenter via Nickel(II)/Trisoxazoline-Catalyzed Sulfa-Michael/Aldol Cascade Reaction: Divergent Access to Chiral Thionucleosides. *Organic Letters*, *23*(1), 81–86. DOI: 10.1021/acs.orglett.0c03747 PMID: 33332122

Huang, X., Li, T., Jiang, X., Wang, Z., Wang, M., Wu, X., Li, J., & Shi, J. (2023). Coassembled supramolecular hydrogel of salvianolic acid B and A phosphopeptide for enhanced wound healing. *ACS Applied Materials & Interfaces*, *15*(39), 45606–45615. DOI: 10.1021/acsami.3c09219 PMID: 37733024

Hu, D., Li, Y., Teng, W., Li, F., Li, L., Li, M., Tan, L., & Wu, Z. (2024). pH-responsive mesoporous silica nanorod for high load and oral delivery of insulin. *Journal of Drug Delivery Science and Technology*, *91*, 105256. DOI: 10.1016/j.jddst.2023.105256

Hu, H., Zhang, J., & Liu, Y. (2015). Metal-free graphene-based catalyst—Insight into the catalytic activity: A short review. *Applied Catalysis A, General*, *492*, 1–9. DOI: 10.1016/j.apcata.2014.11.041

Hu, R., Fan, H., Zhou, Y., Tao, K., Tian, Z., & Ma, H. (2024). Electrostatic interactions dominate thermal conductivity and anisotropy in three-dimensional hydrogen-bonded organic frameworks. *International Journal of Heat and Mass Transfer*, *221*, 125071. DOI: 10.1016/j.ijheatmasstransfer.2023.125071

Huston, M., DeBella, M., & DiCarmine, M. (2021). Green synthesis of nanomaterials. *Nanomaterials (Basel, Switzerland)*, *11*(8), 2130. DOI: 10.3390/nano11082130 PMID: 34443960

Hu, T., Zhang, Z., & Reches, M. (2024). A self-standing superhydrophobic material formed by the self-assembly of an individual amino acid. *Journal of Colloid and Interface Science*, *655*, 899–908. DOI: 10.1016/j.jcis.2023.11.062 PMID: 37979295

Ibrayev, M. K., Nurkenov, O. A., Rakhimberlinova, Z. B., Takibayeva, A. T., Palamarchuk, I. V., Turdybekov, D. M., Kelmyalene, A. A., & Kulakov, I. V. (2022). Synthesis, Structure and Molecular Docking of New 4,5-Dihydrothiazole Derivatives Based on 3,5-Dimethylpyrazole and Cytisine and Salsoline Alkaloids. *Molecules (Basel, Switzerland)*, *27*(21), 7598. Advance online publication. DOI: 10.3390/molecules27217598 PMID: 36364423

Idumah, C. I., Nwuzor, I. C., Odera, S. R., Timothy, U. J., Ngenegbo, U., & Tanjung, F. A. (2024). Recent advances in polymeric hydrogel nanoarchitectures for drug delivery applications. *International Journal of Polymeric Materials*, *73*(1), 1–32. DOI: 10.1080/00914037.2022.2120875

Ingleson, M. J., Heck, R., Gould, J. A., & Rosseinsky, M. J. (2009). Nitric oxide chemisorption in a postsynthetically modified metal– organic framework. *Inorganic Chemistry, 48*(21), 9986–9988. DOI: 10.1021/ic9015977 PMID: 19795833

Jain, A., Maji, S., Shukla, K., Kumari, A., Garg, S., Metre, R. K., Bhattacharyya, S., & Rana, N. K. (2022). Stereoselective synthesis of tri-substituted tetrahydrothiophenes and their in silico binding against mycobacterial protein tyrosine phosphatase B. *Organic & Biomolecular Chemistry, 20*(15), 3124–3135. DOI: 10.1039/D2OB00052K PMID: 35343552

Jakhar, R., Dangi, M., Khichi, A., & Chhillar, A. K. (2020). Relevance of molecular docking studies in drug designing. *Current Bioinformatics, 15*(4), 270–278. DOI: 10.2174/1574893615666191219094216

Jang, E., & Jang, H. (2023). Quantum dot light-emitting diodes. *Chemical Reviews, 123*(8), 4663–4692. DOI: 10.1021/acs.chemrev.2c00695 PMID: 36795794

Jeon, H., Kim, D., Lee, J. H., Song, J., Lee, W. S., Kang, D. W., Kang, S., Lee, S. B., Choi, S., & Hong, K. B. (2018). Hypervalent Iodine-Mediated Alkene Functionalization: Oxazoline and Thiazoline Synthesis via Inter-/Intramolecular Aminohydroxylation and Thioamination. *Advanced Synthesis & Catalysis, 360*(4), 779–783. DOI: 10.1002/adsc.201701087

Jeyadurai, J. A., Muthuselvan, P., Prabhu, A. A. M., Rajendiran, N., & Dasan, A. (2023). Formation of inclusion complexes of 2-hydroxypropyl β-cyclodextrin with 2-hydroxychalcone: Experimental and theoretical study. *Polycyclic Aromatic Compounds, xx*, xx–xx. DOI: 10.1080/10406638.2023.2259565

Jha, A., Kumar, M., & Mishra, B. (2022). Marine Biopolymer-Based Anticancer Drug Delivery Systems. In Jana, S., & Jana, S. (Eds.), *Marine Biomaterials: Drug Delivery and Therapeutic Applications* (pp. 351–401). Springer Nature Singapore. DOI: 10.1007/978-981-16-4787-1_11

Jindal, A. B., Bhide, A. R., Salave, S., Rana, D., & Benival, D. (2023). Long-acting parenteral drug delivery systems for the treatment of chronic diseases. *Advanced Drug Delivery Reviews, 198*, 114862. DOI: 10.1016/j.addr.2023.114862 PMID: 37160247

Jin, S., & Ye, K. (2007). Nanoparticle-mediated drug delivery and gene therapy. *Biotechnology Progress, 23*(1), 32–41. DOI: 10.1021/bp060348j PMID: 17269667

Joshi, N., Azizi Machekposhti, S., & Narayan, R. J. (2023). Evolution of transdermal drug delivery devices and novel microneedle technologies: A historical perspective and review. *JID Innovations : Skin Science from Molecules to Population Health, 3*(6), 100225. DOI: 10.1016/j.xjidi.2023.100225 PMID: 37744689

Jug, M. (2020). Cyclodextrin-based drug delivery systems. In *Nanomaterials for Clinical Applications* (pp. 29–69). Case Studies in Nanomedicines. DOI: 10.1016/B978-0-12-816705-2.00002-3

Kabir, E., & Uzzaman, M. (2022). A review on biological and medicinal impact of heterocyclic compounds. *Results in Chemistry*, *4*, 100606. DOI: 10.1016/j.rechem.2022.100606

Kalantari, K., Fahimi-Kashani, N., & Hormozi-Nezhada, M. R. (2021). Development of a colorimetric sensor array based on monometallic and bimetallic nanoparticles for discrimination of triazole fungicides. *Analytical and Bioanalytical Chemistry*, *413*(7), 1907–1917. PMID: 33855603

Kali, G., Haddadzadegan, S., & Bernkop-Schnürch, A. (2024). Cyclodextrins and derivatives in drug delivery: New developments, relevant clinical trials, and advanced products. *Carbohydrate Polymers*, *324*, 121500. DOI: 10.1016/j.carbpol.2023.121500 PMID: 37985088

Kamal, I. M., & Chakrabarti, S. (2023). MetaDOCK: A combinatorial molecular docking approach. *ACS Omega*, *8*(6), 5850–5860. DOI: 10.1021/acsomega.2c07619 PMID: 36816658

Kang, J. K., Kim, J. C., Shin, Y., Han, S. M., Won, W. R., Her, J., Park, J. Y., & Oh, K. T. (2020). Principles and applications of nanomaterial-based hyperthermia in cancer therapy. *Archives of Pharmacal Research*, *43*(1), 46–57. DOI: 10.1007/s12272-020-01206-5 PMID: 31993968

Karale, U. B., Krishna, V. S., Krishna, E. V., Choudhari, A. S., Shukla, M., Gaikwad, V. R., Mahizhaveni, B., Chopra, S., Misra, S., Sarkar, D., Sriram, D., Dusthackeer, V. N. A., & Rode, H. B. (2019). Synthesis and biological evaluation of 2,4,5-trisubstituted thiazoles as antituberculosis agents effective against drug-resistant tuberculosis. *European Journal of Medicinal Chemistry*, *178*, 315–328. DOI: 10.1016/j.ejmech.2019.05.082 PMID: 31195172

Karim, A. A., Dou, Q., Li, Z., & Loh, X. J. (2016). Emerging Supramolecular Therapeutic Carriers Based on Host-Guest Interactions. *Chemistry, an Asian Journal*, *11*(9), 1300–1321. DOI: 10.1002/asia.201501434 PMID: 26833861

Kaur, A., Singh, N., Kaur, H., Kakoty, V., Sharma, D. S., Khursheed, R., Babu, M. R., Harish, V., Gupta, G., Gulati, M., Kumar, P., Dureja, H., Alharthi, N. S., Khan, F. R., Rehman, Z., Hakami, M. A., Patel, M., Patel, R., Zandi, M., & Singh, S. K. (2023). Neurodegenerative diseases and brain delivery of therapeutics: Bridging the gap using dendrimers. *Journal of Drug Delivery Science and Technology*, *87*, 104868. DOI: 10.1016/j.jddst.2023.104868

Kaur, R., Sharma, S., & Rathore, K. (2023). Stimuli responsive ferrocene-based supramolecular systems: New route for the design of smart devices. *Journal of Organometallic Chemistry, 997*, 122787. DOI: 10.1016/j.jorganchem.2023.122787

Kaur, T., Madgulkar, A., Bhalekar, M., & Asgaonkar, K. (2019). Molecular docking in formulation and development. *Current Drug Discovery Technologies, 16*(1), 30–39. DOI: 10.2174/1570163815666180219112421 PMID: 29468973

Kaushik, A. C., Bharadwaj, S., Kumar, S., & Wei, D. Q. (2018). Nanoparticle mediated inhibition of Parkinson's disease using computational biology approach. *Scientific Reports, 8*(1), 9169. DOI: 10.1038/s41598-018-27580-1 PMID: 29907754

Kelly, K. L., Coronado, E., & Liz, L. M. (2003). The optical properties of metal nanoparticles: The influence of size, shape, and dielectric environment. *The Journal of Physical Chemistry B, 107*(3), 668–677. DOI: 10.1021/jp026731y

Keyur, D. B., Vyas, D. J., & Modi, K. M. (2016). Emerging trends based on calix protected metal nanoparticles. In *Advanced Materials* (Vol. 1, pp. 145–148). TechConnect Briefs.

Khatik, A. S., Kurdhane, S., Batheja, S., & Gupta, U. (2024). Dendrimers: promises and challenges in drug delivery. In Gupta, U., & Goyal, A. K. (Eds.), *Molecular Pharmaceutics and Nano Drug Delivery* (pp. 237–267). Academic Press. DOI: 10.1016/B978-0-323-91924-1.00010-1

Khemnar, M., Galave, V., Kulkarni, V., Chandrakant, M., & Otari, K. (2021). A review on molecular docking. *International Research Journal of Pure and Applied Chemistry, 22*(3), 60–68. DOI: 10.9734/irjpac/2021/v22i330396

Khidre, R. E., & Radini, I. A. M. (2021). Design, synthesis and docking studies of novel thiazole derivatives incorporating pyridine moiety and assessment as antimicrobial agents. *Scientific Reports, 11*(1), 7846. DOI: 10.1038/s41598-021-86424-7 PMID: 33846389

Khin, M. M., Nair, A. S., Babu, V. J., Murugan, R., & Ramakrishna, S. (2012). A review on nanomaterials for environmental remediation. *Energy & Environmental Science, 5*(8), 8075–8109. DOI: 10.1039/c2ee21818f

Kim, B. Y. S., Rutka, J. T., & Chan, W. C. W. (2010). Nanomedicine. *The New England Journal of Medicine, 363*(25), 2434–2443. DOI: 10.1056/NEJMra0912273 PMID: 21158659

Kirla, H., Henry, D. J., Jansen, S., Thompson, P. L., & Hamzah, J. (2023). Use of silica nanoparticles for drug delivery in cardiovascular disease. *Clinical Therapeutics, 45*(11), 1060–1068. DOI: 10.1016/j.clinthera.2023.08.017 PMID: 37783646

Kitagawa, S. (2014). Metal–organic frameworks (MOFs). *Chemical Society Reviews*, *43*(16), 5415–5418. DOI: 10.1039/C4CS90059F PMID: 25011480

Kongor, A. R., Vyas, V. K., & Jain, V. K. (2016). Calix-based nanoparticles: A review. *Topics in Current Chemistry*, *374*(3), 28. DOI: 10.1007/s41061-016-0029-z PMID: 27573268

Kotov, N. A. (2010). Inorganic Nanoparticles as Protein Mimics. *Science*, *330*(6001), 188–189. DOI: 10.1126/science.1190094 PMID: 20929766

Kowalczyk, B., Lagzi, I., & Grzybowski, B. A. (2011). Nanoseparations: Strategies for Size and/or Shape-Selective Purification of Nanoparticles. *Current Opinion in Colloid & Interface Science*, *16*(2), 135–148. DOI: 10.1016/j.cocis.2011.01.004

Krieg, E., Bastings, M. M. C., Besenius, P., & Rybtchinski, B. (2016). Supramolecular Polymers in Aqueous Media. *Chemical Reviews*, *116*(4), 2414–2477. DOI: 10.1021/acs.chemrev.5b00369 PMID: 26727633

Krstić, L., Jarho, P., Ruponen, M., Urtti, A., González-García, M. J., & Diebold, Y. (2022). Improved ocular delivery of quercetin and resveratrol: A comparative study between binary and ternary cyclodextrin complexes. *International Journal of Pharmaceutics*, *624*, 122028. DOI: 10.1016/j.ijpharm.2022.122028 PMID: 35853567

Kumar, A., Jena, P. K., Behera, S., Lockey, R. F., Mohapatra, S., & Mohapatra, S. (2010). Multifunctional magnetic nanoparticles for targeted delivery. *Nanomedicine (London)*, *6*(1), 64–69. DOI: 10.1016/j.nano.2009.04.002 PMID: 19446653

Kumar, P., Bhardwaj, V. K., & Purohit, R. (2024). Highly robust quantum mechanics and umbrella sampling studies on inclusion complexes of curcumin and β-cyclodextrin. *Carbohydrate Polymers*, *323*, 121432. DOI: 10.1016/j.carbpol.2023.121432 PMID: 37940299

Kumar, R., Aadil, K. R., Ranjan, S., & Kumar, V. B. (2020). Advances in nanotechnology and nanomaterials based strategies for neural tissue engineering. *Journal of Drug Delivery Science and Technology*, *57*, 101617. DOI: 10.1016/j.jddst.2020.101617

Kumar, R., Mondal, K., Panda, P. K., Kaushik, A., Abolhassani, R., Ahuja, R., Rubahn, H.-G., & Mishra, Y. K. (2020). Core–shell nanostructures: Perspectives towards drug delivery applications. *Journal of Materials Chemistry. B*, *8*(39), 8992–9027. DOI: 10.1039/D0TB01559H PMID: 32902559

Kumar, R., Nehra, M., Kumar, D., Saharan, B. S., Chawla, P., Sadh, P. K., Manuja, A., & Duhan, J. S. (2023). Evaluation of cytotoxicity, release behavior and phytopathogens control by mancozeb-loaded guar gum nanoemulsions for sustainable agriculture. *Journal of Xenobiotics*, *13*(2), 270–283. DOI: 10.3390/jox13020020 PMID: 37367496

Kumar, R., Soni, P., & Siril, P. F. (2019). Engineering the morphology and particle size of high energetic compounds using drop-by-drop and drop-to-drop solvent–antisolvent interaction methods. *ACS Omega*, *4*(3), 5424–5433. DOI: 10.1021/acsomega.8b03214 PMID: 31459707

Kumar, V. B., Kumar, R., Gedanken, A., & Shefi, O. (2019). Fluorescent metal-doped carbon dots for neuronal manipulations. *Ultrasonics Sonochemistry*, *52*, 205–213. DOI: 10.1016/j.ultsonch.2018.11.017 PMID: 30522849

Kunjiappan, S., Govindaraj, S., Parasuraman, P., Sankaranarayanan, M., Arunachalam, S., Palanisamy, P., Mohan, U. P., Babkiewicz, E., Maszczyk, P., Vellaisamy, S., & Panneerselvam, T. (2020). Design, in silico modeling, and functionality theory of folate-receptor-targeted myricetin-loaded bovine serum albumin nanoparticle formulation for cancer treatment. *Nanotechnology*, *31*(15), 155102. DOI: 10.1088/1361-6528/ab5c56 PMID: 31775133

Kurapati, K. R. V., Atluri, V. S., Samikkannu, T., Garcia, G., & Nair, M. P. N. (2016). Natural Products as Anti-HIV Agents and Role in HIV-Associated Neurocognitive Disorders (HAND): A Brief Overview. *Frontiers in Microbiology*, *6*, 6. DOI: 10.3389/fmicb.2015.01444 PMID: 26793166

Lago, A. B., Pino-Cuevas, A., Carballo, R., & Vázquez-López, E. M. (2016). A new metal–organic polymeric system capable of stimuli-responsive controlled release of the drug ibuprofen. *Dalton Transactions (Cambridge, England)*, *45*(4), 1614–1621. DOI: 10.1039/C5DT04031K PMID: 26692254

Lamberth, C., & Dinges, J. (2012). The Significance of Heterocycles for Pharmaceuticals and Agrochemicals. In *Bioactive Heterocyclic Compound Classes* (pp. 1-20). Academic Press.

Lamey, C. A., Moussa, N., Helmy, M. W., Haroun, M., & Sabra, S. A. (2023). Simultaneous encapsulation of dasatinib and celecoxib into caseinate micelles toward improved in vivo anti-breast cancer efficacy with reduced drug toxicity. *Journal of Drug Delivery Science and Technology*, *87*, 104807. DOI: 10.1016/j.jddst.2023.104807

Langer, R., & Weissleder, R. (2015). Nanotechnology. *Journal of the American Medical Association*, *313*(2), 135–136. DOI: 10.1001/jama.2014.16315 PMID: 25585320

Lanza, G. M., Trousil, R. L., Wallace, K. D., Rose, J. H., Hall, C. S., Scott, M. J., Miller, J. G., Eisenberg, P. R., Gaffney, P. J., & Wickline, S. A. (1998). In vitro characterization of a novel, tissue-targeted ultrasonic contrast system with acoustic microscopy. *The Journal of the Acoustical Society of America, 104*(6), 3665–3672. DOI: 10.1121/1.423948 PMID: 9857523

Lanza, G. M., Wallace, K. D., Scott, M. J., Cacheris, W. P., Abendschein, D. R., Christy, D. H., Sharkey, A. M., Miller, J. G., Gaffney, P. J., & Wickline, S. A. (1996). A novel site-targeted ultrasonic contrast agent with broad biomedical application. *Circulation, 94*(12), 3334–3340. DOI: 10.1161/01.CIR.94.12.3334 PMID: 8989148

Laxmikeshav, K., Kumari, P., & Shankaraiah, N. (2022). Expedition of sulfur-containing heterocyclic derivatives as cytotoxic agents in medicinal chemistry: A decade update. *Medicinal Research Reviews, 42*(1), 513–575. DOI: 10.1002/med.21852 PMID: 34453452

Ledesma-Amaro, R., Santos, M. A., Jiménez, A., & Revuelta, J. L. (2013). 21 - Microbial production of vitamins. In McNeil, B., Archer, D., Giavasis, I., & Harvey, L. (Eds.), *Microbial Production of Food Ingredients, Enzymes and Nutraceuticals* (pp. 571–594). Woodhead Publishing. DOI: 10.1533/9780857093547.2.571

Lee, C. S., Kim, T. W., Kang, Y., Ju, Y., Ryu, J., Kong, H., Jang, Y.-S., Oh, D. E., Jang, S. J., Cho, H., Jeon, S., Kim, J., & Kim, T. H. (2022). Targeted drug delivery nanocarriers based on hyaluronic acid-decorated dendrimer encapsulating gold nanoparticles for ovarian cancer therapy. *Materials Today. Chemistry, 26*, 101083. DOI: 10.1016/j.mtchem.2022.101083

Lee, E., Zhang, X., Noda, T., Miyamoto, J., Kimura, I., Tanaka, T., Sakurai, K., Hatano, R., & Miki, T. (2021). Lecithin inclusion by α-cyclodextrin activates SREBP2 signaling in the gut and ameliorates postprandial hyperglycemia. *International Journal of Molecular Sciences, 22*(19), 10796. DOI: 10.3390/ijms221910796 PMID: 34639136

Lee, Y., Fukushima, S., Bae, Y., Hiki, S., Ishii, T., & Kataoka, K. A. (2007). Protein Nanocarrier from Charge-Conversion Polymer in Response to Endosomal pH. *Journal of the American Chemical Society, 129*(17), 5362–5363. DOI: 10.1021/ja071090b PMID: 17408272

Liang, Y., Mao, G., Dai, J., & Ma, Y. (2023). Biofunctionalized semiconductor quantum dots for virus detection. *Journal of Semiconductors, 44*(2), 023101. DOI: 10.1088/1674-4926/44/2/023101

Li, C., Zhang, Y., & Wang, J. (2023). Recent progress in metal–organic frameworks (MOFs) for electrocatalysis. *Industrial Chemistry & Materials, 1*(1), 9–38. DOI: 10.1039/D2IM00063F

Li, J., Li, X., Ni, X., Wang, X., Li, H., & Leong, K. W. (2006). Self-assembled supramolecular hydrogels formed by biodegradable PEO–PHB–PEO triblock copolymers and α-cyclodextrin for controlled drug delivery. *Biomaterials, 27*(22), 4132–4140. DOI: 10.1016/j.biomaterials.2006.03.025 PMID: 16584769

Li, N., Zhang, T., Xue, W., Zhao, Y., Zhu, B., Pei, X., & Xu, Z. (2024). Tuning composite nanofiltration membranes with γ-cyclodextrin for improved Mg2+/Li+ selectivity. *Separation and Purification Technology, 330*, 125206. DOI: 10.1016/j.seppur.2023.125206

Ling, X., Shen, Y., Sun, R., Zhang, M., Li, C., Mao, J., Xing, J., Sun, C., & Tu, J. (2015). Tumor-targeting delivery of hyaluronic acid–platinum (iv) nanoconjugate to reduce toxicity and improve survival. *Polymer Chemistry, 6*(9), 1541–1552. DOI: 10.1039/C4PY01592D

Lin, S., Chen, W., Alqahtani, M. S., Elkamchouchi, D. H., Ge, Y., Lu, Y., Zhang, G., & Wang, M. (2024). Exploring the therapeutic potential of layered double hydroxides and transition metal dichalcogenides through the convergence of rheumatology and nanotechnology using generative adversarial network. *Environmental Research, 241*, 117262. DOI: 10.1016/j.envres.2023.117262 PMID: 37839531

Li, P., Liu, Q., Xiang, Z., Wang, J., Wu, W.-X., & Yi, W.-J. (2023). ROS-responsive core crosslinked micelles by combination of enzymatic polymerization and thiol-ene click chemistry for anticancer drug delivery. *European Polymer Journal, 199*, 112473. DOI: 10.1016/j.eurpolymj.2023.112473

Liu, C., & Zhang, N. (2011). Nanoparticles in gene therapy: Principles, prospects, and challenges. In Villaverde, A. (Ed.), *Progress in molecular biology and translational science* (Vol. 104, pp. 509–562). Academic Press., DOI: 10.1016/B978-0-12-416020-0.00013-9

Liu, P., Chen, G., & Zhang, J. (2022). A review of liposomes as a drug delivery system: Current status of approved products, regulatory environments, and future perspectives. *Molecules (Basel, Switzerland), 27*(4), 1372. DOI: 10.3390/molecules27041372 PMID: 35209162

Liu, W., Li, Z., Wang, Z., Huang, Z., Sun, C., Liu, S., Jiang, Y., & Yang, H. (2023). Functional system based on glycyrrhizic acid supramolecular hydrogel: Toward polymorph control, stabilization, and controlled release. *ACS Applied Materials & Interfaces, 15*(6), 7767–7776. DOI: 10.1021/acsami.2c19903 PMID: 36732699

Liu, X., Niu, W., Liu, J., Cui, Z., Li, J., Zhang, Z., & Ju, J. (2023). Preparation of β-cyclodextrin and hydroxypropyl-β-cyclodextrin inclusion complexes of baicalein and evaluation of their effects on dextran sulfate sodium-induced acute ulcerative colitis in mice. *Journal of Drug Delivery Science and Technology*, *87*, 104714. DOI: 10.1016/j.jddst.2023.104714

Liu, X.-Y., Zhang, X., Yang, J.-B., Wu, C.-Y., Wang, Q., Lu, Z.-L., & Tang, Q. (2022). Multifunctional amphiphilic peptide dendrimer as nonviral gene vectors for effective cancer therapy via combined gene/photodynamic therapies. *Colloids and Surfaces. B, Biointerfaces*, *217*, 112651. DOI: 10.1016/j.colsurfb.2022.112651 PMID: 35759892

Liu, Y., Liu, J., Chen, S., Lei, T., Kim, Y., Niu, S., & Tok, J. B.-H. (2019). Soft and elastic hydrogel-based microelectronics for localized low-voltage neuromodulation. *Nature Biomedical Engineering*, *3*(1), 58–68. DOI: 10.1038/s41551-018-0335-6 PMID: 30932073

Liu, Z., Lin, W., & Liu, Y. (2022). Macrocyclic supramolecular assemblies based on hyaluronic acid and their biological applications. *Accounts of Chemical Research*, *55*(23), 3417–3429. DOI: 10.1021/acs.accounts.2c00462 PMID: 36380600

Liu, Z., Wang, H., Bu, Y., Wu, T., Chen, X., Yan, H., & Lin, Q. (2024). Fabrication of self-assembled micelles based on amphiphilic oxidized sodium alginate grafted oleoamine derivatives via Schiff base reduction amination reaction for delivery of hydrophobic food active ingredients. *International Journal of Biological Macromolecules*, *257*, 128653. DOI: 10.1016/j.ijbiomac.2023.128653 PMID: 38072345

Li, X., Bian, S., Zhao, M., Han, X., Liang, J., Wang, K., Jiang, Q., Sun, Y., Fan, Y., & Zhang, X. (2021). Stimuli-responsive biphenyl-tripeptide supramolecular hydrogels as biomimetic extracellular matrix scaffolds for cartilage tissue engineering. *Acta Biomaterialia*, *131*, 128–137. DOI: 10.1016/j.actbio.2021.07.007 PMID: 34245894

Li, Y., Liu, F., Abdiryim, T., & Liu, X. (2024). Cyclodextrin-derived materials: From design to promising applications in water treatment. *Coordination Chemistry Reviews*, *502*, 215613. DOI: 10.1016/j.ccr.2023.215613

Lobatto, M. E., Fuster, V., Fayad, Z. A., & Mulder, W. J. M. (2011). Perspectives and Opportunities for Nanomedicine in the Management of Atherosclerosis. *Nature Reviews. Cancer*, *10*(12), 963–963. DOI: 10.1038/nrd3614 PMID: 22015921

Loftsson, T., & Stefánsson, E. (2023). Topical dexamethasone delivery to the retina: An aqueous cyclodextrin-based microsuspension. *Journal of Drug Delivery Science and Technology*, *81*, 104281. DOI: 10.1016/j.jddst.2023.104281

Lue, J. T. (2007). Physical properties of nanomaterials. In *Encyclopedia of nanoscience and nanotechnology* (Vol. 10, pp. 1–46). American Scientific Publishers.

Lu, M., Chen, F., Noy, J. M., Lu, H., & Stenzel, M. H. (2017). Enhanced Antimetastatic Activity of the Ruthenium Anticancer Drug RAPTA-C Delivered in Fructose-Coated Micelles. *Macromolecular Bioscience, 17*(10), 1600513. DOI: 10.1002/mabi.201600513 PMID: 28234423

Luo, D., & Saltzman, W. M. (2000). Enhancement of transfection by physical concentration of DNA at the cell surface. *Nature Biotechnology, 18*(8), 893–895. DOI: 10.1038/78523 PMID: 10932162

Luo, W., Zheng, K., Kuang, H., Li, Z., Wang, J., & Mei, J. (2022). The potential of articaine as new generation of local anesthesia in dental clinics. *Revista de Medicina (São Paulo), 101*(48). PMID: 36482533

Luo, X., Li, Y., Chen, X., Song, Z., Liang, J., Liao, C., Zhu, Z., & Chen, L. (2019). (Z)-Tetrahydrothiophene and (Z)-tetrahydrothiopyran synthesis through nucleophilic substitution and intramolecular cycloaddition of alkynyl halides and EtOCS2K. *Organic & Biomolecular Chemistry, 17*(31), 7315–7319. DOI: 10.1039/C9OB01370A PMID: 31342046

Lu, Y., Li, Y., & Wu, W. (2016). Injected nanocrystals for targeted drug delivery. *Acta Pharmaceutica Sinica. B, 6*(2), 106–113. DOI: 10.1016/j.apsb.2015.11.005 PMID: 27006893

Lyu, Y., Wu, X., Yang, J., Wang, X., & Li, J. (2023). Protocol for preparing dynamic covalent macrocycles for codelivering genes and drugs to cancer cell lines. *STAR Protocols, 4*(3), 102350. DOI: 10.1016/j.xpro.2023.102350 PMID: 37314921

Macedo, V. S., Rodrigues, K. G. O., Ciribelli, N. N., Pereira, A. K. S., Cavallini, G. S., & Pereira, D. H. (2023). DFT study of bismuth subsalicylate adsorption on bow-tie dendrimer nanocarrier as drug delivery agents. *Chemical Physics Impact, 7*, 100381. DOI: 10.1016/j.chphi.2023.100381

Madanayake, S. N., Manipura, A., Thakuria, R., & Adassooriya, N. M. (2023). Opportunities and challenges in mechanochemical cocrystallization toward scaled-up pharmaceutical manufacturing. *Organic Process Research & Development, 27*(3), 409–422. DOI: 10.1021/acs.oprd.2c00314

Mahadev Madgule, N. (2023). Mechanical properties and microstructure of activated TIG welded similar joints of Inconel alloys by desirability approaches. *Materials Today: Proceedings, 77*, 528–533. DOI: 10.1016/j.matpr.2022.12.250

Mak, J. Y. W., Xu, W., & Fairlie, D. P. (2017). Thiazoles in Peptides and Peptidomimetics. In Lubell, W. D. (Ed.), *Peptidomimetics I* (pp. 235–266). Springer International Publishing.

Malik, S., Muhammad, K., & Waheed, Y. (2023). Emerging applications of nanotechnology in healthcare and medicine. *Molecules (Basel, Switzerland), 28*(18), 6624. DOI: 10.3390/molecules28186624 PMID: 37764400

Manimaran, V., Nivetha, R. P., Tamilanban, T., Narayanan, J., Vetriselvan, S., Fuloria, N. K., Chinni, S. V., Sekar, M., Fuloria, S., Wong, L. S., Biswas, A., Ramachawolran, G., & Selvaraj, S. (2023). Nanogels are novel drug nanocarriers for CNS drug delivery. *Frontiers in Molecular Biosciences, 10*, 1232109. DOI: 10.3389/fmolb.2023.1232109 PMID: 37621994

Manjit, & Mishra, B. (2022). Research and Development of Supramolecules as Anticancer Drugs. In N. Goel & N. Kumar (Eds.), *Pharmaceutical Applications of Supramolecules* (pp. 55-87). Springer International Publishing. DOI: 10.1007/978-3-031-21900-9_3

Manna, K., Dey, S., Phukan, A., Dey, S. K., & Pal, S. (2024). Alginic acid-based pH and thermo responsive reversible switched polymeric micelle via RAFT polymerization. *Polymer, 291*, 126581. DOI: 10.1016/j.polymer.2023.126581

Marczylo, E. L., Jacobs, M. N., & Gant, T. W. (2016). Environmentally induced epigenetic toxicity: Potential public health concerns. *Critical Reviews in Toxicology, 2016*(8), 676–700. DOI: 10.1080/10408444.2016.1175417 PMID: 27278298

Markitanov, Y. M., Timoshenko, V. M., Rudenko, T. V., Rusanov, E. B., & Shermolovich, Y. G. (2019). 3-Functional substituted 4-trifluoromethyl tetrahydrothiophenes via [3 + 2]-cycloaddition reactions. *Journal of Sulfur Chemistry, 40*(6), 629–640. DOI: 10.1080/17415993.2019.1633326

Marquet, F., Hagen, H., Stanchieri, M., Beinier, V. S., Grasso, G., Danani, A., Patrulea, V., & Borchard, G. (2024). Clickable polyethyleneimine incorporated into triblock copolymeric micelles as an efficient platform in the delivery of siRNA to NSCLC cells. *International Journal of Pharmaceutics, 649*, 123632. DOI: 10.1016/j.ijpharm.2023.123632 PMID: 38000648

Matea, C. T., Mocan, T., Tabaran, F., Pop, T., Mosteanu, O., Puia, C., Iancu, C., & Mocan, L. (2017). Quantum dots in imaging, drug delivery and sensor applications. *International Journal of Nanomedicine, 12*, 5421–5431. DOI: 10.2147/IJN.S138624 PMID: 28814860

Matos, M. P. S. R., Correia, A. A. S., & Rasteiro, M. G. (2017). Application of carbon nanotubes to immobilize heavy metals in contaminated soils. *Journal of Nanoparticle Research*, *19*(4), 126. DOI: 10.1007/s11051-017-3830-x

Matsumura, Y., & Maeda, H. (1986). A New Concept for Macromolecular Therapeutics in Cancer Chemotherapy: Mechanism of Tumoritropic Accumulation of Proteins and the Antitumor Agent Smancs. *Cancer Research*, *46*, 6387–6392. PMID: 2946403

Mauri, E., Giannitelli, S. M., Trombetta, M., & Rainer, A. (2021). Synthesis of nanogels: Current trends and future outlook. *Gels (Basel, Switzerland)*, *7*(1), 36. DOI: 10.3390/gels7020036 PMID: 33805279

Mauter, M. S., & Elimelech, M. (2008). Environmental applications of carbon-based nanomaterials. *Environmental Science & Technology*, *42*(16), 5843–5859. DOI: 10.1021/es8006904 PMID: 18767635

Ma, X., Yu, X., Huang, H., Zhou, Y., & Song, Q. (2020). Synthesis of Thiazoles and Isothiazoles via Three-Component Reaction of Enaminoesters, Sulfur, and Bromodifluoroacetamides/Esters. *Organic Letters*, *22*(14), 5284–5288. DOI: 10.1021/acs.orglett.0c01275 PMID: 32603128

Ma, X., & Zhao, Y. (2015). Biomedical Applications of Supramolecular Systems Based on Host-Guest Interactions. *Chemical Reviews*, *115*(15), 7794–7839. DOI: 10.1021/cr500392w PMID: 25415447

Ma, Z., & Moulton, B. (2011). Recent advances of discrete coordination complexes and coordination polymers in drug delivery. *Coordination Chemistry Reviews*, *255*(15-16), 1623–1641. DOI: 10.1016/j.ccr.2011.01.031

McKinlay, A. C., Xiao, B., Wragg, D. S., Wheatley, P. S., Megson, I. L., & Morris, R. E. (2008). Exceptional behavior over the whole adsorption– storage– delivery cycle for NO in porous metal organic frameworks. *Journal of the American Chemical Society*, *130*(31), 10440–10444. DOI: 10.1021/ja801997r PMID: 18627150

Meenatchi, V., Sood, A., Bhaskar, R., & Soo Han, S. (2024). Electrospun hydroxypropyl-β-cyclodextrin nanofibers loaded with anti-inflammatory drugs for potential therapeutic wound dressings. *Journal of Molecular Liquids*, *393*, 123558. DOI: 10.1016/j.molliq.2023.123558

Mehata, A. K., Setia, A., Vikas, V., Malik, A. K., Hassani, R., Dailah, H. G., Alhazmi, H. A., Albarraq, A. A., Mohan, S., & Muthu, M. S. (2023). Vitamin E TPGS-based nanomedicine, nanotheranostics, and targeted drug delivery: Past, present, and future. *Pharmaceutics*, *15*(2), 722. DOI: 10.3390/pharmaceutics15030722 PMID: 36986583

Mekuye, B., & Abera, B. (2023). Nanomaterials: An overview of synthesis, classification, characterization, and applications. *Nano Select*, *4*(8), 486–501. DOI: 10.1002/nano.202300038

Melamed, J. R., Riley, R. S., Valcourt, D. M., & Day, E. S. (2016). Using Gold Nanoparticles To DIsrupt the Tumor Microenvironment: An Emerging Therapeutic Strategy. *ACS Nano*, *10*(12), 10631–10635. DOI: 10.1021/acsnano.6b07673 PMID: 28024339

Michlewska, S., Garaiova, Z., Šubjakova, V., Hołota, M., Kubczak, M., Grodzicka, M., Okła, E., Naziris, N., Balcerzak, Ł., Ortega, P., de la Mata, F. J., Hianik, T., Waczulikova, I., Bryszewska, M., & Ionov, M. (2023). Lipid-coated ruthenium dendrimer conjugated with doxorubicin in anticancer drug delivery: Introducing protocols. *Colloids and Surfaces. B, Biointerfaces*, *227*, 113371. DOI: 10.1016/j.colsurfb.2023.113371 PMID: 37244201

Mie, G. (1908). Beiträge zur Optik trüber Medien, speziell kolloidaler Metalllösungen. *Annalen der Physik*, *330*(3), 377–445. DOI: 10.1002/andp.19083300302

Miller, M., & Megson, I. (2007). Recent developments in nitric oxide donor drugs. *British Journal of Pharmacology*, *151*(3), 305–321. DOI: 10.1038/sj.bjp.0707224 PMID: 17401442

Miller, S. R., Heurtaux, D., Baati, T., Horcajada, P., Grenèche, J.-M., & Serre, C. (2010). Biodegradable therapeutic MOFs for the delivery of bioactive molecules. *Chemical Communications*, *46*(25), 4526–4528. DOI: 10.1039/c001181a PMID: 20467672

Mironov, V., Kasyanov, V., & Markwald, R. R. (2008). Nanotechnology in vascular tissue engineering: From nanoscaffolding towards rapid vessel biofabrication. *Trends in Biotechnology*, *26*(6), 338–344. DOI: 10.1016/j.tibtech.2008.03.001 PMID: 18423666

Mirza, S., Ahmad, M. S., & Shah, M. I. A. (2020). Magnetic nanoparticles: Drug delivery and bioimaging applications. In Shah, M. R., Imran, M., & Ullah, S. (Eds.), *Metal nanoparticles for drug delivery and diagnostic applications* (pp. 189–213). Elsevier. DOI: 10.1016/B978-0-12-816960-5.00011-2

Mishra, D. R., Darjee, S. M., Bhatt, K. D., Modi, K. M., & Jain, V. K. (2015). Calix protected gold nanobeacon as turn-off fluorescent sensor for phenylalanine. *Journal of Inclusion Phenomena and Macrocyclic Chemistry*, *82*(3–4), 425–436. DOI: 10.1007/s10847-015-0509-8

Miyamae, K., Nakahata, M., Takashima, Y., & Harada, A. (2015). Self-Healing, Expansion-Contraction, and Shape-Memory Properties of a Preorganized Supramolecular Hydrogel through Host-Guest Interactions. *Angewandte Chemie International Edition in English, 54*(31), 8984–8987. DOI: 10.1002/anie.201502957 PMID: 26080301

Modan, E. M., & Plăia u, A. G. (2020). Advantages and disadvantages of chemical methods in the elaboration of nanomaterials. *The Annals of "Dunarea de Jos" University of Galati. Fascicle IX. Metallurgy and Materials Science, 43*(1), 53–60. DOI: 10.35219/mms.2020.1.06

Moghimi, S. M., Hunter, A. C., & Murray, J. C. (2005). Nanomedicine: Current status and prospects. *The FASEB Journal, 19*(3), 311–330. DOI: 10.1096/fj.04-2747rev PMID: 15746175

Mohammad Asghari, B., Samadi Zadeh, M., Ahmad Panahi, H., Hesami Tackallou, S., & Safaeijavan, R. (2023). Surface modification of nanodiamond with pH/thermo dual responsive polymer and hyperbranched dendrimer as a near-infrared photothermal-triggered drug delivery for cancer therapy. *Journal of Molecular Liquids, 390*, 123155. DOI: 10.1016/j.molliq.2023.123155

Mohammad, A., Singh, S., Swain, S., & Ghose, D. (2024). Quality by design enabled β-cyclodextrin complexes of lisinopril by kneading method: Improved solubility and bioavailability [Article]. *Current Materials Science, 17*(2), 135–147. DOI: 10.2174/2666145416666230407100318

Mohammadjani, N., Karimi, S., Moetasam Zorab, M., Ashengroph, M., & Alavi, M. (2024). Comparative molecular docking and toxicity between carbon-capped metal oxide nanoparticles and standard drugs in cancer and bacterial infections. *BioImpacts, 14*(3), 27778. DOI: 10.34172/bi.2023.27778 PMID: 38505671

Molineux, G. (2002). Pegylation: Engineering improved pharmaceuticals for enhanced therapy. *Cancer Treatment Reviews, 28*, 13–16. DOI: 10.1016/S0305-7372(02)80004-4 PMID: 12173407

Monajati, M., Tamaddon, A. M., Abolmaali, S. S., Yousefi, G., Javanmardi, S., Borandeh, S., Heidari, R., Azarpira, N., & Dinarvand, R. (2023). L-asparaginase immobilization in supramolecular nanogels of PEG-grafted poly HPMA and bis(α-cyclodextrin) to enhance pharmacokinetics and lower enzyme antigenicity. *Colloids and Surfaces. B, Biointerfaces, 225*, 113234. DOI: 10.1016/j.colsurfb.2023.113234 PMID: 36934612

Morbioli, G. G., Mazzoni, T. S., & Furlan, A. P. (2017). Technical aspects and challenges of colorimetric detection with microfluidic paper-based analytical devices (μPADs)—A review. *Analytica Chimica Acta*, *970*, 1–22. DOI: 10.1016/j.aca.2017.03.037 PMID: 28433054

Mourdikoudis, S., Pallares, R. M., & Thanh, N. T. K. (2018). Characterization techniques for nanoparticles: Comparison and complementarity upon studying nanoparticle properties. *Nanoscale*, *10*(27), 12871–12934. DOI: 10.1039/C8NR02278J PMID: 29926865

Muhammed, M. T., & Aki-Yalcin, E. (2024). Molecular docking: Principles, advances, and its applications in drug discovery. *Letters in Drug Design & Discovery*, *21*(3), 480–495. DOI: 10.2174/1570180819666220922103109

Mulvaney, P., Parak, W. J., Caruso, F., & Weiss, P. S. (2016). Standardizing Nanomaterials. *ACS Nano*, *10*(11), 9763–9764. DOI: 10.1021/acsnano.6b07629 PMID: 27934085

Mura, S., Nicolas, J., & Couvreur, P. (2013). Stimuli-responsive nanocarriers for drug delivery. *Nature Materials*, *12*(11), 991–1003. DOI: 10.1038/nmat3776 PMID: 24150417

Musazzi, U. M., Franzè, S., Condorelli, F., Minghetti, P., & Caliceti, P. (2023). Feeding next-generation nanomedicines to Europe: Regulatory and quality challenges. *Advanced Healthcare Materials*, *xx*(30), 2301956. DOI: 10.1002/adhm.202301956 PMID: 37718353

Mustafa, W. W., Khoder, M., Abdelkader, H., Singer, R., & Alany, R. G. (2024). Interactions of cyclodextrins and their hydroxyl derivatives with etodolac: Solubility and dissolution enhancement. *Current Drug Delivery*, *21*(1), 126–139. DOI: 10.2174/1567201820666230320164210 PMID: 36959150

Nagaraju, S., Sathish, K., Satyanarayana, N., Paplal, B., & Kashinath, D. (2020). Regioselective synthesis of spiro isoxazole-oxindole-tetrahydrothiophene hybrids via cascade reactions under catalyst-free conditions. *Journal of Heterocyclic Chemistry*, *57*(1), 469–476. DOI: 10.1002/jhet.3803

Nagavarma, B., Yadav, H. K., Ayaz, A., Vasudha, L., & Shivakumar, H. (2012). Different techniques for preparation of polymeric nanoparticles—A review. *Asian Journal of Pharmaceutical and Clinical Research*, *5*(3), 16–23.

Narenderan, S. T., Meyyanathan, S. N., & Babu, B. (2020). Review of pesticide residue analysis in fruits and vegetables. Pre-treatment, extraction and detection techniques. *Food Research International, 133*, 109141. DOI: 10.1016/j.foodres.2020.109141 PMID: 32466907

Newman, A. D., Brown, D. R., Siril, P., Lee, A. F., & Wilson, K. (2006). Structural studies of high dispersion H 3 PW 12 O 40/SiO 2 solid acid catalysts. *Physical Chemistry Chemical Physics, 8*(24), 2893–2902. DOI: 10.1039/b603979k PMID: 16775645

Ngoc Toan, V., & Dinh Thanh, N. (2021). Novel thiazoline–coumarin hybrid compounds containing sugar moieties: Synthesis, biological evaluation and molecular docking study as antiproliferative agents. *New Journal of Chemistry, 45*(24), 10636–10653. DOI: 10.1039/D1NJ00680K

Nichols, J. W., & Bae, Y. H. (2014). EPR: Evidence and fallacy. *Journal of Controlled Release, 190*, 451–464. DOI: 10.1016/j.jconrel.2014.03.057 PMID: 24794900

Nicolopoulou-Stamati, P., Maipas, S., Kotampasi, C., Stamatis, P., & Hens, L. (2016). Chemical pesticides and human health: The urgent need for a new concept in agriculture. *Frontiers in Public Health, 4*, 148. DOI: 10.3389/fpubh.2016.00148 PMID: 27486573

Nie, Q., Wang, C., Xu, H., Mittal, P., Naeem, A., Zhou, P., Li, H., Zhang, Y., Guo, T., Sun, L., & Zhang, J. (2023). Highly efficient pulmonary delivery of levo-tetrahydropalmatine using γ-cyclodextrin metal-organic framework as a drug delivery platform. *Journal of Drug Delivery Science and Technology, 82*, 104296. DOI: 10.1016/j.jddst.2023.104296

Niu, H., Wang, J., Jia, Q., Wang, Q., Zhao, J., & Yan, F. (2024). QSPR models for complexation performance of α-cyclodextrin and β-cyclodextrin complexes by norm indices. *Chemical Engineering Science, 284*, 119484. DOI: 10.1016/j.ces.2023.119484

Nori, Z. Z., Bahadori, M., Moghadam, M., Tangestaninejad, S., Mirkhani, V., Mohammadpoor-Baltork, I., Jafari, S. S., Emamzadeh, R., & Alem, H. (2023). Synthesis and characterization of a new gold-coated magnetic nanoparticle decorated with a thiol-containing dendrimer for targeted drug delivery, hyperthermia treatment and enhancement of MRI contrast agent. *Journal of Drug Delivery Science and Technology, 81*, 104216. DOI: 10.1016/j.jddst.2023.104216

Nowack, B., Baalousha, M., Bornhöft, N., Chaudhry, Q., Cornelis, G., Cotterill, J., Gondikas, A., Hassellöv, M., Lead, J., Mitrano, D. M., von der Kammer, F., & Wontner-Smith, T. (2015). Progress towards the validation of modelled environmental concentrations of engineered nanomaterials by analytical measurements. *Environmental Science. Nano*, *2*(5), 421–428. DOI: 10.1039/C5EN00100E

Nurunnabi, M., Khatun, Z., Reeck, G. R., Lee, D. Y., & Lee, Y. (2014). Photoluminescent Graphene Nanoparticles for Cancer Phototherapy and Imaging. *ACS Applied Materials & Interfaces*, *6*(15), 12413–12421. DOI: 10.1021/am504071z PMID: 25054687

O'Donnell, J. A., Gelone, S. P., & Safdar, A. (2015). 37 - Topical Antibacterials. In Bennett, J. E., Dolin, R., & Blaser, M. J. (Eds.), *Mandell, Douglas, and Bennett's Principles and Practice of Infectious Diseases* (8th ed., pp. 452–462.e452). W.B. Saunders. DOI: 10.1016/B978-1-4557-4801-3.00037-0

Oliveira, A. R., dos Santos, F. A., Ferreira, L. P. L., Pitta, M. G. R., Silva, M. V. O., Cardoso, M. V. O., Pinto, A. F., Marchand, P., de Melo Rêgo, M. J. B., & Leite, A. C. L. (2021). Synthesis, anticancer activity and mechanism of action of new phthalimido-1,3-thiazole derivatives. *Chemico-Biological Interactions*, *347*, 109597. DOI: 10.1016/j.cbi.2021.109597 PMID: 34303695

Onoue, S., Yamada, S., & Chan, H.-K. (2014). Nanodrugs: Pharmacokinetics and safety. *International Journal of Nanomedicine*, *9*, 1025–1037. DOI: 10.2147/IJN.S38378 PMID: 24591825

Ozboyaci, M., Kokh, D. B., Corni, S., & Wade, R. C. (2016). Modeling and simulation of protein–surface interactions: Achievements and challenges. *Quarterly Reviews of Biophysics*, *49*, e4. DOI: 10.1017/S0033583515000256 PMID: 26821792

Panda, S. S., & Hansen, D. W. (2023). Microwave-assisted synthesis of five-membered azaheterocyclic systems using nanocatalysts. In Bhunia, S., Kumar, B., Singh, P., Oraon, R., & Kim, K.-H. (Eds.), *Nanoparticles in Green Organic Synthesis* (pp. 255–278). Elsevier. DOI: 10.1016/B978-0-323-95921-6.00007-X

Pandey, G., Phatale, V., Khairnar, P., Kolipaka, T., Shah, S., Famta, P., Jain, N., Srinivasarao, D. A., Rajinikanth, P. S., Raghuvanshi, R. S., & Srivastava, S. (2024). Supramolecular self-assembled peptide-engineered nanofibers: A propitious proposition for cancer therapy. *International Journal of Biological Macromolecules*, *256*, 128452. DOI: 10.1016/j.ijbiomac.2023.128452 PMID: 38042321

Pandey, M., Lang, H. X., Loh, J. S., Chai, Y. L., Tee, H. L., Mayuren, J., Candasamy, M., Gorain, B., Jain, N., Gupta, G., & Dua, K. (2023). Dendrimer platform against prostate cancer: Recent update on new horizon of treatment. *Journal of Drug Delivery Science and Technology*, *85*, 104589. DOI: 10.1016/j.jddst.2023.104589

Pan, Z., Zhao, X., Li, Q., Zhang, Z., & Liu, Y. (2023). Recent advances in supramolecular-macrocycle-based nanomaterials in cancer treatment. *Molecules (Basel, Switzerland)*, *28*(3), 1241. DOI: 10.3390/molecules28031241 PMID: 36770907

Parveen, N., Sheikh, A., Abourehab, M. A. S., Karwasra, R., Singh, S., & Kesharwani, P. (2023). Self-nanoemulsifying drug delivery system for pancreatic cancer. *European Polymer Journal*, *190*, 111993. DOI: 10.1016/j.eurpolymj.2023.111993

Pelaz, B., Alexiou, C., Alvarez-Puebla, R. A., Alves, F., Andrews, A. M., Ashraf, S., Balogh, L. P., Ballerini, L., Bestetti, A., Brendel, C., Bosi, S., Carril, M., Chan, W. C. W., Chen, C., Chen, X., Chen, X., Cheng, Z., Cui, D., Du, J., & Parak, W. J. (2017). Diverse applications of nanomedicine. *ACS Nano*, *11*(3), 2313–2381. DOI: 10.1021/acsnano.6b06040 PMID: 28290206

Peng, G., Cai, J., Wang, Z., Zhang, W., Xu, J., Zhang, D., & Gong, D. (2024). Facile fabrication of diatomite biosilica-based nasal drug delivery vehicle for enhanced treatment of allergic rhinitis. *Colloids and Surfaces. B, Biointerfaces*, *234*, 113715. DOI: 10.1016/j.colsurfb.2023.113715 PMID: 38134821

Perles, J., Snejko, N., Iglesias, M., & Monge, M. Á. (2009). 3D scandium and yttrium arenedisulfonate MOF materials as highly thermally stable bifunctional heterogeneous catalysts. *Journal of Materials Chemistry*, *19*(36), 6504–6511. DOI: 10.1039/b902954k

Pinelli, F., Ponti, M., Delleani, S., Pizzetti, F., Vanoli, V., Vangosa, F. B., Castiglione, F., Haugen, H., Nogueira, L. P., Rossetti, A., Rossi, F., & Sacchetti, A. (2023). β-Cyclodextrin functionalized agarose-based hydrogels for multiple controlled drug delivery of ibuprofen. *International Journal of Biological Macromolecules*, *252*, 126284. DOI: 10.1016/j.ijbiomac.2023.126284 PMID: 37572821

Pinzi, L., & Rastelli, G. (2019). Molecular docking: Shifting paradigms in drug discovery. *International Journal of Molecular Sciences*, *20*(18), 4331. DOI: 10.3390/ijms20184331 PMID: 31487867

Pitingolo, G., & Nastruzzi, C. (2023). Production of supramolecular aggregates by microfluidic platforms: application to food industries. In Anandharamakrishnan, C., & Dutta, S. (Eds.), *Liposomal Encapsulation in Food Science and Technology* (pp. 169–187). Academic Press. DOI: 10.1016/B978-0-12-823935-3.00009-6

Pitsillides, C. M., Joe, E. K., Wei, X., Anderson, R. R., & Lin, C. P. (2003). Selective cell targeting with light-absorbing microparticles and nanoparticles. *Biophysical Journal, 84*(6), 4023–4032. DOI: 10.1016/S0006-3495(03)75128-5 PMID: 12770906

Pomal, N. C., Vyas, V. K., & Jain, V. K. (2021). Functionalized silver nanoparticles as colorimetric and fluorimetric sensor for environmentally toxic mercury ions: An overview. *Journal of Fluorescence, 31*(3), 635–649. DOI: 10.1007/s10895-021-02699-z PMID: 33609215

Pomerantseva, E., Dunn, B., & Meng, F. (2019). Energy storage: The future enabled by nanomaterials. *Science, 366*(6468), eaan8285. Advance online publication. DOI: 10.1126/science.aan8285 PMID: 31753970

Pourbadiei, B., Adlsadabad, S. Y., Rahbariasr, N., & Pourjavadi, A. (2023). Synthesis and characterization of dual light/temperature-responsive supramolecular injectable hydrogel based on host-guest interaction between azobenzene and starch-grafted β-cyclodextrin: Melanoma therapy with paclitaxel. *Carbohydrate Polymers, 313*, 120667. DOI: 10.1016/j.carbpol.2023.120667 PMID: 37182982

Priya, A. S., Sivakamavalli, J., Vaseeharan, B., Rajamohan, R., Lee, Y. R., & Thambusamy, S. (2023). Interaction of torsemide with native cyclodextrin through inclusion complexation: In vitro drug release, antibacterial and antibiofilm activities. *Journal of Molecular Structure, 1286*, 135624. DOI: 10.1016/j.molstruc.2023.135624

Puhlmann, N., Vidaurre, R., & Kümmerer, K. (2024). Designing greener active pharmaceutical ingredients: Insights from pharmaceutical industry into drug discovery and development. *European Journal of Pharmaceutical Sciences, 192*, 106614. DOI: 10.1016/j.ejps.2023.106614 PMID: 37858896

Qiu, Y., Shang, K., Xu, N., Chen, P., Gao, H., Mu, H., Feng, W., & Duan, J. (2023). Clearance of intracellular bacterial infections by hyaluronic acid-based ROS responsive drug delivery micelles. *International Journal of Biological Macromolecules, 245*, 125506. DOI: 10.1016/j.ijbiomac.2023.125506 PMID: 37356691

Rahman, M., Almalki, W. H., Afzal, O., Alfawaz Altamimi, A. S., Najib Ullah, S. N. M., Abul Barkat, M., & Beg, S. (2023). Chiral-engineered supraparticles: Emerging tools for drug delivery. *Drug Discovery Today, 28*(1), 103420. DOI: 10.1016/j.drudis.2022.103420 PMID: 36309193

Raia, N. R., Partlow, B. P., McGill, M., Kimmerling, E. P., Ghezzi, C. E., & Kaplan, D. L. (2017). Enzymatically crosslinked silk-hyaluronic acid hydrogels. *Biomaterials, 131*, 58–67. DOI: 10.1016/j.biomaterials.2017.03.046 PMID: 28376366

Rajaram, R., Angaiah, S., & Lee, Y. R. (2023). Polymer supported electrospun nanofibers with supramolecular materials for biological applications – a review. *International Journal of Polymeric Materials*, *72*(13), 1042–1058. DOI: 10.1080/00914037.2022.2075871

Rajput, A., Shevalkar, G., Pardeshi, K., & Pingale, P. (2023). Computational nanoscience and technology. *OpenNano*, *12*, 100147. DOI: 10.1016/j.onano.2023.100147

Ramakrishnan, T., Mohan Gift, M. D., Chitradevi, S., Jegan, R., & Subha Hency Jose, P. (2022). Study of Numerous Resins Used in Polymer Matrix Composite Materials. Advances in Materials Science and Engineering.DOI: 10.1155/2022/1088926

Ramanathan, R., Oluwafemi, O. S., & Kalarikkal, N. (2014). Green synthesis of nanomaterials using biological routes. In *Nanomaterials for environmental protection* (pp. 329–348). Wiley. DOI: 10.1002/9781118845530.ch20

Ram, R. N., Gupta, D. K., & Soni, V. K. (2016). Copper(I)-Promoted Synthesis of Highly Substituted and Functionalized Tetrahydrothiophenes. *European Journal of Organic Chemistry*, *2016*(20), 3434–3440. DOI: 10.1002/ejoc.201600497

Rani, K., Chauhan, C., & Kaur, H. (2014). Potential and automotive applications of nanomaterials in combating cancer and stem cell therapy: An informative overview. *Journal of Nanotechnology and Smart Materials*, *1*(1), 1–6.

Rao, C., & Biswas, K. (2009). Characterization of nanomaterials by physical methods. *Annual Review of Analytical Chemistry (Palo Alto, Calif.)*, *2*(1), 435–462. DOI: 10.1146/annurev-anchem-060908-155236 PMID: 20636070

Rassu, G., Sorrenti, M., Catenacci, L., Pavan, B., Ferraro, L., Gavini, E., Bonferoni, M. C., Giunchedi, P., & Dalpiaz, A. (2021). Versatile nasal application of cyclodextrins: Excipients and/or actives? *Pharmaceutics*, *13*(8), 1180. DOI: 10.3390/pharmaceutics13081180 PMID: 34452141

Ravoo, B. J. (2012). Vesicles in Supramolecular Chemistry. In *Supramolecular Chemistry*. DOI: 10.1002/9780470661345.smc032

Riehemann, K., Schneider, S. W., Luger, T. A., Godin, B., Ferrari, M., & Fuchs, H. (2009). Nanomedicine—Challenge and perspectives. *Angewandte Chemie International Edition*, *48*(5), 872–897. DOI: 10.1002/anie.200802585 PMID: 19142939

Rios-Doria, J., Durham, N., Wetzel, L., Rothstein, R., Chesebrough, J., Holoweckyj, N., Zhao, W., Leow, C. C., & Hollingsworth, R. (2015). Doxil synergizes with cancer immunotherapies to enhance antitumor responses in syngeneic mouse models. *Neoplasia : an International Journal for Oncology Research*, *17*(8), 661–670. DOI: 10.1016/j.neo.2015.08.004 PMID: 26408258

Robles-García, M. A., Rodríguez-Félix, F., Marquez-Rios, E., Aguilar, J. A., Barrera-Rodríguez, A., Aguilar, J., Ruiz-Cruz, S., & Del-Toro-Sánchez, C. L. (2016). Applications of nanotechnology in the agriculture, food, and pharmaceuticals. *Journal of Nanoscience and Nanotechnology*, *16*(8), 8188–8207. DOI: 10.1166/jnn.2016.12925

Rosa Rodes Solanes, Herrero, Royo, & Gòmez. (2016). Spain Patent No. EP 3 040 334 A1. European Patent Office: E. P. Application.

Rouhani, M., Samih, M. A., & Kalantari, S. (2012). Insecticide effect of silver and zinc nanoparticles against *Aphis nerii* Boyer de Fonscolombe (Hemiptera: Aphididae). *Chilean Journal of Agricultural Research*, *72*(4), 590–594. DOI: 10.4067/S0718-58392012000400020

Roy, A., Sharma, A., Yadav, S., Jule, L. T., & Krishnaraj, R. (2021). Nanomaterials for remediation of environmental pollutants. *Bioinorganic Chemistry and Applications*, *1764647*, 1–16. Advance online publication. DOI: 10.1155/2021/1764647 PMID: 34992641

Ruan, D., Qin, L., Chen, R., Xu, G., Su, Z., Cheng, J., Xie, S., Cheng, F., & Ko, F. (2020). Transparent PAN:TiO2 and PAN-co-PMA:TiO2 nanofiber composite membranes with high efficiency in particulate matter pollutants filtration. *Nanoscale Research Letters*, *15*(1), 7. DOI: 10.1186/s11671-019-3225-2 PMID: 31933114

Rubim, A. M., Rubenick, J. B., Vendrame, L. O., Zanella, I., Rolim, C. M. B., & Rhoden, C. R. B. (2024). Formulation and characterization of amiodarone-methyl-beta-cyclodextrin inclusion complexes: A molecular modeling perspective. *Journal of Molecular Graphics & Modelling*, *126*, 108639. DOI: 10.1016/j.jmgm.2023.108639 PMID: 37774479

Rujiralai, T., Thongyoy, S., & Kaewkan, P. (2020). Hydrolyzed product mediated aggregation of l-cysteine-modified gold nanoparticles as a colorimetric probe for carbamate residues in chilis. *Analytical Letters*, *53*(4), 574–588. DOI: 10.1080/00032719.2019.1659804

Saad, A. M., Alabdali, A. Y. M., Ebaid, M., Salama, E., El-Saadony, M. T., Selim, S., Safhi, F. A., ALshamrani, S. M., Abdalla, H., Mahdi, A. H. A., & El-Saadony, F. M. A. (2022). Impact of green chitosan nanoparticles fabricated from shrimp processing waste as a source of nano nitrogen fertilizers on the yield quantity and quality of wheat (*Triticum aestivum* L.) cultivars. *Molecules (Basel, Switzerland)*, *27*(17), 5640. DOI: 10.3390/molecules27175640 PMID: 36080408

Sabela, M., Balme, S., & Bisetty, K. (2017). A review of gold and silver nanoparticle-based colorimetric sensing assays. *Advanced Engineering Materials*, *19*(12), 1700270. DOI: 10.1002/adem.201700270

Safari, J. B., Mona, L. B., Sekaleli, B. T., Avudi, B. K.-N., Isamura, B. K., Mukubwa, G. K., Salami, S. A., Mbinze, J. K., Lobb, K. A., Krause, R. W. M., & Nkanga, C. I. (2023). Inclusion complexation and liposomal encapsulation of an isoniazid hydrazone derivative in cyclodextrin for pH-dependent controlled release. *Journal of Drug Delivery Science and Technology*, *81*, 104302. DOI: 10.1016/j.jddst.2023.104302

Saha, S., Ali, M. R., Khaleque, M. A., Bacchu, M. S., Aly Saad Aly, M., & Khan, M. Z. H. (2023). Metal oxide nanocarrier for targeted drug delivery toward the treatment of global infectious diseases: A review. *Journal of Drug Delivery Science and Technology*, *86*, 104728. DOI: 10.1016/j.jddst.2023.104728

Sahu, K. M., Patra, S., & Swain, S. K. (2023). Host-guest drug delivery by β-cyclodextrin assisted polysaccharide vehicles: A review. *International Journal of Biological Macromolecules*, *240*, 124338. DOI: 10.1016/j.ijbiomac.2023.124338 PMID: 37030461

Saikia, S., & Bordoloi, M. (2019). Molecular docking: Challenges, advances and its use in drug discovery perspective. *Current Drug Targets*, *20*(5), 501–521. DOI: 10.2174/1389450119666181022153016 PMID: 30360733

Sainz, V., Conniot, J., Matos, A. I., Peres, C., Zupan i , E., Moura, L., Silva, L. C., Florindo, H. F., & Gaspar, R. S. (2015). Regulatory aspects on nanomedicines. *Biochemical and Biophysical Research Communications*, *468*(3), 504–510. DOI: 10.1016/j.bbrc.2015.08.023 PMID: 26260323

Salido-Fortuna, S., Salgado, A., Castro-Puyana, M., & Marina, M. L. (2024). Study on the interactions between RS-licarbazepine and carboxyethylated cyclodextrins by electrokinetic chromatography and nuclear magnetic resonance. *Microchemical Journal*, *196*, 109621. DOI: 10.1016/j.microc.2023.109621

Sanhai, W. R., Sakamoto, J. H., Canady, R., & Ferrari, M. (2008). Seven challenges for nanomedicine. *Nature Nanotechnology*, *3*(5), 242–244. DOI: 10.1038/nnano.2008.114 PMID: 18654511

Sanphui, P., Fayaz, T. K. S., Chanduluru, H. K., Lal, P., Ghosh, A., & Chernyshev, V. V. (2024). Structural analysis of anti-retroviral drug raltegravir and its potential impurity C: Investigation of solubility and stability. *CrystEngComm*, *xx*(4), xx–xx. DOI: 10.1039/D3CE01071F

Schmidt, B. V. K. J., & Barner-Kowollik, C. (2016). A Supramolecular Approach to Macromolecular Self-Assembly: Cyclodextrin Host/Guest Complexes. In *Macromolecular Self-assembly* (pp. 1-32). Academic Press.

Selvakumar, P. (2024). AI-Based Cameras to Detect a Leak of Chemicals or Gas in Real Time. In Industry Applications of Thrust Manufacturing: Convergence with Real-Time Data and AI (pp. 221–236). IGI Global. DOI: 10.4018/979-8-3693-4276-3.ch009

Selvakumar, P. (2024). Prediction of Earthquakes, Volcanic Eruptions, Tornadoes, Wildfires, and Droughts Through Artificial Intelligence. In Utilizing AI and Machine Learning for Natural Disaster Management (pp. 133–145). IGI Global.DOI: 10.4018/979-8-3693-3362-4.ch008

Sengul, A. B., & Asmatulu, E. (2020). Toxicity of metal and metal oxide nanoparticles: A review. *Environmental Chemistry Letters*, *18*(5), 1659–1683. DOI: 10.1007/s10311-020-01033-6

Sepulveda, A. F., Kumpgdee-Vollrath, M., Franco, M. K. K. D., Yokaichiya, F., & de Araujo, D. R. (2023). Supramolecular structure organization and rheological properties modulate the performance of hyaluronic acid-loaded thermosensitive hydrogels as drug-delivery systems. *Journal of Colloid and Interface Science*, *630*, 328–340. DOI: 10.1016/j.jcis.2022.10.064 PMID: 36327735

Services, N. J. D. o. H. a. S. (1999). Tetrahydrothiophene. *Hazardous Substance Fact Sheet*, 1-6.

Shafiq, M., Rafique, M., Cui, Y., Pan, L., Do, C.-W., & Ho, E. A. (2023). An insight on ophthalmic drug delivery systems: Focus on polymeric biomaterials-based carriers. *Journal of Controlled Release*, *362*, 446–467. DOI: 10.1016/j.jconrel.2023.08.041 PMID: 37640109

Shah, P., Zhu, X., & Li, C.-Z. (2013). Development of paper-based analytical kit for point-of-care testing. *Expert Review of Molecular Diagnostics*, *13*(1), 83–91. DOI: 10.1586/erm.12.130 PMID: 23256705

Shamraiz, U., Riaz, B., & Badshah, A. (2016). Functional metal sulfides and selenides for the removal of hazardous dyes from water. *Journal of Photochemistry and Photobiology. B, Biology*, *159*, 33–41. DOI: 10.1016/j.jphotobiol.2016.03.013 PMID: 27010842

Shang, Q., Su, Y., Leslie, F., Sun, M., & Wang, F. (2022). Advances in peptide-drug conjugate-based supramolecular hydrogel systems for local drug delivery. *Medicine in Drug Discovery*, *14*, 100125. DOI: 10.1016/j.medidd.2022.100125

Sharma, S., Naman, S., & Baldi, A. (2023). Cyclodextrin-Based Molecular Inclusion by Grinding: Quality by Design in Green Chemistry. *Sustainable Approaches in Pharmaceutical Sciences*, 217-236. DOI: 10.1002/9781119889878.ch9

Sharma, A., Garg, T., Aman, A., Panchal, K., Sharma, R., Kumar, S., & Markandeywar, T. (2016). Nanogel—An advanced drug delivery tool: Current and future. *Artificial Cells, Nanomedicine, and Biotechnology*, *44*(1), 165–177. DOI: 10.3109/21691401.2014.930745 PMID: 25053442

Sharma, A., Kumar, V., & Kumar, P. (2019). Worldwide pesticide usage and its impacts on ecosystem. *SN Applied Sciences*, *1*(11), 1446. DOI: 10.1007/s42452-019-1485-1

Sharma, H., Kalita, B. K., Pathak, D., & Sarma, B. (2023). Low molecular weight supramolecular gels as a crystallization matrix. *Crystal Growth & Design*, *xx*, xx–xx. DOI: 10.1021/acs.cgd.3c01211

Shehzad, M. T., Imran, A., Hameed, A., Rashida, M., Bibi, M., Uroos, M., Asari, A., Iftikhar, S., Mohamad, H., Tahir, M. N., Shafiq, Z., & Iqbal, J. (2021). Exploring synthetic and therapeutic prospects of new thiazoline derivatives as aldose reductase (ALR2) inhibitors. *RSC Advances*, *11*(28), 17259–17282. DOI: 10.1039/D1RA01716K PMID: 35479726

Shen, J., Zhang, W., Qi, R., Mao, Z.-W., & Shen, H. (2018). Engineering functional inorganic–organic hybrid systems: Advances in siRNA therapeutics. *Chemical Society Reviews*, *47*(6), 1969–1995. DOI: 10.1039/C7CS00479F PMID: 29417968

Shi, W., Zhang, D., Han, L., Shao, W., Liu, Q., Song, B., Yan, G., Tang, R., & Yang, X. (2024). Supramolecular chitin-based hydrogels with self-adapting and fast-degradation properties for enhancing wound healing. *Carbohydrate Polymers*, *323*, 121374. DOI: 10.1016/j.carbpol.2023.121374 PMID: 37940271

Shi, Y., & Lammers, T. (2019). Combining nanomedicine and immunotherapy. *Accounts of Chemical Research*, *52*(6), 1543–1554. DOI: 10.1021/acs.accounts.9b00148 PMID: 31120725

Shi, Z., Hu, Y., & Li, X. (2024). Polymer mechanochemistry in drug delivery: From controlled release to precise activation. *Journal of Controlled Release*, *365*, 259–273. DOI: 10.1016/j.jconrel.2023.10.042 PMID: 39491171

Shuai, H., Wu, Y., & Chen, J. (2023). Recent advances of transition metal sulfides/selenides cathodes for aqueous zinc-ion batteries. *Advanced Energy Materials*, *13*(4), 2202992. DOI: 10.1002/aenm.202202992

Singh, R. P., Singh, P., & Srivastava, A. (Eds.). (2023). *Heavy metal toxicity: Environmental concerns, remediation and opportunities*. Springer. DOI: 10.1007/978-981-99-0397-9

Singh, R., Kumar, V., & Sharma, P. (2022). Colorimetric assay for visual determination of imidacloprid in water and fruit samples using asparagine modified gold nanoparticles. *Journal of the Iranian Chemical Society, 19*(2), 599–607. DOI: 10.1007/s13738-021-02334-5

Singh, R., & Lillard, J. W.Jr. (2009). Nanoparticle-Based Targeted Drug Delivery. *Experimental and Molecular Pathology, 86*(3), 215–223. DOI: 10.1016/j.yexmp.2008.12.004 PMID: 19186176

Singh, S. K., Kulkarni, P. P., & Dash, D. (2013). Biomedical applications of nanomaterials: An overview. In *Bio-nanotechnology: A revolution in food, biomedical and health sciences* (pp. 1–32). Wiley.DOI: 10.1002/9781118451915.ch1

Singh, S., & Verma, N. (2015). Fabrication of Ni nanoparticles-dispersed carbon micro-nanofibers as the electrodes of a microbial fuel cell for bio-energy production. *International Journal of Hydrogen Energy, 40*(2), 1145–1153. DOI: 10.1016/j.ijhydene.2014.11.073

Sivaram, A. J., Rajitha, P., Maya, S., Jayakumar, R., & Sabitha, M. (2015). Nanogels for delivery, imaging and therapy. *Wiley Interdisciplinary Reviews. Nanomedicine and Nanobiotechnology, 7*(4), 509–533. DOI: 10.1002/wnan.1328 PMID: 25581024

Soltany, P., Miralinaghi, M., & Pajoum Shariati, F. (2024). Folic acid conjugated poly (amidoamine) dendrimer grafted magnetic chitosan as a smart drug delivery platform for doxorubicin: In vitro drug release and cytotoxicity studies. *International Journal of Biological Macromolecules, 257*(127564), 127564. Advance online publication. DOI: 10.1016/j.ijbiomac.2023.127564 PMID: 37865361

Somavarapu, S., Sornsute, A., Trill, J., Abdelhakim, H. E., Goonatilaka, D., Pannala, A. S., & Taylor, K. M. G. (2023). Engineering artesunate-loaded micelles using spray drying for pulmonary drug delivery. *Journal of Drug Delivery Science and Technology, 86*, 104641. DOI: 10.1016/j.jddst.2023.104641

Sosnovik, D. E., Nahrendorf, M., & Weissleder, R. (2008). Magnetic nanoparticles for MR imaging: Agents, techniques and cardiovascular applications. *Basic Research in Cardiology, 103*(2), 122–130. DOI: 10.1007/s00395-008-0710-7 PMID: 18324368

Soussan, E., Cassel, S., Blanzat, M., & Rico-Lattes, I. (2009). Drug delivery by soft matter: Matrix and vesicular carriers. *Angewandte Chemie International Edition in English, 48*(2), 274–288. DOI: 10.1002/anie.200802453 PMID: 19072808

Srilatha, B. (2011). Nanotechnology in agriculture. *Journal of Nanomedicine & Nanotechnology, 2*(7), 123. DOI: 10.4172/2157-7439.1000123

Stater, E. P., Sonay, A. Y., Hart, C., & Grimm, J. (2021). The ancillary effects of nanoparticles and their implications for nanomedicine. *Nature Nanotechnology*, *16*(11), 1180–1194. DOI: 10.1038/s41565-021-01017-9 PMID: 34759355

Suárez, D., & Díaz, N. (2022). Amphiphilic cyclodextrins: Dimerization and diazepam binding explored by molecular dynamics simulations. *Journal of Molecular Liquids*, *349*, 118457. DOI: 10.1016/j.molliq.2021.118457

Sultana, T., Van Hai, H., Park, M., Lee, S.-Y., & Lee, B.-T. (2020). Controlled release of Mitomycin C from modified cellulose based thermo-gel prevents post-operative de novo peritoneal adhesion. *Carbohydrate Polymers*, *229*, 115552. DOI: 10.1016/j.carbpol.2019.115552 PMID: 31826495

Sun, S., Wang, Y. H., Gao, X., Wang, H. Y., Zhang, L., Wang, N., Li, C., & Xiong, S. Q. (2023). Current perspectives and trends in nanoparticle drug delivery systems in breast cancer: Bibliometric analysis and review. *Frontiers in Bioengineering and Biotechnology*, *11*, 1253048. DOI: 10.3389/fbioe.2023.1253048 PMID: 37771575

Su, Q., Su, W., Xing, S., & Tan, M. (2024). Enhanced stability of anthocyanins by cyclodextrin–metal organic frameworks: Encapsulation mechanism and application as protecting agent for grape preservation. *Carbohydrate Polymers*, *326*, 121645. DOI: 10.1016/j.carbpol.2023.121645 PMID: 38142106

Talreja, S., & Tiwari, S. (2023). Supramolecular chemistry: Unveiling the fascinating world of noncovalent interactions and complex assemblies. *Journal of Pharmacy and Pharmacology Research*, *7*(3), 133–139. DOI: 10.26502/fjppr.075

Tang, L., Zheng, Y., Melo, M. B., Mabardi, L., Castaño, A. P., Xie, Y.-Q., Li, N., Kudchodkar, S. B., Wong, H. C., Jeng, E. K., Maus, M. V., & Irvine, D. J. (2018). Enhancing T cell therapy through TCR-signaling-responsive nanoparticle drug delivery. *Nature Biotechnology*, *36*(8), 707–716. DOI: 10.1038/nbt.4181 PMID: 29985479

Tang, X., Wen, Y., Zhang, Z., Zhu, J., Song, X., & Li, J. (2023). Rationally designed multifunctional nanoparticles as GSH-responsive anticancer drug delivery systems based on host-guest polymers derived from dextran and β-cyclodextrin. *Carbohydrate Polymers*, *320*, 121207. DOI: 10.1016/j.carbpol.2023.121207 PMID: 37659810

Tinkle, S., McNeil, S. E., Mühlebach, S., Bawa, R., Borchard, G., Barenholz, Y., Tamarkin, L., & Desai, N. (2014). Nanomedicines: Addressing the scientific and regulatory gap. *Annals of the New York Academy of Sciences*, *1313*(1), 35–56. DOI: 10.1111/nyas.12403 PMID: 24673240

Torchilin, V. P. (2005). Fluorescence microscopy to follow the targeting of liposomes and micelles to cells and their intracellular fate. *Advanced Drug Delivery Reviews*, *57*(1), 95–109. DOI: 10.1016/j.addr.2004.06.002 PMID: 15518923

Torchilin, V. P. (2007). Micellar Nanocarriers: Pharmaceutical Perspectives. *Pharmaceutical Research*, *24*(1), 1–16. DOI: 10.1007/s11095-006-9132-0 PMID: 17109211

Toy, R., Bauer, L., Hoimes, C., Ghaghada, K. B., & Karathanasis, E. (2014). Targeted nanotechnology for cancer imaging. *Advanced Drug Delivery Reviews*, *76*, 79–97. DOI: 10.1016/j.addr.2014.08.002 PMID: 25116445

Tran, C. T. H., Wang, H.-M. D., Anh, L. T. H., Lin, C., Huang, C.-Y., & Kuo, C.-H. (2024). Evaluate the effect of β-cyclodextrin on the sensory and physicochemical properties of bitter gourd extract during thermal processing. *Food Chemistry*, *433*, 137394. DOI: 10.1016/j.foodchem.2023.137394 PMID: 37690136

Trokowski, R., Akine, S., & Nabeshima, T. (2008). Synthesis, characterization and molecular recognition of a bis-platinum terpyridine dimer. *Chemical Communications*, (7), 889–890. DOI: 10.1039/b716459a PMID: 18253538

Tsai, F.-T., Chiou, S.-J., Tsai, M.-C., Tsai, M.-L., Huang, H.-W., Chiang, M.-H., & Liaw, W.-F. (2005). Dinitrosyl iron complexes (DNICs)[L2Fe (NO) 2]-(L= Thiolate): Interconversion among {Fe (NO) 2} 9 DNICs,{Fe (NO) 2} 10 DNICs, and [2Fe-2S] clusters, and the critical role of the thiolate ligands in regulating NO release of DNICs. *Inorganic Chemistry*, *44*(16), 5872–5881. DOI: 10.1021/ic0505044 PMID: 16060642

Tseng, W.-B. H., Chen, C.-H., Chiu, T.-C., & Tseng, W.-L. (2020). Functionalized gold nanoparticles for sensing of pesticides: A review. *Yao Wu Shi Pin Fen Xi*, *28*(4), 517–534. DOI: 10.38212/2224-6614.1092 PMID: 35696144

Tu, Q., Tian, Y., & Li, T. (2019). In situ colorimetric detection of glyphosate on plant tissues using cysteamine-modified gold nanoparticles. *Analyst*, *144*(6), 2017–2025. DOI: 10.1039/C8AN02473A PMID: 30702090

Turan-Zitouni, G., Yurttaş, L., Tabbi, A., Akalın Çiftçi, G., Temel, H. E., & Kaplancıklı, Z. A. (2018). New Thiazoline-Tetralin Derivatives and Biological Activity Evaluation. *Molecules (Basel, Switzerland)*, *23*(1), 135. Advance online publication. DOI: 10.3390/molecules23010135 PMID: 29320423

Turkevich, J., Stevenson, P. C., & Hillier, J. (1951). A study of the nucleation and growth processes in the synthesis of colloidal gold. *Discussions of the Faraday Society*, *11*, 55–75. DOI: 10.1039/df9511100055

Umapathi, R., Ghoreishian, S. M., & Ranjith, K. S. (2021). Colorimetric based on-site sensing strategies for the rapid detection of pesticides in agricultural foods: New horizons, perspectives, and challenges. *Coordination Chemistry Reviews, 446*, 214061. DOI: 10.1016/j.ccr.2021.214061

Vajtai, R. (Ed.). (2013). *Science and engineering of nanomaterials.* In *Springer handbook of nanomaterials* (pp. 1–36). Springer. DOI: 10.1007/978-3-642-20595-8_1

van Rijt, S., & Habibovic, P. (2017). Enhancing regenerative approaches with nanoparticles. *Journal of the Royal Society, Interface*, *14*(129), 20170093. DOI: 10.1098/rsif.2017.0093 PMID: 28404870

Vandera, K. K. A., Pague, C., Omar, J., González-Gaitano, G., Ways, T. M. M., Khutoryanskiy, V. V., & Dreiss, C. A. (2023). Formation of supramolecular gels from host–guest interactions between PEGylated chitosan and α-cyclodextrin. *Macromolecular Materials and Engineering*, *308*(6), 2200646. DOI: 10.1002/mame.202200646

Vardanyan, R. S., & Hruby, V. J. (2006). 38 - Antihelmintic Drugs. In Vardanyan, R. S., & Hruby, V. J. (Eds.), *Synthesis of Essential Drugs* (pp. 583–593). Elsevier. DOI: 10.1016/B978-044452166-8/50038-8

Vardarajan, V., Gu, L., Kanneganti, A., Mohanty, S., & Koymen, A. (2011). Photothermal therapy of cancer cells using magnetic carbon nanoparticles. In *SPIE Proceedings* (Vol. 7897). DOI: 10.1117/12.875971

Verma, S., Domb, A. J., & Kumar, N. (2011). Nanomaterials for regenerative medicine. *Nanomedicine (London)*, *6*(1), 157–181. DOI: 10.2217/nnm.10.146 PMID: 21182426

Volgin, I. V., Larin, S. V., Lyulin, A. V., & Lyulin, S. V. (2018). Coarse-grained molecular-dynamics simulations of nanoparticle diffusion in polymer nanocomposites. *Polymer*, *145*, 80–87. DOI: 10.1016/j.polymer.2018.04.058

Wang, H.-F., He, Y., Ji, T.-R., & Yan, X.-P. (2009). Surface molecular imprinting on Mn-doped ZnS quantum dots for room-temperature phosphorescence optosensing of pentachlorophenol in water. *Analytical Chemistry*, *81*(4), 1615–1621. DOI: 10.1021/ac802375a PMID: 19170523

Wang, J., Li, B., Qiu, L., Qiao, X., & Yang, H. (2022). Dendrimer-based drug delivery systems: History, challenges, and latest developments. *Journal of Biological Engineering*, *16*(1), 18. DOI: 10.1186/s13036-022-00298-5 PMID: 35879774

Wang, J., Liu, W., Luo, G., Li, Z., Zhao, C., Zhang, H., Zhu, M., Xu, Q., Wang, X., Zhao, C., Qu, Y., Yang, Z., Yao, T., Li, Y., Lin, Y., Wu, Y., & Li, Y. (2018). Synergistic effect of well-defined dual sites boosting the oxygen reduction reaction. *Energy & Environmental Science*, *11*(12), 3375–3379. DOI: 10.1039/C8EE02656D

Wang, M., Xin, Y., Cao, H., Li, W., Hua, Y., Webster, T. J., Zhang, C., Tang, W., & Liu, Z. (2021). Recent advances in mesenchymal stem cell membrane-coated nanoparticles for enhanced drug delivery. *Biomaterials Science*, *9*(4), 1088–1103. DOI: 10.1039/D0BM01164A PMID: 33332490

Wang, M., Zhang, Y., Lv, L., Kong, D., & Niu, G. (2022). Biosynthesis and Chemical Synthesis of Albomycin Nucleoside Antibiotics. *Antibiotics (Basel, Switzerland)*, *11*(4), 438. Advance online publication. DOI: 10.3390/antibiotics11040438 PMID: 35453190

Wang, Q., Yan, X.-T., Fan, J.-W., Xu, S.-H., Yao, H., & Yan, C.-G. (2024). Design of a serum albumin sensitive probe for cell imaging and drug delivery by modifying a fluorescent agent. *Journal of Molecular Structure*, *1297*, 136973. DOI: 10.1016/j.molstruc.2023.136973

Wang, Q., Zhang, A., Zhu, L., Yang, X., Fang, G., & Tang, B. (2023). Cyclodextrin-based ocular drug delivery systems: A comprehensive review. *Coordination Chemistry Reviews*, *476*, 214919. DOI: 10.1016/j.ccr.2022.214919

Wang, X., Wu, Y., Shang, H., Sun, X., An, K., Zhang, Q., & Qiao, N. (2023). Preparation of pH/Light dual-responsive biocompatible polymer micelles: Application to curcumin delivery. *Journal of Drug Delivery Science and Technology*, *86*, 104652. DOI: 10.1016/j.jddst.2023.104652

Wang, X., Zhang, M., Li, Y., Cong, H., Yu, B., & Shen, Y. (2023). Research status of dendrimer micelles in tumor therapy for drug delivery. *Small*, *19*(50), 2304006. DOI: 10.1002/smll.202304006 PMID: 37635114

Wang, Z., Chen, J., Gao, R., Jiang, L., Zhang, G., Zhao, Y., Miao, Y.-B., & Shi, Y. (2024). Spatiotemporal manipulation metal–organic frameworks as oral drug delivery systems for precision medicine. *Coordination Chemistry Reviews*, *502*, 215615. DOI: 10.1016/j.ccr.2023.215615

Wang, Z., Ma, Q., & Wang, J. (2020). Multicolor visual screening of total dithiocarbamate pesticides in foods based on sulfydryl-mediated growth of gold nanobipyramids. *Analytica Chimica Acta*, *1139*, 59–67. DOI: 10.1016/j.aca.2020.09.032 PMID: 33190710

Wei, A., Mehtala, J. G., & Patri, A. K. (2012). Challenges and opportunities in the advancement of nanomedicines. *Journal of Controlled Release*, *164*(2), 236–246. DOI: 10.1016/j.jconrel.2012.10.007 PMID: 23064314

Wei, M., Yang, Z., Li, S., & Le, W. (2023). Nanotherapeutic and stem cell therapeutic strategies in neurodegenerative diseases: A promising therapeutic approach. *International Journal of Nanomedicine*, *18*, 611–626. DOI: 10.2147/IJN.S395010 PMID: 36760756

Wen, Y., Zhang, Z., & Li, J. (2014). Highly Efficient Multifunctional Supramolecular Gene Carrier System Self-Assembled from Redox-Sensitive and Zwitterionic Polymer Blocks. *Advanced Functional Materials*, *24*(25), 3874–3884. DOI: 10.1002/adfm.201303687

Wickline, S. A., & Lanza, G. M. (2003). Nanotechnology for molecular imaging and targeted therapy. *Circulation*, *107*(8), 1092–1095. DOI: 10.1161/01.CIR.0000059651.17045.77 PMID: 12615782

Willstätter, R., & Wirth, T. (1909). Über Thioformamid. *Berichte der Deutschen Chemischen Gesellschaft*, *42*(2), 1908–1922. DOI: 10.1002/cber.19090420267

Wilson, R. L., Stephens, K. K., & Jones, H. N. (2023). Placental nanoparticle gene therapy normalizes gene expression changes in the fetal liver associated with fetal growth restriction in a fetal sex-specific manner. *Journal of Developmental Origins of Health and Disease*, *14*(3), 325–332. DOI: 10.1017/S2040174423000016 PMID: 36794386

Winter, N., Rupcic, Z., Stadler, M., & Trauner, D. (2019). Synthesis and biological evaluation of (±)-hippolachnin and analogs. *The Journal of Antibiotics*, *72*(6), 375–383. DOI: 10.1038/s41429-019-0176-x PMID: 30979942

Wlodarczyk, M. T., Dragulska, S. A., Camacho-Vanegas, O., Dottino, P. R., Jarzęcki, A. A., Martignetti, J. A., & Mieszawska, A. J. (2018). Platinum (II) complex-nuclear localization sequence peptide hybrid for overcoming platinum resistance in cancer therapy. *ACS Biomaterials Science & Engineering*, *4*(2), 463–467. DOI: 10.1021/acsbiomaterials.7b00921 PMID: 32042890

Wolf, L. K. (2012). Sweating the Small Stuff. *Chemical and Engineering News*, *90*(22), 48–50. DOI: 10.1021/cen-09022-scitech1

Wu, L.-P., Wang, D., & Li, Z. (2020). Grand challenges in nanomedicine. *Materials Science and Engineering C*, *106*, 110302. DOI: 10.1016/j.msec.2019.110302 PMID: 31753337

Wu, X., Chen, X., Wang, X., He, H., Chen, J., & Wu, W. (2024). Paclitaxel-lipid prodrug liposomes for improved drug delivery and breast carcinoma therapy. *Chinese Chemical Letters, 35*(2), 108756. DOI: 10.1016/j.cclet.2023.108756

Xavier-Jr, F., Gueutin, C., Chacun, H., Vauthier, C., & Egito, E. (2019). Mucoadhesive paclitaxel-loaded chitosan-poly (isobutyl cyanoacrylate) core-shell nanocapsules containing copaiba oil designed for oral drug delivery. *Journal of Drug Delivery Science and Technology, 53*, 101194. DOI: 10.1016/j.jddst.2019.101194

Xiang, S., Guilbaud-Chéreau, C., Hoschtettler, P., Stefan, L., Bianco, A., & Ménard-Moyon, C. (2024). Preparation and optimization of agarose or polyacrylamide/amino acid-based double network hydrogels for photocontrolled drug release. *International Journal of Biological Macromolecules, 255*, 127919. DOI: 10.1016/j.ijbiomac.2023.127919 PMID: 37944737

Xiao, B., Wheatley, P. S., Zhao, X., Fletcher, A. J., Fox, S., Rossi, A. G., & Thomas, K. M. (2007). High-capacity hydrogen and nitric oxide adsorption and storage in a metal– organic framework. *Journal of the American Chemical Society, 129*(5), 1203–1209. DOI: 10.1021/ja066098k PMID: 17263402

Xiao, T., Zhong, W., Xu, L., Sun, X. Q., Hu, X. Y., & Wang, L. (2019). Supramolecular vesicles based on pillar[n]arenes: Design, construction, and applications. *Organic & Biomolecular Chemistry, 17*(6), 1336–1350. DOI: 10.1039/C8OB03095B PMID: 30638249

Xie, T.-Z., Guo, K., Li, J.-Y., Zhang, B., Zheng, K., Moorefield, C. N., & Wesdemiotis, C. (2016). Coordination-Driven, Self-Assembly of a Polycyclic, Terpyridine-Based Nanobelt. *Journal of Inorganic and Organometallic Polymers and Materials, 26*(4), 907–913. DOI: 10.1007/s10904-016-0393-8

Xie, X. (2023). Application of nanomedicine in diagnostic technology. *Highlights in Science. Engineering and Technology, 40*, 125–131. DOI: 10.54097/hset.v40i.6577

Xing, P., & Zhao, Y. (2016). Multifunctional Nanoparticles Self-Assembled from Small Organic Building Blocks for Biomedicine. *Advanced Materials, 28*(34), 7304–7339. DOI: 10.1002/adma.201600906 PMID: 27273862

Xiong, L., Yang, T., Yang, Y., Xu, C., & Li, F. (2010). Long-term in vivo biodistribution imaging and toxicity of polyacrylic acid-coated upconversion nanophosphors. *Biomaterials, 31*(27), 7078–7085. DOI: 10.1016/j.biomaterials.2010.05.065 PMID: 20619791

Xue, H., Ju, Y., Ye, X., Dai, M., Tang, C., & Liu, L. (2024). Construction of intelligent drug delivery system based on polysaccharide-derived polymer micelles: A review. *International Journal of Biological Macromolecules, 254*, 128048. DOI: 10.1016/j.ijbiomac.2023.128048 PMID: 37967605

Xu, J., Wang, K., Li, Y., Li, Y., Li, B., Luo, H., Shi, H., Guan, X., Zhang, T., Sun, Y., Chen, F., He, H., Zhang, J., Cai, L., Song, W., Wu, J., & Li, X. (2023). Injectable host-guest supramolecular hydrogel codelivers hydrophobic and hydrophilic agents for enhanced wound healing. *Chemical Engineering Journal, 454*, 140027. DOI: 10.1016/j.cej.2022.140027

Yahya, R., Al-Rajhi, A. M. H., Alzaid, S. Z., Al Abboud, M. A., Almuhayawi, M. S., Al Jaouni, S. K., Selim, S., Ismail, K. S., & Abdelghany, T. M. (2022). Molecular docking and efficacy of Aloe vera gel based on chitosan nanoparticles against Helicobacter pylori and its antioxidant and anti-inflammatory activities. *Polymers, 14*(15), 2994. DOI: 10.3390/polym14152994 PMID: 35893958

Yang, H., Chen, Y., Chen, Z., Geng, Y., Xie, X., Shen, X., Li, T., Li, S., Wu, C., & Liu, Y. (2017). Chemo-photodynamic combined gene therapy and dual-modal cancer imaging achieved by pH-responsive alginate/chitosan multilayer-modified magnetic mesoporous silica nanocomposites. *Biomaterials Science, 5*(5), 1001–1013. DOI: 10.1039/C7BM00043J PMID: 28327716

Yang, H., Yuan, B., Zhang, X., & Scherman, O. A. (2014). Supramolecular Chemistry at Interfaces: Host–Guest Interactions for Fabricating Multifunctional Biointerfaces. *Accounts of Chemical Research, 47*(7), 2106–2115. DOI: 10.1021/ar500105t PMID: 24766328

Yang, J., & Liu, H. (2015). *Metal-based composite nanomaterials*. Springer. DOI: 10.1007/978-3-319-12220-5

Yang, K., Pei, Y., Wen, J., & Pei, Z. (2016). Recent advances in pillar [n] arenes: Synthesis and applications based on host–guest interactions. *Chemical Communications, 52*(60), 9316–9326. DOI: 10.1039/C6CC03641D PMID: 27332040

Yang, Y., Xu, L., Zhu, W., Feng, L., Liu, J., Chen, Q., Dong, Z., Zhao, J., Liu, Z., & Chen, M. (2018). One-pot synthesis of pH-responsive charge-switchable PEGylated nanoscale coordination polymers for improved cancer therapy. *Biomaterials, 156*, 121–133. DOI: 10.1016/j.biomaterials.2017.11.038 PMID: 29195181

Yan, M., Wu, S., Wang, Y., Liang, M., Wang, M., Hu, W., Yu, G., Mao, Z., Huang, F., & Zhou, J. (2023). Recent progress of supramolecular chemotherapy based on host–guest interactions. *Advanced Materials, xx*(21), 2304249. DOI: 10.1002/adma.202304249 PMID: 37478832

Yan, X., Li, H., & Su, X. (2018). Review of optical sensors for pesticides. *Trends in Analytical Chemistry*, *103*, 1–20. DOI: 10.1016/j.trac.2018.03.004

Yan, X., Wang, F., Zheng, B., & Huang, F. (2012). Stimuli-responsive supramolecular polymeric materials. *Chemical Society Reviews*, *41*(18), 6042–6065. DOI: 10.1039/c2cs35091b PMID: 22618080

Yan, Y., Zhang, Y., & Wang, J. (2021). Colorimetric assay based on arginine-functionalized gold nanoparticles for the detection of dibutyl phthalate in Baijiu samples. *Analytical Methods*, *13*(43), 5179–5186. DOI: 10.1039/D1AY01464A PMID: 34672311

Yan, Z., Zheng, X., Wang, E., & Wang, J. (2013). Thermodynamic and kinetic specificities of ligand binding. *Chemical Science (Cambridge)*, *4*(6), 2387–2395. DOI: 10.1039/c3sc50478f

Yilmaz, A. S., Ozturk, S., Salih, B., Ayyala, R. S., & Sahiner, N. (2023). ESI-IM-MS characterization of cyclodextrin complexes and their chemically cross-linked alpha (α-), beta (β-) and gamma (γ-) cyclodextrin particles as promising drug delivery materials with improved bioavailability. *Colloids and Surfaces. B, Biointerfaces*, *230*, 113522. DOI: 10.1016/j.colsurfb.2023.113522 PMID: 37657404

Yin, Y., Hu, B., Yuan, X., Cai, L., Gao, H., & Yang, Q. (2020). Nanogel: A versatile nano-delivery system for biomedical applications. *Pharmaceutics*, *12*(3), 290. DOI: 10.3390/pharmaceutics12030290 PMID: 32210184

Yong Joon Kim, J., Sang, Z., Xiang, Y., Shen, Z., & Shi, Y. (2023). Nanobodies: Robust miniprotein binders in biomedicine. *Advanced Drug Delivery Reviews*, *195*, 114726. DOI: 10.1016/j.addr.2023.114726 PMID: 36754285

Yoon, H.-J., & Jang, W.-D. (2010). Polymeric supramolecular systems for drug delivery. *Journal of Materials Chemistry*, *20*(2), 211–222. DOI: 10.1039/B910948J

Yousfan, A., Al Rahwanji, M. J., Hanano, A., & Al-Obaidi, H. (2024). A comprehensive study on nanoparticle drug delivery to the brain: Application of machine learning techniques. *Molecular Pharmaceutics*, *21*(2), 333–345. DOI: 10.1021/acs.molpharmaceut.3c00880 PMID: 38060692

Yousuf, M., Khan, H. M., Rasool, F., Khan, K. U., Usman, F., Ghalloo, B. A., Umair, M., Babalghith, A. O., Kamran, M., Aadil, R. M., Al Jaouni, S. K., Selim, S., Korma, S. A., & Conte-Junior, C. A. (2022). Chemical profiling, formulation development, in vitro evaluation, and molecular docking of Piper nigrum seed extract loaded with emulgel for anti-ageing. *Molecules (Basel, Switzerland)*, *27*(18), 5990. DOI: 10.3390/molecules27185990 PMID: 36144719

Yuan, W., Ma, J., Zhao, Z., & Liu, S. (2020). Self-Assembly of Supramolecular DNA Amphiphiles through Host–Guest Interaction and Their Stimuli-Responsiveness. *Macromolecular Rapid Communications*, *41*(9), 2000022. DOI: 10.1002/marc.202000022 PMID: 32196823

Yu, G., Jie, K., & Huang, F. (2015). Supramolecular Amphiphiles Based on Host–Guest Molecular Recognition Motifs. *Chemical Reviews*, *115*(15), 7240–7303. DOI: 10.1021/cr5005315 PMID: 25716119

Yu, H., Song, Z., Yu, J., Ren, B., Dong, Y., You, Y., Zhang, Z., Jia, C., Zhao, Y., Zhou, X., Sun, H., & Zhang, X. (2024). Supramolecular self-assembly of EGCG-selenomethionine nanodrug for treating osteoarthritis. *Bioactive Materials*, *32*, 164–176. DOI: 10.1016/j.bioactmat.2023.09.020 PMID: 37822916

Yu, L., Liu, S., Jia, S., & Xu, F. (2023). Emerging frontiers in drug delivery with special focus on novel techniques for targeted therapies. *Biomedicine and Pharmacotherapy*, *165*, 115049. DOI: 10.1016/j.biopha.2023.115049 PMID: 37364480

Yu, M., Chen, Y., Zhang, N., & Liu, Y. (2010). Construction and transmembrane dissociation behavior of supramolecular assembly of quinolinocyclodextrin with porphyrin. *Organic & Biomolecular Chemistry*, *8*(18), 4148–4154. DOI: 10.1039/c0ob00080a PMID: 20657925

Yu, R., Yang, Y., He, J., Li, M., & Guo, B. (2021). Novel supramolecular self-healing silk fibroin-based hydrogel via host–guest interaction as wound dressing to enhance wound healing. *Chemical Engineering Journal*, *417*, 128278. DOI: 10.1016/j.cej.2020.128278

Yu, S. J., Kang, M. W., Chang, H. C., Chen, K. M., & Yu, Y. C. (2005). Bright Fluorescent Nanodiamonds: No Photobleaching and Low Cytotoxicity. *Journal of the American Chemical Society*, *127*(50), 17604–17605. DOI: 10.1021/ja0567081 PMID: 16351080

Zakharova, L. Y., Maganova, F. I., Sinyashin, K. O., Gaynanova, G. A., Mirgorodskaya, A. B., Vasilieva, E. A., & Sinyashin, O. G. (2023). Supramolecular strategy for the design of nanocarriers for drugs and natural bioactives: Current state of the art (a review). *Russian Journal of General Chemistry*, *93*(7), 1867–1899. DOI: 10.1134/S1070363223070253

Zhang, A., Jung, K., Li, A., Liu, J., & Boyer, C. (2019). Recent advances in stimuli-responsive polymer systems for remotely controlled drug release. *Progress in Polymer Science*, *99*, 101164. DOI: 10.1016/j.progpolymsci.2019.101164

Zhang, B., Yan, W., Zhu, Y., Yang, W., Le, W., Chen, B., Zhu, R., & Cheng, L. (2018). Nanomaterials in neural-stem-cell-mediated regenerative medicine: Imaging and treatment of neurological diseases. *Advanced Materials*, *30*(17), 1705694. DOI: 10.1002/adma.201705694 PMID: 29543350

Zhang, C., Niu, J., Li, J., Zhang, H., Yu, Q., Chen, Y., & Liu, Y. (2024). Polysaccharide based supramolecular injectable hydrogels for in situ treatment of bladder cancer. *Chinese Chemical Letters*, *35*(1), 108556. DOI: 10.1016/j.cclet.2023.108556

Zhang, C., Yan, L., Wang, X., Zhu, S., Chen, C., Gu, Z., & Zhao, Y. (2020). Progress, challenges, and future of nanomedicine. *Nano Today*, *35*, 101008. DOI: 10.1016/j.nantod.2020.101008

Zhang, G., Khan, A. A., Wu, H., Chen, L., Gu, Y., & Gu, N. (2018). The application of nanomaterials in stem cell therapy for some neurological diseases. *Current Drug Targets*, *19*(3), 279–298. DOI: 10.2174/1389450118666170328115801 PMID: 28356028

Zhang, L., Gu, F., Chan, J., Wang, A., Langer, R., & Farokhzad, O. (2008). Nanoparticles in medicine: Therapeutic applications and developments. *Clinical Pharmacology and Therapeutics*, *83*(5), 761–769. DOI: 10.1038/sj.clpt.6100400 PMID: 17957183

Zhang, L., Xu, Y., & Wei, W. (2023). Water-soluble organic macrocycles based on dye chromophores and their applications. *Chemical Communications*, *59*(91), 13562–13570. DOI: 10.1039/D3CC04159J PMID: 37901908

Zhang, Q., Re Ko, N., & Kwon Oh, J. (2012). Modulated morphologies and tunable thiol-responsive shedding of aqueous block copolymer aggregates. *RSC Advances*, *2*(21), 8079–8086. DOI: 10.1039/c2ra21209a

Zhang, W., Yang, W., & Zhou, J. (2023). Biphenarenes, versatile synthetic macrocycles for supramolecular chemistry. *Molecules (Basel, Switzerland)*, *28*(11), 4422. DOI: 10.3390/molecules28114422 PMID: 37298898

Zhang, X., Ren, K., Xiao, C., & Chen, X. (2023). Guanosine-driven hyaluronic acid-based supramolecular hydrogels with peroxidase-like activity for chronic diabetic wound treatment. *Acta Biomaterialia*, *172*, 206–217. DOI: 10.1016/j.actbio.2023.10.014 PMID: 37839631

Zhang, Y.-J., Liang, J.-X., Xu, Y.-S., Liu, Y.-X., Cui, Y., Qiao, Z.-Y., & Wang, H. (2024). Covalent drugs based on small molecules and peptides for disease theranostics. *Biomaterials Science*, *xx*(3), xx–xx. DOI: 10.1039/D3BM01138K PMID: 37975197

Zhang, Y., Wang, J., & Li, M. (2020). Colorimetric detection of paraquat in aqueous and fruit juice samples based on functionalized gold nanoparticles. *Journal of Food Composition and Analysis*, *92*, 103574. DOI: 10.1016/j.jfca.2020.103574

Zhang, Z., Cao, P., Fang, M., Zou, T., Han, J., Duan, Y., Xu, H., Yang, X., & Li, Q.-S. (2021). Design, synthesis, and SAR study of novel 4,5-dihydropyrazole-Thiazole derivatives with anti-inflammatory activities for the treatment of sepsis. *European Journal of Medicinal Chemistry*, *225*, 113743. DOI: 10.1016/j.ejmech.2021.113743 PMID: 34403978

Zhang, Z., Xie, B., Lu, X., Xiong, L., Li, X., Zhang, Y., Li, C., & Wang, C. (2024). Intracellular self-aggregation of biomimetic Fe_3O_4 nanoparticles for enhanced ferroptosis-inducing therapy of breast cancer. *Nanoscale*, *xx*(2), xx–xx. DOI: 10.1039/D3NR04305C PMID: 38108145

Zhao, J., Liu, Y., Wang, L., Zhou, Y., Du, J., & Wang, Y. (2018). Functional and modified nanocrystals technology for target drug delivery. *Journal of Nanoscience and Nanotechnology*, *18*(8), 5207–5221. DOI: 10.1166/jnn.2018.15421 PMID: 29458570

Zhao, M.-X., & Zhu, B.-J. (2016). The research and applications of quantum dots as nanocarriers for targeted drug delivery and cancer therapy. *Nanoscale Research Letters*, *11*(1), 207. DOI: 10.1186/s11671-016-1394-9 PMID: 27090658

Zhao, Q., Cheng, N., Sun, X., Yan, L., & Li, W. (2023). The application of nano-medicine in clinical settings. *Frontiers in Bioengineering and Biotechnology*, *11*, 1219054. DOI: 10.3389/fbioe.2023.1219054 PMID: 37441195

Zhao, T., Zhou, M., Wu, R., Wang, H., Zouboulis, C. C., Zhu, M., & Lee, M. (2023). Dendrimer-conjugated isotretinoin for controlled transdermal drug delivery. *Journal of Nanobiotechnology*, *21*(1), 285. DOI: 10.1186/s12951-023-02052-5 PMID: 37605256

Zhao, Y., Zheng, Z., Yu, C., & Wei, H. (2024). Engineered cyclodextrin-based supramolecular hydrogels for biomedical applications. *Journal of Materials Chemistry. B*, *12*(1), 39–63. DOI: 10.1039/D3TB02101G PMID: 38078497

Zharov, V. P., Letfullin, R. R., & Galitovskaya, E. N. (2005). Microbubbles-overlapping mode for laser killing of cancer cells with absorbing nanoparticle clusters. *Journal of Physics. D, Applied Physics*, *38*(15), 2571–2581. DOI: 10.1088/0022-3727/38/15/007

Zheng, H.-y., Liu, Y., Zhang, Y., Shi, Q., Hou, X., & An, L. (2024). Camptothecin-loaded supramolecular nanodelivery system based on amphiphilic calix[4]arene for targeted tumor therapy. *New Journal of Chemistry*, *xx*(3), xx. DOI: 10.1039/D3NJ03289B

Zheng, M., Mei, T., Deng, C., Xin, J., Aras, O., An, F., & Chen, N. (2024). The impact of PEG chain length on the in vivo tumor accumulation of self-assembled amphiphilic drug-PEG polymeric micelles. *Colloids and Surfaces. A, Physicochemical and Engineering Aspects*, *682*, 132823. DOI: 10.1016/j.colsurfa.2023.132823

Zhou, J., Li, Y., Wang, L., Lv, P., Chen, M., Xiao, F., Si, T., Tao, J., & Yang, B. (2024). Bifunctional drug delivery system with carbonic anhydrase IX targeting and glutathione-responsivity driven by host-guest amphiphiles for effective tumor therapy. *Carbohydrate Polymers*, *326*, 121577. DOI: 10.1016/j.carbpol.2023.121577 PMID: 38142063

Zhou, Y., Jiang, Y., & Chen, S. J. (2022). RNA–ligand molecular docking: Advances and challenges. *Wiley Interdisciplinary Reviews. Computational Molecular Science*, *12*(3), e1571. DOI: 10.1002/wcms.1571 PMID: 37293430

Zhu, J., Wang, J., & Zhang, X. (2022). Colorimetric measurement of deltamethrin pesticide using a paper sensor based on aggregation of gold nanoparticles. *Coatings*, *12*(1), 63. DOI: 10.3390/coatings12010063

Zhuo, S., Liang, Y., Wu, Z., Zhao, X., Han, Y., & Guo, B. (2024). Supramolecular hydrogels for wound repair and hemostasis. *Materials Horizons*, *xx*(1), xx–xx. DOI: 10.1039/D3MH01403G PMID: 38018225

Zielińska, A., Carreiró, F., Oliveira, A. M., Neves, A., Pires, B., Venkatesh, D. N., Durazzo, A., Lucarini, M., Eder, P., Silva, A. M., Santini, A., & Souto, E. B. (2020). Polymeric nanoparticles: Production, characterization, toxicology and ecotoxicology. *Molecules (Basel, Switzerland)*, *25*(16), 3731. DOI: 10.3390/molecules25163731 PMID: 32824172

About the Contributors

Arumugam Akila has 13 years of research experience in the field of Organic Chemistry and 7.5 years of teaching experience in the field of Chemistry. She is working for Avinashilingam Institute for Home Science and Higher Education for Women as Assistant Professor of Chemistry. She obtained her Masters of Philosophy from Government Arts College affiliated to Bharathiar University, Coimbatore, Tamilnadu, India in the year 2012. She obtained her Doctor of Philosophy from Government Arts College affiliated to Bharathiar University, Coimbatore, Tamilnadu, India in the year 2016. Her area of interest includes Synthetic Organic Chemistry, Stereochemistry, Spectroscopy. https://orcid.org/0000-0001-7580-7093.

Sumanta Bhattacharya from Salkia, Howrah, West Bengal is an Expert commissioner member IUCN Environmental, economics and social policy, Individual Policy Analyst (ADB) . He is among the top 2% world scientist (AD Index), Special civilian member Ministry of Defence, (USI) Member at Ministry of Culture (AS). Member of Ministry of Defence {[Defence and strategic analysis] IDSA}, Member of World Economic Forum, littérateur at National Economic Research Bureau, Member of Ministry of Personnel, Pension and Public grievance (IIPA) He completed his B.Tech, M.Tech in Textile Technology along with that he has MA in Development studies, M.Sc in Environmental Science, LLB, MA in security and defence law from NALSAR, MA in Economics, MBA, Post Graduate Diploma in Environment and Sustainable Development, DIA&D, DS&DS, MPI(Oxford University). Being a research scholar he has 216 research papers published in International and Scopus index journal, Wiley, Springer, Hindawi in various sectors like law, economic, Development studies, Textiles, Climate Change, SDGs, Public policy, Nanotechnology, Biotechnology, cancer, environment studies, 57 Book chapters published, Edited 2 books, wrote 1 book, 80 patents International and National, 4 copyrights, attended 93 International and National conference and presented papers and won 25 awards for his excellent in education, policy making and innovation which includes Best Water Management Solution Award 2022, Young Researcher Award 2020, Bravo International Book of Records, Mahatma Gandhi Ratna Award 2020, Sustainable Technology Award 2022, Global Change Maker Award 2020, Best Water Management Solution Award at the Rural & Urban Development Summit & Awards 2022, New Delhi, The Progress Global Awards 2023 for Excellence in Research, GIS -Person of the year – Water Conservation, Green India award, 2023, Global Legend Award 2023 by GSIL Educational Society . He is also a member of various International and National organization which includes IEI, IICHE, Asiatic Society, IIPA, IDSA, YOUNGO, IUCN, USI and many more.

Aarati Budar is currently pursuing my Masters of Pharmacy with specialization in Pharmaceutics at St. John Institute of Pharmacy and Research. As a dedicated and passionate student, I am committed to advancing the field of pharmaceutics through rigorous academic study and hands-on research. With a strong foundation in pharmaceutical sciences, I have developed a keen interest in the formulation and delivery of therapeutic agents.

Sudha D. is an Assistant Professor (Sl. G) in the Department of Chemistry at KPR Institute of Engineering and Technology, Coimbatore.She has successfully completed her Doctoral studies in Chemistry, specializing in the area of Advanced Materials, at Bharathiar University Coimbatore, Tamil Nadu. With 15 years of teaching experience, she has made significant contributions to the academic community. She has published over 15 papers in peer reviewed international journals, authored 3 book chapters and published 3 patents. She was awarded Best Faculty award, Best Innovative teacher and Best Mentor awards at KPRIET and 4 Best Paper awards for Paper presentation in International conferences. She has been the Resource person in various National and International conferences, reviewer for few scopus and UGC care journals. She is an active member of professional organizations such as ISTE and IEEE. Her research interests include advanced materials, nanocomposites, environmental chemistry and metal-organic complexes.

T.C. Manjunath was born on 6th Feb. 1967 in Bangalore. He did his BE from RV College of Engg., Bangalore Univ. in 1989 with First Class, ME in Robotics and Automation from the Govt. LD College of Engg. in 1992 with First Rank – FCD from Gujarat Univ., Ahmedabad and PhD in Systems and Control Engg. from IIT Bombay in the year 2007. He is currently working as the Dean Research, R and D and as Professor in the Dept. of Computer Science and Engineering with specialization in IoT, Cyber Security and Blockchain Tech. in Rajarajeswari College of Engg., Bengaluru, Karnataka, India. He has got a teaching, academic, research and administrative experience of more than 35+ years with 10 years as Principal, 8 yrs as Professor - HoD and 17 yrs as faculty. He has published more than 1200+ research papers in various national and international journals, conferences, with 5000+ citations, 175+ patents, 50+ book-chapters, 25 awards, published 25 text books in various field of science, engineering and technology, completed funded research projects worth Rs. 9 Crores till date apart from having guided 10 PhD scholars, 25 PG Students, more than 200 UG students towards their project works. He is a member of 30 Professional Society Chapters and is a fellow of IE, IETE, AMIEE and a chartered engineer. He got the best research scholar award in 2006 in IIT Bombay amongst all the research scholars of the institute.

Selvakumar P. is currently an Associate Professor in the Department of Science and Humanities at Nehru Institute of Technology, Coimbatore, Tamil Nadu, India. Since 2023, he has served as the Department IQAC Coordinator at the institute. Prior to joining Nehru Institute of Technology, he held positions as an Assistant Professor in the Department of Science and Humanities various engineering colleges in Coimbatore, Tamil Nadu, India, spanning from 2011 to 2023. Dr.P. Selvakumar's academic and research interests lie primarily in the field of phytochemistry. He has made significant contributions to his field, with an impressive record that includes 30 journal publications, 12 patents, and 20 conference papers. Additionally, he has organized 5 seminars and conferences, further demonstrating his commitment to academic and scientific exchange. Throughout his career, Dr. P.Selvakumar has been recognized for his dedication and excellence in teaching, having been honored with several awards, including the Best Teacher Awards. He holds a prominent position in the academic community, contributing actively to research.

Nihalkumar Pravinbhai Patel is an Assistant Professor in the Department of Chemistry at Ganpat University, Mehsana Urban Institute of Sciences (MUIS), Gujarat, India. He completed his M.Phil. in Chemistry (A+) from Ganpat University in 2019–2020, followed by a Ph.D. in Chemistry at the same institution from 2020 to 2023. His academic journey began with a B.Sc. in Chemistry from Veer Narmad South Gujarat University (2014–2017) Dr. Patel initially joined Ganpat University as a Research Assistant and then Research Associate (2019–2021) before taking on the faculty role in early 2021. His research focuses on nanoscience, supramolecular chemistry, and the design of fluorescent and colorimetric chemosensors, particularly using functionalized calix pyrrole and triazole derivatives. He has co-authored over 20 peer reviewed publications, including computational and experimental studies on novel sensor platforms for pesticide and antibiotic detection. Dr. Patel is actively involved in collaborative interdisciplinary research, working closely with senior colleagues such as Dr. Keyur Bhatt, and is contributing significantly to emerging sensor technologies for environmental and biomedical applications.

Dhruvansh Patni is a faculty member in the Life Sciences / Microbiology and Biotechnology disciplines at Sigma University. He completed his education at Sardar Patel University, Vallabh Vidyanagar, before joining Sigma University as faculty . His research interests include microbiology, nanoparticles, biomedical sensors, and bioengineering applications. He was co-author on a 2025 study (Indian Journal of Science & Technology) focusing on proteolytic activity of Aspergillus terreus isolated from cowshed soil, investigating enzyme production and optimization under laboratory conditions. He is connected with early-career academia, as evidenced by his LinkedIn activity promoting Sigma University's symposium CTOBS 2025, aimed at biotechnology students and professionals.

Nandan C. Pomal served as the Head of the Department of Science and Humanities at Tolani Foundation Gandhidham Polytechnic and is affiliated with Sigma University's academic network in Vadodara. He holds strong expertise in nanotechnology, specializing in supramolecules, calix[4]pyrrole chemistry, and metal nanoparticles. His research includes designing functional nanomaterials—including silver and gold nanoparticle systems—for environmental and biomedical sensing applications. Dr. Pomal has authored several notable publications, such as the 2023 Applied Nanoscience study on palladium nanoparticles and the ChemistrySelect article on fluorescence sensing of fungicide dimoxystrobin, reflecting his ongoing contributions to sensor development and nanoscience innovations. As an educator and researcher associated with Sigma University, he is engaged in evolving areas of functional nanomaterials and supramolecular chemistry.

Bhupendra Prajapati works as Professor in Department of Pharmaceutics, Shree S.K.Patel College of Pharmaceutical Education and Research, Ganpat University, North Gujarat, India. He did his Ph.D. from Hemchandracharya North Gujarat University, Patan. He did his PG and UG from M.S. University, Baroda. He has 19 years of experience in academic/industry (17+2). He was awarded with the Carrier Award for Young Teacher by AICTE, New Delhi in 2013. He was also awarded for Distinguished Associate Professor in TechNExt India 2017 by CSI, Mumbai and President award of Staff Excellence in Research (2019) and Capacity Building (2020) by Ganpat University for consecutive year 2021 and 2022 respectively. He claims on his name more than sixty national and international publications. He fetched grants for Research Projects, Staff Development Programs, Seminars, Conferences and Travel Grants from National and State Government agencies. He is also given his guidance in industrial consultancy projects conducted at institute. His two patents were published and Three application submitted at the Indian Patent office in the field of NDDS. He has delivered expert talks and invited scientific sessions in several national and international conferences and seminars. He is actively working in the field of lipid-based drug delivery and nanotech formulations. He guided 7 Ph.D. and 45 PG research scholars. 6 Ph.D and 3 PG research scholars are currently working under his guidance in the field of Nanoparticulate Drug Delivery and Bioavailability Enhancement. In last three-year 7 Prizes secured by his PG and Ph.D. research scholars work at national and international conferences. He is an active life member of IPA, APTI and CRSIC.

Seenivasan S. is currently working as a Associate Professor Department of Mechanical Engineering at Rathinam Technical Campus, Coimbatore, TamilNadu, India. He completed BE (Mechanical Engineering) in the year of May 2009 at Kumaraguru College of Technology, Coimbatore and ME (Engineering Design) in the year May 2012 at Anna University Regional campus, Coimbatore. Ph.D (Mechanical Engineering) in the Year 2021 at Anna University, Chennai, in the area of Composite Materials, machine learning, deep learning. He has 12 years of teaching & 5 Years of Research Experience. Orcid id; 0000-0002-2155-9781.

Bancha Yingngam, Associate Professor at Ubon Ratchathani University, Thailand, specializes in Pharmaceutical Sciences and Technology. He earned his pharmacy degree in 2005 and his Ph.D. in 2011 from the same institution. A postdoctoral fellowship was held at Karl-Franzens-Universität Graz, Austria, from 2013 to 2021. He has published more than 20 scholarly works with an h-index of 10 according to the public profile in the Scopus database. Dr. Yingngam has also reviewed more than 100 Scopus-listed articles, underscoring his commitment to advancing his field. He has supervised 2 Ph.D.s and Master's theses.

Index

A

B

C

D

F

G

H

I

L

M

www.ingramcontent.com/pod-product-compliance
Lightning Source LLC
LaVergne TN
LVHW061541290926
849162LV00004B/868